Fermented Foods: Biochemistry and Biotechnology

Fermented Foods: Biochemistry and Biotechnology

Contributors

Stephanie N. Chilton, Jeremy P. Burton et al.

AURIS
Reference

www.aurisreference.com

Fermented Foods: Biochemistry and Biotechnology

Contributors: Stephanie N. Chilton , Jeremy P. Burton et al.

Published by Auris Reference Limited

www.aurisreference.com

United Kingdom

Fermented Foods: Biochemistry and Biotechnology

ISBN: 978-1-78154-807-3

British Library Cataloguing in Publication Data
A CIP record for this book is available from the British Library

Printed in the United Kingdom

Exclusively distributed by CBS Publishers & Distributors Pvt. Ltd.

Sales & Distribution Rights only for India, Pakistan, Bangladesh, Sri Lanka, Nepal and Bhutan.This book is not to be sold outside these territories.

Contents

List of Abbreviations

AR	Antibiotic resistance
ATR	Acid Tolerance Response
Aw	Water activity
BCAAs	Branched chain amino acids
BLIS	Bacteriocin-like inhibitory substances
CFU	Colony forming units
CIPs	Cold Induced Proteins
CNS	Chinese Nutrition Society
DLSU	De La Salle University
EMP	Embden Meyerhof Parnas
ETEC	Enterotoxigenic Escherichia coli
FOSHU	Food for specified health uses
FPLC	Fast Protein Liquid Chromatography
GABA	γ-amino butyric acid
GAD	Glutamate decarboxylase
GDH	Glutamate dehydrogenase
GMOs	Genetically modified microorganisms
GRAS	Generally regarded as safe
HIV	Human immunodeficiency virus
HPH	High pressure homogenization
HPLC	Normal chromatographic technique
HPP	High-pressure processing
IBD	Inflammatory bowel disease
IMTECH	Institute of Microbial Technology
LAB	Lactic acid bacteria
LDL	low-density lipoproteins
MBC	Mozzarella di Bufala Campana
MPE	Monascus purpureus extract
MRS	Man Rogosa Sharpe
MTCC	Microbial Type Culture Collection
NO	Nitric oxide
NWSC	Natural whey starter cultures
ORAC	Oxygen radical absorbance capacity
ROS	Reactive oxygen species
SIM	Selected ion mode
SOD	Superoxide dismutase
UPM	University of the Philippines Manila
UST	University of Santo Tomas
VIP	Vasoactive intestinal peptide
VSU	Visayas State University
WHO	World Health Organization

List of Contributors

Stephanie N. Chilton
Canadian Centre for Human Microbiome and Probiotic Research, Lawson Health Research Institute, 268 Grosvenor St., N6A 4V2, London, ON, N6A 4V2, Canada
Department of Physiology and Pharmacology, University of Western Ontario, London, ON, N6A 3K7, Canada

Jeremy P. Burton
Canadian Centre for Human Microbiome and Probiotic Research, Lawson Health Research Institute, 268 Grosvenor St., N6A 4V2, London, ON, N6A 4V2, Canada
Department of Physiology and Pharmacology, University of Western Ontario, London, ON, N6A 3K7, Canada
Department of Microbiology and Immunology, University of Western Ontario, London, ON, N6A 3K7, Canada
Department of Surgery, University of Western Ontario, London, ON, N6A 3K7, Canada
Division of Urology, University of Western Ontario, London, ON, N6A 3K7, Canada

Gregor Reid
Canadian Centre for Human Microbiome and Probiotic Research, Lawson Health Research Institute, 268 Grosvenor St., N6A 4V2, London, ON, N6A 4V2, Canada
Department of Physiology and Pharmacology, University of Western Ontario, London, ON, N6A 3K7, Canada
Department of Microbiology and Immunology, University of Western Ontario, London, ON, N6A 3K7, Canada
Department of Surgery, University of Western Ontario, London, ON, N6A 3K7, Canada

Charina Gracia B. Banaay
Institute of Biological Sciences, College of Arts and Sciences, University of the Philippines Los Baños, Laguna, Philippines

Marilen P. Balolong
Department of Biology, University of the Philippines Manila, Padre Faura Street, Ermita, Manila, Philippines
Institute of Molecular Biology and Biotechnology, National Institutes of

Health, University of the Philippines Manila, Pedro Gil Street, Ermita, Manila, Philippines

Francisco B. Elegado
National Institute of Molecular Biology and Biotechnology, University of the Philippines Los Baños, Laguna, Philippines

Mohan-Kumari Honganoor Puttananjaiah
Department of Food Microbiology, Central Food Technological Research Institute, Council of Scientific and Industrial Research, Mysore, India;

Mohan Appasaheb Dhale
Biological Oceanography Division, National Institute of Oceanography, Council of Scientific and Industrial Research, Dona Paula, India.

Vijayalakshmi Govindaswamy
Department of Food Microbiology, Central Food Technological Research Institute, Council of Scientific and Industrial Research, Mysore, India;

Olaoluwa Oyedeji
Department of Microbiology, Obafemi Awolowo University, Ile-Ife, Nigeria;

Samuel Temitope Ogunbanwo
[2]Department of Microbiology, University of Ibadan, Ibadan, Nigeria.

Anthony Abiodun Onilude
Department of Microbiology, University of Ibadan, Ibadan, Nigeria.

Chiara Devirgiliis
CRA-NUT, Food & Nutrition Research Center, Agricultural Research Council, Via Ardeatina 546, 00178 Rome,

Paola Zinno
CRA-NUT, Food & Nutrition Research Center, Agricultural Research Council, Via Ardeatina 546, 00178 Rome,

Mariarita Stirpe
CRA-NUT, Food & Nutrition Research Center, Agricultural Research Council, Via Ardeatina 546, 00178 Rome, Italy
Department of Biology and Biotechnology Charles Darwin, Sapienza University of Rome, Piazzale Aldo Moro 5, 00185 Rome, Italy

Simona Barile
CRA-NUT, Food & Nutrition Research Center, Agricultural Research Council, Via Ardeatina 546, 00178 Rome,

Giuditta Perozzi
CRA-NUT, Food & Nutrition Research Center, Agricultural Research Council, Via Ardeatina 546, 00178 Rome, Italy

Grazina Juodeikiene
Kaunas University of Technology,, Lithuania

Elena Bartkiene
Veterinary Academy, Lithuanian University of Health Sciences, Lithuania

Pranas Viskelis[3]
Institute of Horticulture, Lithuanian Research Centre for Agriculture and Forestry, Lithuania

Dalia Urbonaviciene
Institute of Horticulture, Lithuanian Research Centre for Agriculture and Forestry, Lithuania

Dalia Eidukonyte
Kaunas University of Technology,, Lithuania

Ceslovas Bobinas
Institute of Horticulture, Lithuanian Research Centre for Agriculture and Forestry,, Lithuania

Lavinia Claudia Buruleanu
Department of Food Engineering, Faculty of Environmental Engineering and Biotechnology, Valahia University of Targoviste, Romania

Magda Gabriela Bratu
Department of Food Engineering, Faculty of Environmental Engineering and Biotechnology, Valahia University of Targoviste, Romania

Iuliana Manea
Department of Food Engineering, Faculty of Environmental Engineering and Biotechnology, Valahia University of Targoviste, Romania

Daniela Avram
Department of Food Engineering, Faculty of Environmental Engineering and Biotechnology, Valahia University of Targoviste, Romania

Carmen Leane Nicolescu
Department of Food Engineering, Faculty of Environmental Engineering and Biotechnology, Valahia University of Targoviste, Romania

Diana I. Serrazanetti
Department of Food Science, Alma Mater Studiorum, University of Bologna,Bologna, Italy
Inter-Departmental Center of Industrial Agri-Food Research (CIRI Agroalimentare), Cesena, Italy

Davide Gottardi
Department of Food Science, Alma Mater Studiorum, University of Bologna,Bologna, Italy
Inter-Departmental Center of Industrial Agri-Food Research (CIRI Agroalimentare), Cesena, Italy

Chiara Montanari
Department of Food Science, Alma Mater Studiorum, University of Bologna,Bologna, Italy
Inter-Departmental Center of Industrial Agri-Food Research (CIRI Agroalimentare), Cesena, Italy

Andrea Gianotti
Department of Food Science, Alma Mater Studiorum, University of Bologna,Bologna, Italy
Inter-Departmental Center of Industrial Agri-Food Research (CIRI Agroalimentare), Cesena, Italy

Evans C. Egwim
Biochemistry Department, Federal University of Technology, Minna, Niger State, Nigeria

Amanabo Musa
Biochemistry Department, Federal University of Technology, Minna, Niger State, Nigeria

Yahaya Abubaka

Biochemistry Department, Federal University of Technology, Minna, Niger State, Nigeria

Bello Mainuna
Biochemistry Department, Federal University of Technology, Minna, Niger State, Nigeria

Huayi Suo
College of Food Science, Southwest University, Chongqing 400715, China
Chongqing Engineering Research Center of Regional Food, Chongqing 400715, China

Xia Feng
Department of Biological and Chemical Engineering, Chongqing University of Education, Chongqing 400067, China
Chongqing Collaborative Innovation Center of Functional Food, Chongqing University of Education, Chongqing 400067, China

Kai Zhu
Department of Biological and Chemical Engineering, Chongqing University of Education, Chongqing 400067, China
Chongqing Collaborative Innovation Center of Functional Food, Chongqing University of Education, Chongqing 400067, China

Cun Wang
Department of Biological and Chemical Engineering, Chongqing University of Education, Chongqing 400067, China
Chongqing Collaborative Innovation Center of Functional Food, Chongqing University of Education, Chongqing 400067, China

Xin Zhao
Department of Biological and Chemical Engineering, Chongqing University of Education, Chongqing 400067, China
Chongqing Collaborative Innovation Center of Functional Food, Chongqing University of Education, Chongqing 400067, China

Jianquan Kan
College of Food Science, Southwest University, Chongqing 400715, China
Chongqing Engineering Research Center of Regional Food, Chongqing 400715, China

Mohsen Zareian
Faculty of Food Science and Technology, Universiti Putra Malaysia, Serdang 43400, Selangor, Malaysia

Afshin Ebrahimpour
Faculty of Food Science and Technology, Universiti Putra Malaysia, Serdang 43400, Selangor, Malaysia

Fatimah Abu Bakar
Faculty of Food Science and Technology, Universiti Putra Malaysia, Serdang 43400, Selangor, Malaysia

Abdul Karim Sabo Mohamed
Faculty of Food Science and Technology, Universiti Putra Malaysia, Serdang 43400, Selangor, Malaysia

Bita Forghani
Faculty of Food Science and Technology, Universiti Putra Malaysia, Serdang 43400, Selangor, Malaysia

Mohd Safuan B. Ab-Kadir
Faculty of Food Science and Technology, Universiti Putra Malaysia, Serdang 43400, Selangor, Malaysia

Nazamid Saari
Faculty of Food Science and Technology, Universiti Putra Malaysia, Serdang 43400, Selangor, Malaysia

Changhyun Roh
Division of Biotechnology, Advanced Radiation Technology Institute (ARTI), Korea Atomic Energy Research Institute (KAERI), 1266, Sinjeong-dong, Jeongeup, Jeonbuk 580-185, Korea;

Preface

Traditional fermented foods are not only the staple food for most of developing countries but also the key healthy food for developed countries. As the healthy functions of these foods are gradually discovered, more high throughput biotechnologies are being used to promote the fermented food industries. As a result, the microorganisms, process biochemistry, manufacturing, and down-streaming processing, as well as the bioactive metabolites released by the fermenting organisms and, above all, the healthy functions of these foods were extensively researched. Fermented Foods Biochemistry and Biotechnology discusses the general aspects of biochemistry and biotechnological application of fermented foods involving acetic acid bacteria, lactic acid bacteria, ethanolic yeasts, and fungi in accelerating the many and variable functional factors in the fermented foods as well as metagenomics of fermented foods. First chapter focuses on the inclusion of fermented foods in food guides around the World. Lactic acid bacteria in Philippine traditional fermented foods is presented in second chapter. In third chapter, we have identified the non-toxic effect of MPE on lactic acid bacteria (LAB) and biotransformation of isoflavone glycosides by LAB in culture medium supplemented with MPE. The objective of fourth chapter is to isolate and characterize predominant lactic acid bacteria species during the course of traditional fufu and ogi fermentations. Functional screening of antibiotic resistance genes from a representative metagenomic library of food fermenting microbiota is presented in fifth chapter. Sixth chapter reviews on fermentation processes using lactic acid bacteria producing bacteriocins for preservation and improving functional properties of food products. Seventh chapter explores on fermentation of vegetable juices by lactobacillus acidophilusla-5. Eighth chapter highlights on dynamic stresses of lactic acid bacteria associated to fermentation processes. Ninth chapter explores on processes and prospects of Nigerian indigenous fermented foods. Tenth chapter aims to study the anti-gastric mucosal injury effects of Shuidouchi fermented in different containers and explain the mechanism of the anti-oxidation effects of Shuidouchi. Last chapter presents an approach on glutamic acid-producing lactic acid bacteria isolated from Malaysian fermented foods.

Chapter 1

INCLUSION OF FERMENTED FOODS IN FOOD GUIDES AROUND THE WORLD

Stephanie N. Chilton [1,2], Jeremy P. Burton [1,2,3,4,5] and Gregor Reid [1,2,3,4]

[1]Canadian Centre for Human Microbiome and Probiotic Research, Lawson Health Research Institute, 268 Grosvenor St., N6A 4V2, London, ON, N6A 4V2, Canada

[2]Department of Physiology and Pharmacology, University of Western Ontario, London, ON, N6A 3K7, Canada

[3]Department of Microbiology and Immunology, University of Western Ontario, London, ON, N6A 3K7, Canada

[4]Department of Surgery, University of Western Ontario, London, ON, N6A 3K7, Canada

[5]Division of Urology, University of Western Ontario, London, ON, N6A 3K7, Canada

ABSTRACT

Fermented foods have been a well-established part of the human diet for thousands of years, without much of an appreciation for, or an understanding of, their underlying microbial functionality, until recently. The use of many organisms derived from these foods, and their applications in probiotics, have further illustrated their impact on gastrointestinal wellbeing and diseases affecting other sites in the body. However, despite the many benefits of fermented foods, their recommended consumption has not been widely translated to global inclusion in food guides. Here, we present the case for such inclusion, and challenge health authorities around the world to consider advocating for the many benefits of these foods.

INTRODUCTION

Recommendations for the consumption of certain nutritious foods date back to the Hippocratic Corpus of Ancient Greece. More recently, the United States Department of Agriculture first created nutrition guidelines in 1894 which advocated variety, proportionality and moderation, calorie measuring, nutrient-rich foods and consumption of less fat, sugar and starch [1]. Canada's first Food Guide was introduced in July 1942, to provide guidance to Canadians on proper nutrition during a period when wartime rationing was common [2].

While such guidelines result from consultation with knowledge providers, they need not reflect traditions practiced by populations nor do they appreciate the benefits of foods consumed by generations of ethnic groups. Foods that are prepared by fermentation (the slow decomposition process of organic substances induced by microorganisms, or by complex proteinaceous substances (enzymes) of plant or animal origin [3]), occurs due to biochemical changes brought about by the anaerobic or partially anaerobic oxidation of carbohydrates. This process has long been shown to help retain shelf-life and prevent food spoilage. The absence of fermented foods from some food guides, as will be discussed later, should not be interpreted as suggesting these foods are not beneficial. Rather, they may not have had a history of use in a particular country, and may be made at home instead of being purchased from a commercial enterprise. The aim of the present article is to examine the history of fermented foods, their health benefits and the basis for why they are, or should be, included in the food guides of different countries across the continents. Such a review with evidence of the effectiveness of fermented foods, is one of the means that regulatory agencies, such as Health Canada, use to evaluate whether or not certain foods are worthy of inclusion in a revised food guide.

FERMENTED FOODS

What exactly are fermented foods? Fermentation is a process that has been used by humans for thousands of years, with major roles in food preservation and alcohol production. Fermentation is primarily an anaerobic process converting sugars, such as glucose, to other compounds like alcohol, while producing energy for the microorganism or cell. Bacteria and yeast are microorganisms with the enzymatic capacity for fermentation, specifically, lactic acid fermentation in the former and ethanol fermentation in the latter. Many different products around the world are a result of fermentation, either occurring naturally or through addition of a starter culture. Different bacterial and yeast species are present in each case, which contribute to the unique flavors and textures present in fermented foods (Table 1). These bacteria and yeasts are referred to as "probiotic" when they adhere to the following World Health Organization (WHO) definition: "live microorganisms which, when administered in adequate amounts, confer a health benefit on the host" [4].

During lactic acid fermentation, the pyruvate molecules from glycolysis are converted into lactate. Lactic acid bacteria (LAB) consist of homo and hetero-lactic acid organisms, and are a broad category of bacteria, including *Lactobacillus, Streptococcus, Enterococcus, Lactococcus* and *Bifidobacterium* [5], with the ability to produce lactate primarily from sugars. They are among the most commercially used bacteria today [6,7], contributing to yogurt, sauerkraut, kimchi [8] and kefir production [9], the pickling of vegetables, curing of fish, and many other traditional dishes around the world [10,11].

In comparison, ethanol fermentation produces carbon dioxide and ethanol from pyruvate molecules, mainly through the actions of various yeasts. *Saccharomyces cerevisiae* is used in bread making, helping the dough rise through the production of carbon dioxide. A separate strain of *S. cerevisiae* is also used in alcohol production, including beer and wine, in combination with other yeast species [12].

Table 1: Examples of fermented foods and countries in which they are believed to originate and remain particularly popular.

Fermented Food and Main Constituents	Country
Yogurt—milk, *L. bulgaricus, S. thermophilus*	Greece, Turkey
Kefir—milk, kefir grains, *Saccharomyces cerevisiae* and *L. plantarum*	Russia
Sauerkraut—green cabbage, *L. plantarum*	Germany
Kimchi—cabbage, *Leuconostoc mesenteroides*	South Korea
Cortido—cabbage, onions, carrots	El Salvador
Sourdough—flour, water, *L. reuteri, Saccharomyces cerevisiae*	Egypt
Kvass—beverage from black or rye bread, *Lactobacillus*	Russia
Kombucha—black, green, white, pekoe, oolong, or darjeeling tea, water, sugar, *Gluconacetobacter* and *Zygosaccharomyces*	Russia and China
Pulque—beverage from agave plant sap, *Zymomonas mobilis*	Mexico
Kaffir beer—beverage from kaffir maize, *Lactobacillus sp.*	South Africa
Ogi—cereal, *Lactobacillus sp., Saccharomyces sp., Candida sp.*	Africa
Igunaq—fermented walrus	Canada
Miso—soybeans, *Aspergillus oryzae, Zygosaccharomyces, Pediococcus sp.*	Japan
Tepa—Stinkhead fermented fish	USA
Dosa—fermented rice batter and lentils, *L. plantarum*	India
Cheddar and stilton cheeses—*Penicillium roquefort, Yarrowia lipolytica, Debaryomyces hansenii, Trichosporon ovoides*	United Kingdom
Surströmming—fermented herring, brine, *Haloanaerobium praevalens, Haloanaerobium alcaliphilum*	Sweden
Crème fraiche—soured dessert cream, *L. cremoris, L. lactis*	France
Fermented sausage—Lactobacillus, Pediococcus, or Micrococcus	Greece and Italy
Wine—various organisms particularly *Saccharomyces cerevisiae*	Georgia

EXAMPLES OF FERMENTED FOODS FROM AROUND THE WORLD

The ability to create tasty food using microbes reflects human culinary innovation at its best. The use of microbial fermenters has been instrumental in making a large range of foods, popular around the world. Examples of these are given in Table 1, illustrating diversity and opportunism by the originators of the food formulae.

These traditional foods have been consumed in some cases for thousands of years, with recipes being passed down through generations, as well-documented elsewhere [13]. Initially, many foods underwent fermentation naturally, but today, a number of them are made with the addition of a starter culture and the process has become automated and more reproducible and reliable. There are clearly types of fermented foods consumed across countries and continents, such as sauerkraut, kimchi and cortido, all products of fermented cabbage. Likewise, some foods remain quite limited in the scope of who consumes them.

A trend in the past 20 or so years has been in the globalization of foods, aided by shipping and airline delivery, and a desire by consumers to gain access to products. Thus, in the depths of winter in Canada, consumers can still purchase "fresh" fruit and vegetables from countries in the southern hemisphere. However, for the most part, global distribution is not required for fermented foods. Instead, they tend to be made locally with outside temperature not being an issue. Often, immigrants will introduce these foods for their own use, then their popularity grows and consumption becomes widespread. The net result is that fermented foods are widely consumed across the globe (Table 2) [14,15].

Table 2: Widely consumed fermented foods, the country they are consumed in, and the average amount of consumption per person annually.

Food	Country	Average Annual Consumption (per Person)
Beer	Germany	106 L
Cheese	UK	10 kg
Kimchi	Korea	22 kg
Miso	Japan	7 kg
Soy Sauce	Japan	10 L
Tempeh	Indonesia	18 kg
Wine	Italy, Portugal	90 L
	Argentina	70 L
	Finland	40 L
Yogurt	Netherlands	25 L

NUTRITIONAL GUIDES FROM AROUND THE WORLD

Nutritional guides around the world come in many different formats, illustrated as pyramids, pie charts, text and tables, yet they are similar in terms of content. Japan, like most countries, states the importance of every food group taken daily in moderation in order to achieve a well-balanced diet, but it emphasizes more carbohydrates than proteins and does not specifically highlight fermented foods as a category (Figure 1). In Canada and the USA, food guides have yogurt and kefir as recommended items listed under the dairy products section, but there is no emphasis on them being fermented foods, nor is there inclusion of fermented foods as a healthy category. The United Kingdom presents their Food Guide as a plate, with emphasis on carbohydrates, fruits and vegetables (Figure 2). The Swedish model for healthy eating, also in the form of a plate, has no section allotted to dairy products or any fermented foods. They only stress the importance of consuming foods that are low in fat and high in fiber at every meal.

Given the history of fermented foods in Asia, it is surprising that Japan and China, for example, do not recommend this as a category in their Food Guides. China's "food pagoda" stresses the importance of dietary balance and places crude wheat, rice, corn and sorghum cereals as "Level 1" for energy sources. The Chinese Nutrition Society (CNS) does suggest the use of yogurt for those who do not tolerate milk [16]. The one exception in Asia is India, whose Guide explicitly encourages the consumption of fermented foods. The National Institute of Nutrition's 2010 "Dietary Guidelines for Indians" document suggests specifically to pregnant women that they should: "eat more whole grains, sprouted grams and fermented foods" [17]. The document also describes the enhanced digestibility of fermented foods and increased nutritional value, through greater production of vitamins B and C. Again, fermented foods are encouraged later in the document when discussing various methods of food preparation. In Japan, probiotics are listed in "Food for specified health uses" (FOSHU), allowing labelling of health-promoting functions.

With few exceptions, fermented foods are generally absent as a recommended category of food for daily intake, in Food Guides. We believe this reflects a failure to appreciate the benefits resulting from the process of fermentation, which have been supported by numerous studies. When individual fermented foods such as yogurt are included, it is because of their nutritional value, such as high calcium levels.

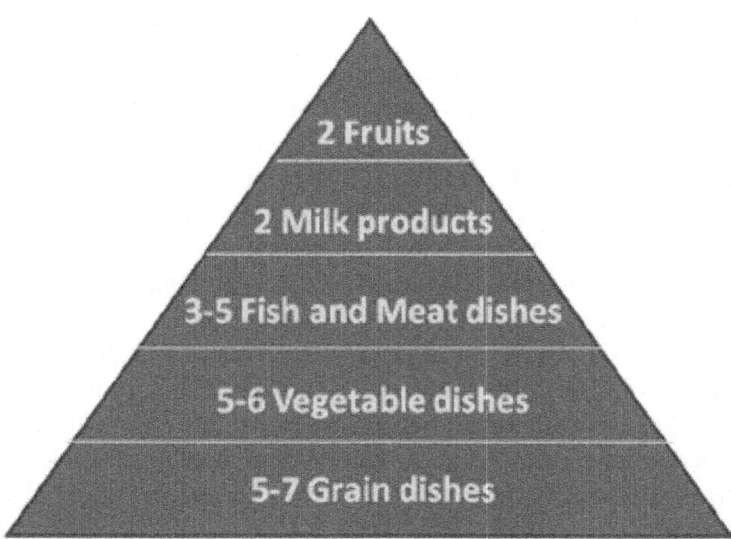

Figure 1: Breakdown of the food groups in the Japanese Food Guide pyramid, with portions per day.

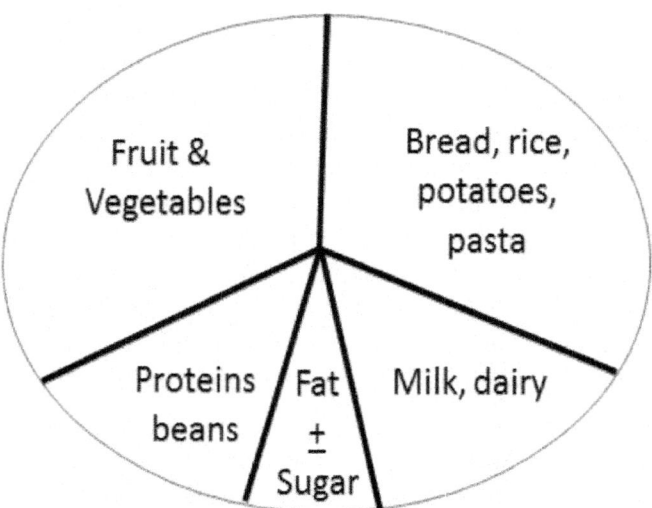

Figure 2: The pie or plate design illustrates the British Food Chart with emphasis on carbohydrates, fruit and vegetables, and no category of fermented food.

BENEFITS OF FERMENTED FOODS

Benefits of Fermented Dairy Products

Fermented foods and the microorganisms that contribute to the fermentation process have been associated with many beneficial effects on human health. Recent large cohort studies in the Netherlands and Sweden have examined the effects of regular consumption of fermented dairy products on the risk of bladder cancer [18] and cardiovascular disease [19]. Dairy products were divided according to those fermented and those that were not. In both studies, only fermented milk products were significantly associated with decreasing disease prevalence. In another large study on Danish participants, the effects of dairy products on periodontitis were examined [20]. It was reported that calcium intake specifically associated with fermented foods was inversely and significantly correlated with periodontitis, while calcium from other dairy foods was not. These findings emphasize the need to differentiate the types of dairy products, fermented and non-fermented, with regards to their health benefits, instead of promoting all dairy products, as is the case with many food guides.

The basis for fermented dairy products conferring health effects, in addition to the nutritional value of non-fermented milk, is multi-fold. The proteolysis that occurs in fermenting milk results in a higher content of peptides and free amino acids, especially cystine, histidine, and asparagine [21], which play various roles in health, and produce a more digestible food than milk *per se*. The breakdown of lactose concentration by the bacteria containing β-galactosidase, not only in the fermentation process but also in the stomach when the bacteria die and release this enzyme, then allows many lactose-intolerant individuals to consume the milk product. Lactose-free products are available for particularly sensitive individuals. Although the level of thermostable vitamins, niacin and pantothenic acid, are not destroyed by milk pasteurization that occurs prior to fermentation, some LAB can resynthesize folates, which are destroyed by the heat and have been shown to confer many health benefits [22,23]. Although there is some evidence that fermented foods alleviate constipation [24], studies using probiotic *B. lactis* DN-173 010 [25] or *L. casei* Shirota fermented milk [26] were no better than control products in showing a difference in constipation severity and stool frequency over a three or four-week period.

The two main health effects from fermented dairy consumption are immune and metabolic, especially with the addition of probiotic organisms. Fermented milk supplemented with probiotics can improve intestinal health, humoral and cell-mediated immunity [27], and salivary and fecal antibodies [28,29].

Some evidence suggests that this can lead to reduced incidence, or duration, of respiratory infections [30,31,32] presumably because the priming at the intestinal mucosal level impacts the lung's immune response. This is just one example of distant site health effects of probiotic fermented foods, but many others occur, including improvements to vaginal [33], bladder [34], bone [35], liver [36,37], body mass and blood pressure indices [38] and skin health [39]. Interestingly, a study of children from a low socioeconomic area of Argentina had no elevation of antibodies irrespective of vaccinations, suggesting that their system was already primed by exposure to pathogens [40].

The modulation of immune parameters is particularly challenging in two extreme conditions: inflammatory bowel disease (IBD) and human immunodeficiency virus (HIV) infection. The safe use of fermented foods in these types of extreme cases is important if they are to be recommended as part of a national food guide. As IBD is a Th1 immune response, treatment requires administration of anti-inflammatory therapy, such as antibodies to TNFα, or probiotic yogurt, which increases Treg cells [41]. Overall, the data are limited for remediation of IBD using probiotics, but some studies are supportive [42]. The Th2 immune response in HIV-infected patients associated with depletion of CD4+ and dendritic cells can lead to compromised epithelial repair mechanisms and enhanced epithelial permeability. Probiotic fermented foods can help maintain epithelial layer integrity [43,44], and reduce the loss of CD4+ cells in HIV patients [45,46]. In the latter group, maintenance of gut barrier function also helps reduce bacterial translocation that can cause serious infectious complications to the immune-suppressed host.

Benefits of Fermented Foods in Vulnerable Populations

Food guides are designed for the general public, not hospitalized patients. Nevertheless, the studies showing benefits of probiotic food to aid recovery from organ transplantation and abdominal surgery [47,48,49,50] further demonstrate safety and effectiveness. An important safety study of infants randomly assigned to receive probiotics or placebo for a total of five months, starting two months prior to vaccination, showed no adverse interference with the immune response to mumps, measles, rubella and varicella vaccine [51].

The high prevalence of allergies affecting skin, gut and respiratory tract, have led to probiotics being tested in humans as a means of prevention or treatment. In a nested unmatched case-control study, 237 infants were given probiotics prenatally and at 6 months of age. By age 2, the risk of eczema, food allergy, asthma, and rhinitis was assessed [52]. In infants with high fecal IgA concentration at 6 months, the risk of having any allergic disease tended to reduce (odds ratio (OR): 0.52), as did the risk for any IgE-associated (atopic) disease (OR: 0.49). High fecal calprotectin at the age of 6 months was associated with lower risk for IgE-associated diseases (OR: 0.49). This study showed the broad potential of probiotic food against allergy. In a study of almost 200 infants aged 4 to 13 months, use of probiotic cereal showed signs of preventing early manifestation of allergy, and the higher Th1/Th2 ratio suggested an effect on the T-cell-mediated immune response [53]. It is not unexpected that foods are less able to treat conditions like atopic dermatitis [54], but their demonstrated safety in such infants is reassuring. Likewise, in adults prone to seasonal grass and ragweed allergies, a pilot study showed some markers of success and no adverse responses [55].

The increasing diabetes rates amongst pregnant women and children have led to consideration of probiotic intervention. In a study undertaken in Iran, daily consumption of probiotic yogurt for 9 weeks maintained serum insulin levels, potentially preventing pregnant women from developing insulin resistance [56]. Preservation of insulin sensitivity [57] and controlling glycemic index are very important for patients at risk of, or suffering from, diabetes [58,59]. A recent study with an impressive nested case-cohort and a random sub-cohort of 4000 subjects followed-up for 11 years showed that greater low-fat fermented dairy product intake was associated with a decreased risk of type 2 diabetes development [60]. Thus, with pre-diabetes and type II diabetes affecting several million Canadians, the inclusion of fermented foods in their diet could be significant.

Considering that arthritis is even more prevalent, affecting almost five million Canadians, and the influence that the microbiome has on arthritis [61], it is worth investigating whether fermented foods and probiotics could alleviate pain, swelling and discomfort. Reactive arthritis is known to be triggered in some patients by Gram-negative gastrointestinal infection. A mouse study showed that consumption of probiotic fermented milk prevented *Salmonella*-induced synovitis by altering the intestinal milieu necessary for differentiation of cells involved in the generation of joint inflammation [62]. A small human study on rheumatoid arthritis, in patients with at least four swollen and four tender joints and stable medications with no steroids for at least one month prior to and during the study, showed a significant improvement in the Health Assessment Questionnaire score after three months of probiotic treatment [63]. This illustrates that probiotics recommended in a food guide, will not cause harm even in subjects who might consume such products while suffering from common arthritic disease.

A large portion of populations in developed countries take a range of prescription medications, most of which confer adverse effects of one sort or another. Indeed, large numbers of patients admitted to emergency clinics do so following drug complications, with agents used to manage cardiovascular disease being the main culprits [64,65]. As fermented foods are known to provide benefits to managing some cardiovascular risk factors [66] and potentially even improve recovery post-infarction [67], the reduction by probiotic food intake of drug side effects [68,69], especially antibiotics [70,71], further suggests benefits, and no harm, if taken by a general population, some of whom are receiving drug therapy.

The intake of LAB can help to reduce the load of pathogens, even in the nasal cavity [72]. This may be an unusual attribute with respect to a food, but since *Staphylococcus aureus*, *Streptococcus pneumoniae*, and β-hemolytic streptococci are major causes of disease, the ability of probiotic yogurt to deplete them should be seen as a positive outcome, and therefore something aligned with the expectations from any recommended food type.

Other Fermented Foods and Benefits

Not every consumer eats dairy products, and not every fermented food requires use of milk. Kimchi, fermented vegetables, is a popular side dish originating in Korea that has been associated with numerous health benefits, including prevention of cancer and obesity, reduction in cholesterol levels and immune system promotion [73]. A study that looked at the preventative effects of kimchi, fresh and fermented, in pre-diabetic individuals showed promising results. Insulin resistance decreased while insulin sensitivity increased, and overall, glucose tolerance improved by 33%, compared to only a 9.5% increase in those receiving fresh non-fermented control [74].

The long history and wide diversity of fermented foods across African countries attests to the benefits they have accrued over many generations [75]. Studies have attributed benefits to include prevention of diarrhea and constipation [76]. Researchers there, and in other developing countries with a history of fermented food production, have been examining the properties of strains in their products. A *Lactobacillus plantarum* strain isolated from the common Indian fermented food Dosa, has been shown to inhibit the growth of a range of food-borne pathogens [77].

Adverse Effects of Fermented Foods

In parts of Asia, fermented fish sauce is widely consumed. A study performed in the Chaoshan area of China, showed an increased risk of squamous cell carcinoma of the esophagus in habitual consumers of fermented fish sauce [75]. Another Chinese study showed that *N*-nitroso compounds and genotoxins present before and after nitrosation, appear to be responsible for the cancer risk

[76]. An Egyptian study also found high levels of histamine in fermented fish [77].

CONCLUSIONS AND RECOMMENDATIONS

The expansive use of, and benefits gained from, fermented foods supports their greater inclusion in Food Guides around the world. They have long been a part of the human diet, and with further supplementation of probiotic microbes in some cases, they offer nutritional and health attributes worthy of recommendation of regular consumption. It is hoped that this review contributes to policy changes and increases the inclusion of fermented foods when Food Guides are next revised. This might, for now, exclude fermented fish consumed in parts of Asia. It would be a great detriment to human health if fermented food use were to decline, as has been noted in parts of Africa, through lack of generational transfer of knowledge, poor availability and affordability of probiotics [78,79,80,81], and introduction of food and drink products high in certain sugars [82,83,84].

AUTHOR CONTRIBUTIONS

Stephanie N. Chilton drafted and helped revise the manuscript. Jeremy P. Burton provided input during the writing process. Gregor Reid conceived the idea, reviewed and revised the manuscript.

CONFLICTS OF INTEREST

The authors declare no conflict of interest.

REFERENCES

1. Various, A. Regimen for Health. In *Hippocratic Writings*; Lloyd, G.E.R., Chadwick, J., Mann, W.N., Eds.; Penguin Books Ltd.: London, UK, 1983; pp. 154–196.

2. Canada's Food Guides from 1942 to 1992. Available online: http://www.hc-sc.gc.ca/fn-an/food-guide-aliment/context/fg_history-histoire_ga-eng.php#fnb1 (accessed on 12 October 2014).

3. Walker, P.M.B. *Chambers Science and Technology Dictionary*; Cambridge University Press: Cambridge, UK, 1988.

4. Health and Nutritional Properties of Probiotics in Food Including Powder Milk with Live Lactic Acid Bacteria. Available online: ftp://ftp.fao.org/docrep/fao/009/a0512e/a0512e00.pdf (accessed on 23 September 2014).

5. Masood, M.I.; Qadir, M.I.; Shirazi, J.H.; Khan, I.U. Beneficial effects of lactic acid bacteria on human beings. *Crit. Rev. Microbiol.* **2011**, *37*, 91–98.

6. Leroy, F.; de Vuyst, L. Lactic acid bacteria as functional starter cultures for the food fermentation industry. *Trends Food Sci. Technol.* **2004**, *15*, 67–78.

7. Parvez, S.; Malik, K.A.; Kang, S.A.; Kim, H.-Y. Probiotics and their fermented food products are beneficial for health.*J. Appl. Microbiol.* **2006**, *100*, 1171–1185.

8. Zeng, Z.; Lin, J.; Gong, D. Identification of lactic acid bacterial strains with high conjugated linoleic acid-producing ability from natural sauerkraut fermentations. *J. Food Sci.* **2009**, *74*, 154–158.

9. Carasi, P.; Diaz, M.; Racedo, S.M.; de Antoni, G.; Urdaci, M.C.; Serradell Mde, L. Safety characterization and antimicrobial properties of kefir-isolated *Lactobacillus kefiri*. *Biomed. Res. Int.* **2014**, *2014*.

10. Anukam, K.C.; Reid, G. African traditional fermented foods and probiotics. *J. Med. Food* **2009**, *12*, 1177–1184.

11. Rhee, S.J.; Lee, J.E.; Lee, C.H. Importance of lactic acid bacteria in Asian fermented foods. *Microb. Cell Fact.* **2011**, *10*.

12. Sicard, D.; Legras, J.L. Bread, beer and wine: Yeast domestication in the *Saccharomyces sensu stricto* complex. *C. R. Biol.***2011**, *334*, 229–236.

13. Farnworth, E.R. *Handbook of Fermented Functional Foods*; CRC Press: Boca Raton, FL, USA, 2008; pp. 1–602. [Google Scholar]

14. Fermented Fruits and Vegetables: A Global Perspective. Available online: http://www.fao.org/docrep/x0560e/x0560e00.htm (accessed on 23 September 2014).

15. Fermented Foods. Available online: http://www.nzifst.org.nz/myfiles/ Expt_6.ppt (accessed on 23 September 2014).

16. Pagoda Illustration. Available online: http://www.cnsoc.org/en/nutrition. asp?s=9&nid=806 (accessed on 23 September 2014).

17. Dietary Guidelines for Indians: A Manual. Available online: http:// nininindia.org/DietaryguidelinesforIndians-Finaldraft.pdf (accessed on 23 September 2014).

18. Keszei, A.P.; Schouten, L.J.; Goldbohm, A.; van den Brandt, P.A. Dairy intake and the risk of bladder cancer in the Netherlands cohort study on diet and cancer. *Am. J. Epidemiol.* **2009**, *171*, 436–446.

19. Sonedstedt, E.; Wirfält, E.; Wallstrom, P.; Gullberg, B.; Orho-Melander, M.; Hedblad, B. Dairy products and its association with incidence

of cardiovascular disease: The Malmö diet and cancer cohort. *Eur. J. Epidemiol.* **2011**, *26*, 609–618.

20. Adegboye, A.R.; Christensen, L.B.; Holm-Pedersen, P.; Avlund, K.; Boucher, B.J.; Heitmann, B.L. Intake of dairy products in relation to periodontitis in older Danish adults. *Nutrients* **2012**, *4*, 1219–1229.

21. Ghosh, D.; Chattorai, D.K.; Chattopadhyay, P. Studies on changes in microstructure and proteolysis in cow and soy milk curd during fermentation using lactic cultures for improving protein bioavailability. *J. Food Sci. Technol.* **2013**, *50*, 979–985.

22. Chen, P.; Li, C.; Li, X.; Li, J.; Chu, R.; Wang, H. Higher dietary folate intake reduces the breast cancer risk: A systematic review and meta-analysis. *Br. J. Cancer* **2014**, *110*, 2327–2338.

23. Morse, N.L. Benefits of docosahexaenoic acid, folic acid, vitamin D and iodine on foetal and infant brain development and function following maternal supplementation during pregnancy and lactation. *Nutrients* **2012**, *4*, 799–840.

24. Pitkala, K.H.; Strandberg, T.E.; Finne Soveri, U.H.; Ouwehand, A.C.; Poussa, T.; Salminen, S. Fermented cereal with specific bifidobacteria normalizes bowel movements in elderly nursing home residents. A randomized, controlled trial. *J. Nutr. Health Aging* **2007**, *11*, 305–311.

25. Tabbers, M.M.; Chmielewska, A.; Roseboom, M.G.; Crastes, N.; Perrin, C.; Reitsma, J.B.; Norbruis, O.; Szajewska, H.; Benninga, M.A. Fermented milk containing *Bifidobacterium lactis* DN-173 010 in childhood constipation: A randomized, double-blind, controlled trial. *Pediatrics* **2011**, *127*, 1392–1399.

26. Mazlyn, M.M.; Nagarajah, L.H.; Fatimah, A.; Norimah, A.K.; Goh, K.L. Effects of a probiotic fermented milk on functional constipation: A randomized, double-blind, placebo-controlled study. *J. Gastroenterol. Hepatol.* **2013**, *28*, 1141–1147.

27. Wang, S.; Zhu, H.; Lu, C.; Kang, Z.; Luo, Y.; Feng, L.; Lu, X. Fermented milk supplemented with probiotics and prebiotics can effectively alter the intestinal microbiota and immunity of host animals. *J. Dairy Sci.* **2012**, *95*, 4813–4822.

28. Surono, I.S.; Koestomo, F.P.; Novitasari, N.; Zakaria, F.R. Novel probiotic *Enterococcus faecium* IS-27526 supplementation increased total salivary sIgA level and bodyweight of pre-school children: A pilot study. *Anaerobe* **2011**, *17*, 496–500.

29. Campeotto, F.; Suau, A.; Kapel, N.; Magne, F.; Viallon, V.; Ferraris, L.; Waligora-Dupriet, A.J.; Soulaines, P.; Leroux, B.; Kalach, N.; *et al.* A fermented formula in pre-term infants: Clinical tolerance, gut microbiota, down-regulation of faecal calprotectin and up-regulation of faecal secretory IgA. *Br. J. Nutr.* **2011**, *105*, 1843–1851.

30. Guillemard, E.; Tondu, F.; Lacoin, F.; Schrezenmeir, J. Consumption of a fermented dairy product containing the probiotic *Lactobacillus casei* DN-114001 reduces the duration of respiratory infections in the elderly in a randomised controlled trial. *Br. J. Nutr.* **2010**, *103*, 58–68.

31. Makino, S.; Ikegami, S.; Kume, A.; Horiuchi, H.; Sasaki, H.; Orii, N. Reducing the risk of infection in the elderly by dietary intake of yoghurt fermented with *Lactobacillus delbrueckii* ssp. bulgaricus OLL1073R-1. *Br. J. Nutr.* **2010**, *104*, 998–1006.

32. De Vrese, M.; Winkler, P.; Rautenberg, P.; Harder, T.; Noah, C.; Laue, C.; Ott, S.; Hampe, J.; Schreiber, S.; Heller, K.; *et al.* Probiotic bacteria reduced duration and severity but not the incidence of common cold episodes in a double blind, randomized, controlled trial. *Vaccine* **2006**, *24*, 6670–6674.

33. Gardiner, G.; Heinemann, C.; Baroja, M.L.; Bruce, A.W.; Beuerman, D.; Madrenas, J.; Reid, G. Oral administration of the probiotic combination *Lactobacillus rhamnosus* GR-1 and *L. fermentum* RC-14 for human intestinal applications. *Int. Dairy J.* **2002**, *12*, 191–196.

34. Ohashi, Y.; Nakai, S.; Tsukamoto, T.; Masumori, N.; Akaza, H.; Miyanaga, N.; Kitamura, T.; Kawabe, K.; Kotake, T.; Kuroda, M.; *et al.* Habitual intake of lactic acid bacteria and risk reduction of bladder cancer. *Urol. Int.* **2002**, *68*, 273–280.

35. Narva, M.; Nevala, R.; Poussa, T.; Korpela, R. The effect of *Lactobacillus helveticus* fermented milk on acute changes in calcium metabolism in postmenopausal women. *Eur. J. Nutr.* **2004**, *43*, 61–68.

36. Higashikawa, F.; Noda, M.; Awaya, T.; Nomura, K.; Oku, H.; Sugiyama, M. Improvement of constipation and liver function by plant-derived lactic acid bacteria: A double-blind, randomized trial. *Nutrition* **2010**, *26*, 367–374.

37. Moroti, C.; Souza Magri, L.F.; de Rezende Costa, M.; Cavallini, D.C.; Sivieri, K. Effect of the consumption of a new symbiotic shake on glycemia and cholesterol levels in elderly people with type 2 diabetes mellitus. *Lipids Health Dis.* **2012**, *11*, 29.

38. Sharafedtinov, K.K.; Plotnikova, O.A.; Alexeeva, R.I.; Sentsova, T.B.; Songisepp, E.; Stsepetova, J.; Smidt, I.; Mikelsaar, M. Hypocaloric diet

supplemented with probiotic cheese improves body mass index and blood pressure indices of obese hypertensive patients—A randomized double-blind placebo-controlled pilot study. *Nutr. J.* **2013**, *12*, 138.

39. Peguet-Navarro, J.; Dezutter-Dambuyant, C.; Buetler, T.; Leclaire, J.; Smola, H.; Blum, S.; Bastien, P.; Breton, L.; Gueniche, A. Supplementation with oral probiotic bacteria protects human cutaneous immune homeostasis after UV exposure-double blind, randomized, placebo controlled clinical trial. *Eur. J. Dermatol.* **2008**, *18*, 504–511.

40. Pérez, N.; Iannicelli, J.C.; Girard-Bosch, C.; González, S.; Varea, A.; Disalvo, L.; Apezteguia, M.; Pernas, J.; Vicentin, D.; Cravero, R.; *et al.* Effect of probiotic supplementation on immunoglobulins, isoagglutinins and antibody response in children of low socio-economic status. *Eur. J. Nutr.* **2010**, *49*, 173–179.

41. Baroja, M.L.; Kirjavainen, P.V.; Hekmat, S.; Reid, G. Anti-inflammatory effects of probiotic-yogurt in inflammatory bowel disease patients. *Clin. Exp. Immunol.* **2007**, *149*, 470–479.

42. Jonkers, D.; Penders, J.; Masclee, A.; Pierik, M. Probiotics in the management of inflammatory bowel disease: A systematic review of intervention studies in adult patients. *Drugs* **2012**, *72*, 803–823.

43. Orlando, A.; Linsalata, M.; Notarnicola, M.; Tutino, V.; Russo, F. *Lactobacillus* GG restoration of the gliadin induced epithelial barrier disruption: The role of cellular polyamines. *BMC Microbiol.* **2014**, *14*, 19.

44. Persborn, M.; Gerritsen, J.; Wallon, C.; Carlsson, A.; Akkermans, L.M.; Söderholm, J.D. The effects of probiotics on barrier function and mucosal pouch microbiota during maintenance treatment for severe pouchitis in patients with ulcerative colitis. *Aliment. Pharmacol. Ther.* **2013**, *38*, 772–783.

45. Anukam, K.C.; Osazuwa, E.O.; Osadolor, B.E.; Bruce, A.W.; Reid, G. Yogurt containing probiotic *Lactobacillusrhamnosus* GR-1 and *L. reuteri* RC-14 helps resolve moderate diarrhea and increases CD4 count in HIV/AIDS patients.*J. Clin. Gastroenterol.* **2008**, *42*, 239–243.

46. Irvine, S.L.; Hummelen, R.B.S.; Hekmat, S.; Looman, C.; Changalucha, J.; Habbema, D.F.; Reid, G. Probiotic yogurt consumption is associated with an increase of CD4 count among people living with HIV/AIDS. *J. Clin. Gastroenterol.***2010**, *44*, 201–205.

47. Rayes, N.; Seehofer, D.; Hansen, S.; Boucsein, K.; Müller, A.R.; Serke, S.; Bengmark, S.; Neuhaus, P. Early enteral supply of *Lactobacillus* and

fiber *versus* selective bowel decontamination: A controlled trial in liver transplant recipients. *Transplantation* **2002**, *74*, 123–127.

48. Rayes, N.; Seehofer, D.; Müller, A.R.; Hansen, S.; Bengmark, S.; Neuhaus, P. Influence of probiotics and fibre on the incidence of bacterial infections following major abdominal surgery—Results of a prospective trial. *Z. Gastroenterol.* **2002**, *40*, 869–876.

49. Rayes, N.; Hansen, S.; Seehofer, D.; Müller, A.R.; Serke, S.; Bengmark, S.; Neuhaus, P. Early enteral supply of fiber and lactobacilli *versus* conventional nutrition: A controlled trial in patients with major abdominal surgery. *Nutrition* **2002**, *18*, 609–615.

50. Liu, Z.H.; Huang, M.J.; Zhang, X.W.; Wang, L.; Huang, N.Q.; Peng, H.; Lan, P.; Peng, J.S.; Yang, Z.; Xia, Y.; *et al.* The effects of perioperative probiotic treatment on serum zonulin concentration and subsequent postoperative infectious complications after colorectal cancer surgery: A double-center and double-blind randomized clinical trial. *Am. J. Clin. Nutr.* **2013**, *97*, 117–126.

51. Youngster, I.; Kozer, E.; Lazarovitch, Z.; Broide, E.; Goldman, M. Probiotics and the immunological response to infant vaccinations: A prospective, placebo controlled pilot study. *Arch. Dis. Child.* **2011**, *96*, 345–349.

52. Kukkonen, K.; Kuitunen, M.; Haahtela, T.; Korpela, R.; Poussa, T.; Savilahti, E. High intestinal IgA associates with reduced risk of IgE-associated allergic diseases. *Pediatr. Allergy Immunol.* **2010**, *21*, 67–73.

53. West, C.E.; Hammarström, M.L.; Hernell, O. Probiotics during weaning reduce the incidence of eczema. *Pediatr. Allergy Immunol.* **2009**, *20*, 430–437.

54. Grüber, C.; Wendt, M.; Sulser, C.; Lau, S.; Kulig, M.; Wahn, U.; Werfel, T.; Niggemann, B. Randomized, placebo-controlled trial of *Lactobacillus rhamnosus* GG as treatment of atopic dermatitis in infancy. *Allergy* **2007**, *62*, 1270–1276.

55. Koyama, T.; Kirjavainen, P.V.; Fisher, C.; Anukam, K.; Summers, K.; Hekmat, S.; Reid, G. Development and pilot evaluation of a novel probiotic mixture for the management of seasonal allergic rhinitis. *Can. J. Microbiol.* **2010**, *56*, 730–738.

56. Asemi, Z.; Samimi, M.; Tabassi, Z.; Naghibi Rad, M.; Rahimi Foroushani, A.; Khorammian, H.; Esmaillzadeh, A. Effect of daily consumption of probiotic yoghurt on insulin resistance in pregnant women: A randomized controlled trial. *Eur. J. Clin. Nutr.* **2013**, *67*, 71–74.

57. Andreasen, A.S.; Larsen, N.; Pedersen-Skovsgaard, T.; Berg, R.M.; Møller, K.; Svendsen, K.D.; Jakobsen, M.; Pedersen, B.K. Effects of *Lactobacillus acidophilus* NCFM on insulin sensitivity and the systemic inflammatory response in human subjects. *Br. J. Nutr.* **2010**, *104*, 1831–1838.

58. Granfeldt, Y.E.; Björck, I.M. A bilberry drink with fermented oatmeal decreases postprandial insulin demand in young healthy adults. *Nutr. J.* **2011**, *10*, 57.

59. De Angelis, M.; Rizzello, C.G.; Alfonsi, G.; Arnault, P.; Cappelle, S.; di Cagno, R.; Gobbetti, M. Use of sourdough lactobacilli and oat fibre to decrease the glycaemic index of white wheat bread. *Br. J. Nutr.* **2007**, *98*, 1196–1205.

60. O'Connor, L.M.; Lentjes, M.A.; Luben, R.N.; Khaw, K.T.; Wareham, N.J.; Forouhi, N.G. Dietary dairy product intake and incident type 2 diabetes: A prospective study using dietary data from a 7-day food diary. *Diabetologia* **2014**, *57*, 909–917.

61. Yeoh, N.; Burton, J.P.; Suppiah, P.; Reid, G.; Stebbings, S. The role of the microbiome in rheumatic diseases. *Curr. Rheumatol. Rep.* **2013**, *15*, 314.

62. Noto Llana, M.; Sarnacki, S.H.; Aya Castañeda Mdel, R.; Bernal, M.I.; Giacomodonato, M.N.; Cerquetti, M.C. Consumption of *Lactobacillus casei* fermented milk prevents *Salmonella* reactive arthritis by modulating IL-23/IL-17 expression. *PLoS One* **2013**, *8*.

63. Pineda Mde, L.; Thompson, S.F.; Summers, K.; de Leon, F.; Pope, J.; Reid, G. A randomized, double-blinded, placebo-controlled pilot study of probiotics in active rheumatoid arthritis. *Med. Sci. Monit.* **2011**, *17*, 347–354.

64. Dechanont, S.; Maphanta, S.; Butthum, B.; Kongkaew, C. Hospital admissions/visits associated with drug-drug interactions: A systematic review and meta-analysis. *Pharmacoepidemiol. Drug Saf.* **2014**, *23*, 489–497.

65. Budnitz, D.S.; Lovegrove, M.C.; Shehab, N.; Richards, C.L. Emergency hospitalizations for adverse drug events in older Americans. *N. Engl. J. Med.* **2011**, *365*, 2002–2012.

66. Plana, N.; Nicolle, C.; Ferre, R.; Camps, J.; Cos, R.; Villoria, J.; Masana, L.; DANACOL Group. Plant sterol-enriched fermented milk enhances the attainment of LDL-cholesterol goal in hypercholesterolemic subjects. *Eur. J. Nutr.* **2008**, *47*, 32–39.

67. Gan, X.T.; Ettinger, G.; Huang, C.X.; Burton, J.P.; Haist, J.V.; Rajapurohitam, V.; Sidaway, J.E.; Martin, G.; Gloor, G.B.; Swann, J.R.; *et al*. Probiotic administration attenuates myocardial hypertrophy and heart failure after myocardial infarction in the rat. *Circ. Heart Fail.* **2014**, *7*, 491–499.

68. Li, S.; Huang, X.L.; Sui, J.Z.; Chen, S.Y.; Xie, Y.T.; Deng, Y.; Wang, J.; Xie, L.; Li, T.J.; He, Y.; *et al*. Meta-analysis of randomized controlled trials on the efficacy of probiotics in *Helicobacter pylori* eradication therapy in children. *Eur. J. Pediatr.* **2014**, *173*, 153–161.

69. Yamashiro, Y.; Nagata, S. Beneficial microbes for premature infants, and children with malignancy undergoing chemotherapy. *Benef. Microbes* **2010**, *1*, 357–365.

70. Beausoleil, M.; Fortier, N.; Guénette, S.; L'ecuyer, A.; Savoie, M.; Franco, M.; Lachaine, J.; Weiss, K. Effect of a fermented milk combining *Lactobacillus acidophilus* Cl1285 and *Lactobacillus casei* in the prevention of antibiotic-associated diarrhea: A randomized, double-blind, placebo-controlled trial. *Can. J. Gastroenterol.* **2007**, *21*, 732–736.

71. Wenus, C.; Goll, R.; Loken, E.B.; Biong, A.S.; Halvorsen, D.S.; Florholmen, J. Prevention of antibiotic-associated diarrhoea by a fermented probiotic milk drink. *Eur. J. Clin. Nutr.* **2008**, *62*, 299–301.

72. Glück, U.; Gebbers, J.O. Ingested probiotics reduce nasal colonization with pathogenic bacteria (*Staphylococcus aureus,Streptococcus pneumoniae*, and beta-hemolytic streptococci). *Am. J. Clin. Nutr.* **2003**, *77*, 517–520.

73. Park, K.Y.; Jeong, J.K.; Lee, Y.E.; Daily, J.W. Health benefits of kimchi (Korean fermented vegetables) as a probiotic food. *J. Med. Food* **2014**, *17*, 6–20.

74. An, S.Y.; Lee, M.S.; Jeon, J.Y.; Ha, E.S.; Kim, T.H.; Yoon, J.Y.; Ok, C.O.; Lee, H.K.; Hwang, W.S.; Choe, S.J.; *et al*. Beneficial effects of fresh and fermented kimchi in prediabetic individuals. *Ann. Nutr. MeTable* **2013**, *63*, 111–119.

75. Ke, L.; Yu, P.; Zhang, Z.X. Novel epidemiologic evidence for the association between fermented fish sauce and esophageal cancer in South China. *Int. J. Cancer* **2002**, *99*, 424–426.

76. Chen, C.S.; Pignatelli, B.; Malaveille, C.; Bouvier, G.; Shuker, D.; Hautefeuille, A.; Zhang, R.F.; Bartsch, H. Levels of direct-acting mutagens, total *N*-nitroso compounds in nitrosated fermented fish products, consumed in a high-risk area for gastric cancer in southern China. *Mutat. Res.* **1992**, *265*, 211–221.

77. Rabie, M.A.; Elsaidy, S.; el-Badawy, A.A.; Siliha, H.; Malcata, F.X. Biogenic amine contents in selected Egyptian fermented foods as determined by ion-exchange chromatography. *J. Food Prot.* **2011**, *74*, 681–685.

78. Franz, C.M.; Huch, M.; Mathara, J.M.; Abriouel, H.; Benomar, N.; Reid, G.; Galvez, A.; Holzapfel, W.H. African fermented foods and probiotics. *Int. J. Food Microbiol.* **2014**, *190*, 84–96.

79. Mathara, J.M.; Schillinger, U.; Kutima, P.M.; Mbugua, S.A.; Holzapfel, W.H. Isolation, Identification and characterization of the dominant microorganisms of *kule naoto*: The Maasai traditional fermented milk in Kenya. *Int. J. Food Microbiol.* **2004**, *94*, 269–278.

80. Gupta, A.; Tiwari, S.K. Probiotic potential of *Lactobacillus plantarum* LD1 isolated from batter of Dosa, a South Indian fermented food. *Probiotics Antimicrob. Proteins* **2014**, *6*, 73–81.

81. Watson, F.E.; Ngesa, A.; Onyang'o, J.; Alnwick, D.; Tomkins, A.M. Fermentation—A traditional anti-diarrhoeal practice lost? The use of fermented foods in urban and rural Kenya. *Int. J. Food Sci. Nutr.* **1996**, *47*, 171–179.

82. Kelishadi, R.; Mansourian, M.; Heidari-Beni, M. Association of fructose consumption and components of metabolic syndrome in human studies: A systematic review and meta-analysis. *Nutrition* **2014**, *30*, 503–510.

83. Chan, T.F.; Lin, W.T.; Chen, Y.L.; Huang, H.L.; Yang, W.Z.; Lee, C.Y.; Chen, M.H.; Wang, T.N.; Huang, M.C.; Chiu, Y.W.; *et al.* Elevated serum triglyceride and retinol-binding protein 4 levels associated with fructose-sweetened beverages in adolescents. *PLoS One* **2014**, *9*.

84. Walker, R.W.; Dumke, K.A.; Goran, M.I. Fructose content in popular beverages made with and without high-fructose corn syrup. *Nutrition* **2014**, *30*, 928–935.

Chapter 2

LACTIC ACID BACTERIA IN PHILIPPINE TRADITIONAL FERMENTED FOODS

Charina Gracia B. Banaay[1], Marilen P. Balolong[2,3] and Francisco B. Elegado[4]

[1]Institute of Biological Sciences, College of Arts and Sciences, University of the Philippines Los Baños, Laguna, Philippines

[2]Department of Biology, University of the Philippines Manila, Padre Faura Street, Ermita, Manila, Philippines

[3]Institute of Molecular Biology and Biotechnology, National Institutes of Health, University of the Philippines Manila, Pedro Gil Street, Ermita, Manila, Philippines

[4]National Institute of Molecular Biology and Biotechnology, University of the Philippines Los Baños, Laguna, Philippines

INTRODUCTION

The Philippine archipelago is home to a diverse array of ecosystems, organisms, peoples, and cultures. Filipino cuisine is no exception as distinct regional flavors stem from the unique food preparation techniques and culinary traditions of each region. Although Philippine indigenous foods are reminiscent of various foreign influences, local processes are adapted to indigenous ingredients and in accordance with local tastes. Pervasive throughout the numerous islands of the Philippines is the use of fermentation to enhance the organoleptic qualities as well as extend the shelf-life of food.

Traditional or indigenous fermented foods are part and parcel of Filipino culture since these are intimately entwined with the life of local people. The three main island-groups of the Philippines, namely – Luzon, Visayas, and Mindanao, each have their own fermented food products that cater to the local palate. Fermentation processes employed in the production of these indigenous fermented foods often rely entirely on natural microflora of the raw material and the surrounding environment; and procedures are handed down from one generation to the next as a village-art process. Because traditional food fermentation industries are commonly home-based and highly reliant on indigenous materials without the benefit of using commercial starter cultures, microbial assemblages are unique and highly variable per product and per region. Hence the possibility of discovering novel organisms, products, and interactions are likely.

Various microorganisms are involved in common food fermentation processes. In particular, lactic acid bacteria (LAB) in food is a type of biopreservation system. They not only contribute to the flavor of the food but LAB are also able to control pathogenic and spoilage microorganisms through various ways that include, but are not limited to, production of peroxidases, organic acids, and bacteriocins. Traditionally, identification of LAB in foods is largely dependent on culture-based methods; and properties of each isolate are evaluated under controlled conditions. However, with the advent of molecular techniques, the enumeration of microorganisms missed by culture-dependent methods is now possible. Also, as more LAB metabolites, such as bacteriocins, are being reported, a wider database for identification and comparison with potential novel products are now available.

As the production and consumption of traditional fermented food products become increasingly relevant in the face of rapidly increasing population and food insecurity, more research and development to ensure the safety and nutritional quality of these fermented products is warranted. For a more extensive discussion of the principles and technology of Philippine fermented foods, the readers are directed toSanchez (2008). This book is a detailed reference based on decades of research. Some data from the book will be presented again here in addition to other data from more recent studies. It is not the intention of this present paper to repeat what has been presented in the book, especially regarding fermentation processes, but only to present, as complete as possible, the data that are available regarding LAB present in indigenous/traditional fermented foods.

This paper aims to briefly review the various lactic acid-fermented indigenous fermented specialties in the different regions of the Philippines. Majority of the discussion will focus on recent data gathered from bacteriocin research and metagenomics studies of Philippine fermented specialties. Lastly, the health applications of the different fermented food products and their development as functional foods will be evaluated.

REGIONAL FERMENTED SPECIALTIES IN THE PHILIPPINES

There are various lactic acid-fermented indigenous food products in the Philippines. Table 1 gives a summary of these different fermented specialties found in the different regions. Although a particular product type can be seen throughout the whole country, the texture, taste, and appearance would vary depending on the local taste, materials used, and process employed. For example, bagoong is a common fermented fish paste found all over the Philippines but the characteristic of the product found in Luzon is different from that found in the Visayas and Mindanao regions. Bagoong also takes on different names; there is bagoong na isda, bagoong alamang, bagoong na sisi, and guinamos (Sanchez, 2008). A product that is processed in a similar manner is dayok; it is made of brined fish entrails. Research indicates that this is also a lactic acid-fermented food but the LAB involved have not been identified yet (Besas and Dizon, 2012). Longanisa is sausage made of beef, pork, or chicken. It also takes on many forms depending on where it is made. The more famous ones are Vigan Longanisa in Northern Luzon, Pampanga Longanisa in Central Luzon, Lucban Longanisa in Southern Luzon, and Cebu Longanisa in the Visayas. The tastes vary from spicy, garlicky, sour, to sweet.

In lactic acid-fermented foods, LAB are important in preventing the growth of spoilage organisms, and altering flavor, aroma, and texture of the product. Although LAB are initially present in low numbers in the raw materials used, they soon proliferate as other organisms are inhibited by the initial addition of salt and as the continuous growth of LAB decreases the pH of the food making it less conducive for growth of other organisms. Recent studies, however, have shown that there are a lot more benefits that can be derived from LAB in traditional fermented foods.

Table 1: Regional Lactic Acid-Fermented Specialties in the Philippines

CATEGORY	PRODUCT NAME	REGION	MAJOR INGREDIENTS	LACTIC ACID BACTERIA INVOLVED (as determined from culture-based methods)	APPEARANCE AND/OR USAGE
Fermented vegetables, fruits	Burong mustasa	Luzon	Mustard leaves, cooked rice and/or rice washings	*Leuconostoc mesenteroides, Enterococcus faecalis, Lactobacillus plantarum*	Side dish
	Burong pipino	Whole Phil	cucumber	*Leu. mesenteroides, L. brevis, Pediococcus cerevisiae, L. plantarum*	Side dish
	Burong mangga	Whole Phil	Immature mango	*Leu. mesenteroides, L. brevis, P. cerevisiae, L. plantarum*	Side dish
	Atchara	Whole Phil	Immature papaya or chayote, or turnip (singkamas)	Unknown	Side dish
Cheese	Kesong puti	Luzon, Visayas	Cow or carabao milk	*Lactococcus lactis*	White soft cheese
Fermented fish and fishery products	Balao-balao	Luzon	Cooked rice, shrimp, salt	*Leu. mesenteroides, P. cerevisiae, L. plantarum*	Side dish, condiment
	Burong-isda	Luzon	Freshwater fish, rice, salt	*Leu. mesenteroides, E. faecalis, P. cerevisiae, L. plantarum, P. acidilactici, Leu. paramesenteroides*	Side dish, condiment
	Tinabal	Visayas	Parrot fish (for tinabal molmol) and frigate fish (for tinabal mangko), salt	*P. pentosaceus, S. equinus, Leuconostoc sp., Lactobacillus sp.*	Side dish, viand
	Burong talangka	Luzon	Small shore crabs (*Varuna litterata*)	*Leu. mesenteroides, E. faecalis, P. cerevisiae, L. plantarum*	Side dish, viand
	Patis	Whole Phil	Small fish, salt	*P. halophilus* (in mixed fermentation)	Fish sauce (patis), fish paste (bagoong), used as condiment, sauce, flavoring agent, viand
	Bagoong isda	Whole Phil	Small fish, salt		
	Bagoong alamang	Whole Phil	Small shrimps, salt		
	Bagoong na sisi	Visayas	Shell fish, salt		
	Guinamos	Bagoong isda in Visayas, Mindanao	Salt water small fish (dilis/belabid – *Stolephorus* sp.), salt		Condiment, viand, side dish
	Dayok	Visayas, Mindanao	Fish entrails, salt	Unidentified LAB	Condiment, viand, side dish
Fermented meat, sausages	Longanisa	Whole Phil	Ground pork, beef, or chicken meat, spices and preservatives	*P. acidilactici, Lactococcus lactis* (together with *Micrococcus aurantiacus*)	Viand
	Agos-os	Visayas	Sweet potato and ground pig's head	*E. faecalis*	Viand
	Burong kalabi	Luzon	Cooked rice, ground carabao meat	*L. plantarum*	Side dish, viand
	Burong babi	Luzon	Cooked rice, ground pork	*L. plantarum*	Side dish, viand

CATEGORY	PRODUCT NAME	REGION	MAJOR INGREDIENTS	LACTIC ACID BACTERIA INVOLVED (as determined from culture-based methods)	APPEARANCE AND/OR USAGE
Fermented vegetables, fruits	Burong mustasa	Luzon	Mustard leaves, cooked rice and/or rice washings	*Leuconostoc mesenteroides, Enterococcus faecalis, Lactobacillus plantarum*	Side dish
	Burong pipino	Whole Phil	cucumber	*Leu. mesenteroides, L. brevis, Pediococcus cerevisiae, L. plantarum*	Side dish
	Burong mangga	Whole Phil	Immature mango	*Leu. mesenteroides, L. brevis, P. cerevisiae, L. plantarum*	Side dish
	Atchara	Whole Phil	Immature papaya or chayote, or turnip (singkamas)	Unknown	Side dish
Cheese	Kesong puti	Luzon, Visayas	Cow or carabao milk	*Lactococcus lactis*	White soft cheese
Fermented fish and fishery products	Balao-balao	Luzon	Cooked rice, shrimp, salt	*Leu. mesenteroides, P. cerevisiae, L. plantarum*	Side dish, condiment
	Burong-isda	Luzon	Freshwater fish, rice, salt	*Leu. mesenteroides, E. faecalis, P. cerevisiae, L. plantarum, P. acidilactici, Leu. paramesenteroides*	Side dish, condiment
	Tinabal	Visayas	Parrot fish (for tinabal molmol) and frigate fish (for tinabal mangko), salt	*P. pentosaceus, S. equinus, Leuconostoc sp., Lactobacillus sp.*	Side dish, viand
	Burong talangka	Luzon	Small shore crabs (*Varuna litterata*)	*Leu. mesenteroides, E. faecalis, P. cerevisiae, L. plantarum*	Side dish, viand
	Patis	Whole Phil	Small fish, salt	*P. halophilus* (in mixed fermentation)	Fish sauce (patis), fish paste (bagoong), used as condiment, sauce, flavoring agent, viand
	Bagoong isda	Whole Phil	Small fish, salt		
	Bagoong alamang	Whole Phil	Small shrimps, salt		
	Bagoong na sisi	Visayas	Shell fish, salt		
	Guinamos	Bagoong isda in Visayas, Mindanao	Salt water small fish (dilis/belabid – *Stolephorus* sp.), salt		Condiment, viand, side dish
	Dayok	Visayas, Mindanao	Fish entrails, salt	Unidentified LAB	Condiment, viand, side dish
Fermented meat, sausages	Longanisa	Whole Phil	Ground pork, beef, or chicken meat, spices and preservatives	*P. acidilactici, Lactococcus lactis* (together with *Micrococcus aurantiacus*)	Viand
	Agos-os	Visayas	Sweet potato and ground pig's head	*E. faecalis*	Viand
	Burong kalabi	Luzon	Cooked rice, ground carabao meat	*L. plantarum*	Side dish, viand
	Burong babi	Luzon	Cooked rice, ground pork	*L. plantarum*	Side dish, viand

	Puto	Whole Phil	Rice, sugar	*L. mesenteroides, E. faecalis, P. cerevisiae* (in mixed fermentation with *Saccharomyces cerevisiae*)	Steamed rice cake
Fermented rice, cassava, sugar cane, coconut, soya	Bibingka	Whole Phil	Rice, sugar		Baked rice cake
	Tapuy	Luzon	Rice, glutinous rice	*Leuconostoc, L. plantarum* (in mixed fermentation with molds and yeasts)	Wine; beer
	Pangasi	Mindanao	Rice	Unknown	Wine
	Landang	Visayas, Mindanao	Cassava, or buli palm flour	Unknown	Dried jelly pellets pellets, rice substitute
	Puto balanghoy	Mindanao	Cassava	Unknown	Steamed cake
	Basi	Luzon	Sugar cane	Unknown	Wine
	Suka	Whole Phil	Sugar cane juice (for sukang Iloco), palm inflorescence sap (for sukang tuba)	*Leuconostoc, Lactobacillus, Streptococcus* in the initial fermentation phase only	Vinegar, condiment, flavoring
	Sinamak	Luzon	Sugar cane juice, spices (chilies, onions, garlic)	Unknown	Spiced vinegar, condiment, flavoring
	Pinakurat	Visayas, Mindanao	Coconut sap, chilies, salt, various spices	Unknown	Spiced vinegar, condiment, flavoring
	Tuba	Whole Phil	Coconut sap	Unknown	Wine
	Lambanog	Whole Phil	Coconut sap	Unknown	Wine
	Toyo	Whole Phil	Soybeans	*P. halophilus, E. faecalis, L. delbrueckii* (in mixed fermentation with *Aspergillus sojae* and *Saccharomyces rouxii*)	Condiment, flavoring agent, seasoning

RESEARCH INITIATIVES ON LAB FROM PHILIPPINE FERMENTED FOODS

Bacteriocin Research

Bacteriocins are antimicrobial proteins or peptides produced by certain bacterial strains. Unlike the peptide antibiotics they usually have a narrow spectrum of antimicrobial activity, usually inhibiting growth of closely related bacterial species or strains and lacking lethality to the producer strain (Riley and Wertz, 2002).

The bacteriocins of LAB are small, cationic, hydrophobic, or amphiphilic peptides or small proteins, composed of 20 to 60 amino acid residues (Chen & Hoover, 2003). The bactericidal mode of action and biochemical properties depend on the protein moiety that could be specific to a particular LAB strain,*i.e.* the N-terminal amino acids as determinant of receptors in the cell wall of the susceptible strains/species and C-terminal amino acids for the biochemical properties. LAB bacteriocin must have the following desirable properties: "(1) not active and nontoxic to eukaryotic cells, (2) become inactivated by digestive proteases, having little influence on the gut microbiota, (3) low pH and heat-tolerant, (4) have a relatively broad antimicrobial spectrum, against many food-borne pathogenic and spoilage bacteria, (5) show a bactericidal mode of action, usually acting on the bacterial cytoplasmic membrane: no cross resistance with antibiotics, and (6) have genetic determinants that are usually plasmid-encoded, facilitating genetic manipulation" (Apaga, 2012 as cited from Abriouel et al., 2007).

LAB bacteriocins have attracted attention in recent years because of their generally regarded as safe (GRAS) status and good value as natural biopreservatives which can find applications in the food and cosmetic industries (Cleveland et al., 2001; Daeschel, 1993; Riley and Wertz, 2002). Nisin, produced by strains of *Lactococcus lactis*, has been used in over 50 countries as anti-listerial and anti-clostridium substance. LAB bacteriocins with selective inhibition on food pathogens such as *Listeria monocytogenes*, but no inhibition on important lactic acid bacterial inocula such as the noted probiotic*Lactobacillus paracasei* or *Lactobacillus rhamnosus*; and yogurt-producing *Lactobacillus delbrueckii*subsp. *bulgaricus* and *Lactococcus thermophilus*, may provide advantage over those that have a wider spectrum of antimicrobial activity and would kill these beneficial organisms, including nisin (De Vos, 1993; Jack and Ray, 1995; Nielsen et al., 1990). Hence, efforts on the search for LAB bacteriocins and elucidation of their properties are actively being pursued by several research laboratories. The future holds a wide array of LAB bacteriocins available for various specific applications.

Isolation and Identification of Bacteriocin-Producing Lab

Some efforts on the isolation of bacteriocin-producing LAB had been started for more than a decade now in two major research institutions in the country namely: University of the Philippines Los Banos (specifically, the National Institutes of Molecular Biology and Biotechnology or BIOTECH-UPLB and the Institute of Biological Sciences or IBS-UPLB) and the Philippine Root Crop Research and Training Center, Visayas State University (VSU). These two institutions branched out knowledge on bacteriocin research through affiliate tutorship, as thesis advisers and as trainors to students and staff from a few other academic institutions which also did bacteriocin researches like University of Santo Tomas (UST), University of the Philippines Manila (UPM), De La Salle University (DLSU) and Ateneo de Manila University (ADMU). BIOTECH-UPLB and IBS-UPLB jointly worked on bacteriocins of*Lactobacillus plantarum* or plantaricins and those of *Pediococcus acidilactici* or pediocins. On the other hand, VSU devoted some efforts on the enterocins of *Enterococcus* spp. (Tan et al., 2001). DLSU also tried isolation of bacteriocin-producing LAB for food applications. UST was able to isolate bacteriocin-like inhibitory substances against medically important pathogens like *K. pneumoniae* (Dedeles et al., 2011). UPM and ADMU worked on human and animal health applications of bacteriocins.

Various fermented food products with proteinaceous components were the major sources of isolated LAB for bacteriocin screening. Such fermented food products are home-grown or produced by small enterprises and are still commercially available from public markets in Luzon, Philippines and some parts of the Visayas like Leyte island. Examples of Philippine indigenous fermented foods that were good sources of bacteriocin-producing LAB are fermented rice and shrimp (*balao-balao*), fermented rice and fish mixture (*burong kanin at isda*), fermented pork (*burong babi*) in Central Luzon (Elegado et al., 2003; Gervasio and Lim, 2007) and fermented pork and sweet potato (*agos-os*) in Eastern Visayan region (Samar and Leyte). On the other hand, pickled vegetables like mustard leaf (*burong mustasa*) and green papaya (*achara*), fermenting fruits like pickled green mango, *bignay* or mango wine (Samnang 2010), fermented salted fish (*bagoong*), spicy sausages (*longganisa*) may contain some LAB but often times they are not bacteriocinogenic (Gervasio and Lim, 2007). The obvious reasons are the presence of inhibitory substances like salt, spices, alcohol or acid and of course the dearth of proteinaceous materials in the food material.

In one of the first isolation studies for bacteriocinogenic LAB, various proteinaceous fermented foods native to Central and Southern, Philippines were screened for bacteriocin-producing bacterial isolates. Seventy one out of several hundreds of colony-forming unit isolated by agar plate streaking were found antagonistic to the indicator microorganism, *Lactobacillus plantarum* ATCC 14917, through direct assay. By "spot-on-lawn" assay by pH-neutralized culture supernatant, nine (9) isolates were confirmed to be bacteriocin producers (Elegado et al., 2003). Banaay et al. in 2004 also reported on the isolation of 1,100 putative LAB from indigenous fermented foods in Luzon, Philippines. A strain of *Lactobacillus plantarum* was selected as the best bacteriocin producer. In another study, out of the 160 putative LAB obtained from 19 fermented food products from public markets in Central Luzon, 32 LAB isolates were found to be bacteriocinogenic (Gervasio and Lim, 2007). Santiago et al. (2008) were also able to find two LAB isolates, *Lactobacillus fermentum* LBA-19 and *Lactobacillus casei* LTI-21, screened from among several LAB isolates from various fermented food products from different regions in the Philippines.

Being pleomorphic, identification of LAB is quite challenging. A combination of various microbiological and molecular biology tools would help in finding the real identity. Banaay et al. (2004) did a thorough identification of the bacteriocinogenic LAB isolate using conventional morphological, biochemical and physiological methods, chemotaxonomic methods, as well as molecular methods. This is especially relevant to the identification of *Lactobacillus plantarum* which is a known pleomorphic bacteria. Most other Philippine LAB researchers often times directly apply 16S rRNA gene sequencing and homology search for LAB purified through repeated agar streaking and putatively identified as LAB just after determining its acid–forming, Gram positive and catalase negative properties. (Elegado et al., 2003; Gervasio and Lim, 2007; Santiago et al., 2008). Aside from 16S rRNA genes, other conserved genes were used for identification such as phenylalanyl-tRNA synthase (*pheS*) gene (Dedeles et al., 2011). Detection of bacteriocin genes through PCR may also be helpful in confirming the identity of the bacteriocinogenic LAB as well as the probability of producing the bacteriocin (Table 2).

Table 2: Identification and bacteriocin gene determination of putative *Pediococcus acidilactici* through 16S rRNA and pediocin gene PCR amplification and sequencing.

ISOLATE/ STRAIN No.	IDENTIFICATION	(primer) HOMOLOGY to *P. acidilactici* type strain	REFERENCE	Bacteriocin gene by PCR; fingerprinting; HOMOLOGY
AA-5a	partial 16S rRNA gene ID: *P. acidilactici*	(1492R)98% *P. acidilactici* UL5; 99% *P. acidilactici* DSM20284 (27F) 99% *P. acidilactici* LAB 001; 99% *P. acidilactici* DSM20284	Elegado et al. 2003	ped⁺; REP and RAPD
4E2	partial 16S rRNA gene ID: *P. acidilactici*	(1492R) 98% *P. acidilactici* UL5; 99% *P. acidilactici* DSM20284 (27F) 99% *P. acidilactici* LAB 001; 99% *P. acidilactici* DSM20284	Apaga (2012)	ped⁺
4E4	partial 16S rRNA gene ID: *P. acidilactici*	(1492R) 97% *P. acidilactici* UL5; 99% *P. acidilactici* DSM20284 (27F) 98% *P. acidilactici* 8D2CCH01MX; 99% *P. acidilactici* DSM20284	Apaga (2012)	ped⁺
4E5	partial 16S rRNA gene ID: *P. acidilactici*	(1492R) 99% *P. acidilactici* DSM20284 (27F) 99% *P. acidilactici* DSM20284	Laxamana et al. (2011)	ped⁺ ; REP
4E6	partial 16S rRNA gene ID: *P. acidilactici*	(1492R) 98% *P. acidilactici* UL5; 99% *P. acidilactici* DSM20284; (27F) 99% *P. acidilactici* 8D2CCH01MX ; 99% *P. acidilactici* DSM20284	Apaga (2012)	ped⁺;[99% *P. acidilactici* bacteriocin genes ; pSMB74]
4E10	partial 16S rRNA gene ID: *P. acidilactici*	(1492R) 96% *P. acidilactici* UL5; 99% *P. lolii* to NGRI0510Q (27F) 99% *P. acidilactici* LAB 001 ; 99% *P. lolii* NGRI 0510Q	Apaga (2012)	ped⁻
4BL7	partial 16S rRNA gene ID: *P. acidilactici*	(1492R) 98% *P. acidilactici* UL5; 99% *P. acidilactici* DSM20284 (27F) 99% *P. acidilactici* 8D2CCH01MX; 99% *P. acidilactici* DSM20284	Apaga (2012)	ped⁺

3G3	API CHL50 ID: *Lactobacillus pentosus*(doubtful) partial 16S rRNA gene ID: *P. acidilactici*	(1492R) 99% *P. acidilactici* IMAU20090 (27F) 98% *P. acidilactici* DSM20284	Elegado and Perez (2012)	ped⁺ ; REP; ped⁺
3G8	partial 16S rRNA gene ID: *P. acidilactici*	(1492R) 99% *P. acidilactici* UL5 (27F) 98% *P. acidilactici* DSM20284	Elegado and Perez (2012)	ped⁺
3F3	partial 16S rRNA gene ID: *P. acidilactici*	(1492R) 95% *P. acidilactici* UL5 (27F) 98% *P. acidilactici* UL5; 99% *P. acidilactici* DSM20284	Apaga (2012)	ped⁺
3F8	partial 16S rRNA gene ID: *P.acidilactici*	(1492R) 98% *P. acidilactici* UL5; 99% *P. acidilactici* DSM20284 (27F) 99% *P. acidilactici* LAB 001; 99% *P. acidilactici* DSM20284	Apaga (2012)	ped⁺
3F10	partial 16S rRNA gene ID: *P. acidilactici*	(1492R) 97% *P. acidilactici* UL5; 99% *P. acidilactici* DSM20284 (27F) 97% *P. acidilactici* LAB 001; 99% *P. acidilactici* DSM20284	Apaga (2012)	ped⁺ [99% *P. acidilactici* genomic scaffold];
IG7	partial 16S rRNA gene ID: *P. acidilactici*	(1492R) 97% *P.acidilactici* UL5 99% *P. acidilactici* 8D2CCH01MX; (27F) 98% *P. acidilactici* DSM20284	Apaga (2012)	ped⁺ [100% pediocin operon;PSMB74];

K₂A₂-3	API: *Pediococcus pentosaceus* (good) partial 16S rRNA gene ID: *P. acidilactici*	(1492R) 97% *P. acidilactici* UL5; 99% *P. acidilactici* DSM20284 (27F) 99% *P. acidilactici* LAB 001; 99% *P. acidilactici* DSM20284	Villarante (2011); Elegado and Perez (2012)	ped⁺ ; plan⁺ ped⁺ ; REP
K₂A₂-1	API: *P. acidilactici* (doubtful)	-	Abuel (2007)	ped⁺ ; plan⁺ped⁺
K₂A₂-5	API: *P. acidilactici* (doubtful); partial 16S rRNA gene ID: *P. acidilactici*	(1492R) 97% *P. acidilactici* UL5; 99% *P. acidilactici* DSM20284 (27F) 99% *P. acidilactici* LAB 001; 99% *P. acidilactici* DSM20284	Apaga (2012)	ped⁺ [99% *P. acidilactici* genomic scaffold]; plan⁺
K₂A₁-1	partial 16S rRNA gene ID: *P. acidilactici*	(1492R) 99% *P. acidilactici* L94; 99% *P. acidilactici* DSM20284 (27F) 98% *P.acidilactici* JS-9-4; 99% *P. acidilactici* DSM20284	Apaga (2012)	ped⁺
K₃A₂-2	API: *Lactococcus lactis* (good) partial 16S rRNA gene ID: *P. pentosaceus*	-		ped⁺ ; plan⁺; ped⁺
K₃A₂-3	partial 16S rRNA gene ID: *P. acidilactici*	100% *P. acidilactici* UL5	Elegado and Perez (2012)	ped⁺
S3	partial 16S rRNA gene ID: *P. acidilactici*	(1492R) 98% *P. acidilactici* UL5; 99% *P. acidilactici* DSM20284 (27F) 97% *P. acidilactici* LAB 001; 99% *P. acidilactici* DSM20284	Apaga (2012)	ped⁺ [99% pediocin operon; pSMB72];

Purification and Characterization of Bacteriocins

Purification of bacteriocin peptides or small proteins into homogeneity is necessary in order to fully characterize them, particularly the determination of molecular mass, the primary structure or amino acid sequence and secondary structure. For pediocin, it was found that a simple and rapid method is effective for its purification. This method involves adsorption of pediocin onto the cell wall of the producer cell at pH 6 and 0.05 M NaCl and then subsequent desorption at pH 2.0 and 1 M NaCl (Elegado et al., 1997;Yang et al., 1992). This method seemed more applicable to pediocin but not with the lactococcin, nisin or plantaricin. The reason is not clear but it could be related to variation in cell wall properties. The pH-adsorption/desorption method was able to provide materials for pH and temperature tolerance assays, estimation of molecular mass through SDS-PAGE, residual activity determination after protease, amylase and other enzyme actions (Laxamana et al., 2011). Enough amount of semi-purified bacteriocin from pediococci using this method was obtained for further purification through preparative reverse phase HPLC for various characterization studies, including the determination of secondary structures by circular dichroism and confirmation of double bonds through trypsin digestion and electrospray mass spectrometry (Elegado and Kwon, 1998). Other preparative purification methods prior to reverse phase HPLC and spectrometry included ion exchange chromatography and gel filtration chromatography (Elegado et al., 2003), and hydrophobic interaction chromatography (Villarante et al., 2011). This method could also be applied with bacteriocins of pediococci and lactobacilli. The properties obtained from well characterized bacteriocinogenic LAB are shown in Table 3.

Table 3: List of purified and characterized bacteriocins from LAB isolated from Philippine indigenous fermented foods.

Isolate	Identity	Bacteriocin	Purification mode	Properties
AA5a	*Pediococcus acidilactici*	pediocin	pH adsorption/desorption Reversed-phase HPLC	Tolerant to pH 2-9 and 121 °C
BS25	*Lactobacillus plantarum*	plantaricin	Gel filtration chromatography Reversed-phase HPLC	MW = 3,830 Da
K2a2-3	*Pediococcus acidilactici*	pediocin	Hydrophobic interaction and ion-exchange chromatographies Reversed-phase HPLC	MW = 4,626 Da
K2a2-1	*Pediococcus acidilactici*	pediocin	pH adsorption/desorption	Optimum pH = 5-7 Resistant to boiling but not to autoclaving

4E5	*Pediococcus acidilactici*	pediocin	pH adsorption/de-sorption	Tolerant to pH 2-9; slight loss of activity at 100 °C; loss of activity at 121 °C; tolerates high salt; est. MW = 6,500 Da by SDS-PAGE

Optimization of Bacteriocin Production through Fermentation Kinetics

Bacteriocin production is largely dependent on the nutrients and nitrogen content of the fermentation medium. For instance, increased yeast extract concentration and polypeptone amount increases bacteriocin production. Molasses, raw sugar and sago hydrolyzates of amylase digestion were found to be good carbon sources. Other possible substrate base and supplements are cheese whey, coconut water and rice bran extract. Initial sugar concentration of usually 2 to 3% and inoculation rate of 3% by volume of at least 10^8 cells/mL provides good bacteriocin production (Elegado et al., 2001).

Bacteriocin production is highly dependent on cell or biomass growth. LAB are microaerophilic and most are either mesophilic or slightly thermophilic. The following conditions are applicable to their production: pH= 5.5 to 6.0; temperature = 35 – 40 °C; agitation = 50 rpm; without aeration. Usually, bacteriocin is optimally produced or secreted in the culture broth during the early stationary phase of growth. For *Pediococcus acidilactici*, culturing at 40 °C promotes earlier optimum bacteriocin production of around 10-12 hours. At 37 °C, bacteriocin production is from 14-16 hours (Sagpao et al., 2007).

Applications

Pediocins and plantaricins are the commonly found bacteriocins in Philippine fermented foods so far studied. Their antimicrobial properties have been investigated in several studies (Banaay et al., 2004;Elegado et al., 2003, 2004, 2007; Marilao et al., 2007). Although pediocins and plantaricins show promise, their applications are limited at present because it is a well-known fact that other bacteriocins aside from nisin are not yet approved for food use. For pediocins and plantaricins, the most practical use for now would be dermatological and animal health care use. But since the bacteriocin-producing LAB are of GRAS status, those with probiotic properties such as tolerance to acidic pH (2.0 -3.0) and bile (0.3%) and adhesion properties to intestinal mucosa would be an advantage when used as adjunct inocula in fermented food products (Gervasio and Lim, 2007).

Perhaps another importance of bacteriocin-producing LAB is their effectiveness in biomedical applications. In one study, for example, partially-purified pediocin K2a2-3, through pH-mediated bacteriocin extraction method, was found cytotoxic against human colon adenocarcinoma (HT29) and human cervical carcinoma (HeLa) cells *in vitro* as determined by MTT [3-(4,5-dimethylthiazol-2-yl)-2,5-diphenyltetrazolium   bromide]   assay (Villarante et al., 2011). Other potential biomedical applications will be discussed in the succeeding section.

PROBIOTICS AND FUNCTIONAL FOODS

An offshoot of the initial research on bacteriocins of LAB isolated from indigenous fermented foods is the emergence of probiotic research towards developing functional foods for biomedical applications. Probiotics refer to microorganisms that, when administered in adequate amounts, confers health benefits to the host. Although there are many microorganisms that can be considered as probiotics, LAB are the most common types because they produce antimicrobial compounds that inhibit other harmful microorganisms, they are able to tolerate acids and bile present in the digestive system, and they are able to adhere and establish themselves in the gut surfaces.

Many benefits have been ascribed to probiotics. For example, *Lactobacillus casei* (Shirota strain in Yakult®) have been shown effective in preventing diarrhea due to enterotoxigenic *Escherichia coli*(ETEC) and choleragenic vibrios (*V. cholerae* biotype E1 Tor and classical *V. cholerae*) using rats (Jacalne et al., 1990). This may be accounted for by its ability to kill the pathogens and inhibit further growth (Consignado et al., 1994). Because the probiotic used in the two studies mentioned is a commercial strain, current research on probiotics progressed to the search for indigenous LAB for use in the development of locally-produced functional food and investigation of their utility for biomedical applications. Metagenomic approaches to investigating LAB present in fermented foods have shown the diversity of potentially beneficial species present other than those that are readily detected by conventional culture-based methods. The development of functional food products shows potential in disease management. Research using metagenomic analysis in searching for microbial markers for use in functional foods to address certain lifestyle diseases as well as malnutrition is on the way.

Metagenomic and Diversity Studies

Traditional culture-based methods have been used for isolating LAB from fermented foods. These studies form the basis for the starter cultures used in food fermentation technologies employed for commercial production. Sanchez

(2008) gives detailed information on the different technologies and cultures used for the production of some traditional as well as developed technologies that have arisen from the culture-based studies conducted in earlier years.

In recent years culture-based approaches in LAB isolation have become more targeted for detection of bacteriocin-producers and those that have potential as probiotics. In one initiative, LAB isolates from fermented foods were screened for bacteriocin production and a PCR-based assay was used to detect specific bacteriocin-encoding genes. Acid and bile tolerance were also determined. Among all the isolates tested, *Lactobacillus fermentum* 4B1 and *Lactobacillus pentosus* 3G3 (later identified as*Pediococcus acidilactici*) have been identified as most promising for the development of new probiotic food products, hence they were chosen for subsequent biomedical application assays (Lim and Gervacio, 2007). In another study, LAB from traditionally fermented wine and vinegar from Visayas and Mindanao were isolated, identified, and tested for inhibitory activity against *Enterococcus faecium*,*Listeria innocua*, and *Staphylococcus aureus*. Five *Lactobacillus paracasei* and one *Lactobacillus brevis* showed antimicrobial properties against the tester strains (Licaros and Bautista, 2009).

With the advent of molecular techniques, the existence of non-culturable microorganisms has been acknowledged especially since the occurrence of culture-bias is already well-accepted. Culture-independent approaches, therefore, have been gaining popularity in microbial diversity studies and this includes researches on microorganisms found in fermented foods. The microbial populations in selected Philippine fermented foods were assessed through Polymerase Chain Reaction followed by Denaturing Gradient Gel Electrophoresis (PCR-DGGE) in two recent studies (Dalmacio et al., 2011; Larcia, 2010). Food samples tested include *burong mustasa* (fermented mustard), *alamang* (fermented shrimp paste),*burong isda* (fermented rice-fish mixture), *balao-balao/burong hipon* (fermented rice-shrimp mixture),*tuba* (sugar cane wine), and *sinamak* (spiced vinegar). Analysis of the 16S rRNA gene sequences revealed the presence of several LAB that have not been reported in these food products before.*Weissella cibaria, Lactobacillus plantarum, Lactobacillus pontis, Lactobacillus panis,* and *Lactobacilus fermentum* were detected in *burong mustasa* (Larcia, 2010). *L. panis* and *L. fermentum* were present in*alamang*; *L. pontis* and *L. plantarum* in *burong isda*; *L. panis, L. pontis*, and *L. fermentum* in burong hipon; and *W. cibaria, L. pontis, L. panis, L. fermentum* and *L. plantarum* in *burong mustasa* (Dalmacio et al., 2011).

The results of the two studies using molecular approaches in defining diversity of LAB in Philippine fermented foods show that culture-independent approaches are efficient tools for the analysis of microbial populations in fermented foods. Majority of the identified bacteria (LAB and other bacterial groups) have not been reported in culture-dependent studies. As such, the isolated bacterial 16S rRNA genes were cloned to have an initial partial 16S rRNA gene library for Philippine fermented foods (Dalmacio et al., 2011).

Biomedical Applications

Anti-Obesity

Obesity is defined as an abnormal or excessive fat accumulation that presents risks to health. Probiotics can help in fighting obesity by reducing lipid absorption through its action on bile acid metabolism, and by assimilation of cholesterol thus eliminating it from the host's system. Several studies were conducted to examine anti-obesity properties of different probiotic strains.

In one study, oral administration of *Lactobacillus paracasei* K3-4C, isolated from a locally fermented food had significant effect on lowering blood glucose levels (by 46%) and body weight (by 13%) in female BALB/c mice induced to be diabetic and obese through a 28-day high-fat diet (Parungao et al., 2006). In another study, orally administered *L. fermentum* 4B1 reduced adipose cell size, and decreased adipose tissue weight and overall body weight of mice fed with a high-fat diet for 49 days (Bautista et al., 2008). Likewise, oral administration of *P. acidilactici* 3G3 reduced body weight in diet-induced obese female Swiss mice (Parungao et al., 2009). In the last two studies described, the effects of the probiotics were determined to be comparable with the effects of the commercial anti-obesity drug Orlistat based on the parameters measured.

Recently, it has been postulated that the development of obesity may be caused by a shift in the composition of the gut microbiota towards the Firmicutes population (Ley et al., 2005). Firmicutes characterize obese versus lean/non-obese individuals together with a drop or no change in Bacteroidetes (Delzenne and Cani, 2010). Interestingly, Ley et al. (2006) found that a low fat diet had an effect to reverse the shift of Firmicutes/Bacteroidetes proportion. Because of this, dietary manipulation has been seen as a potential means of changing bacterial populations in the colonic microbiota and perhaps treating or at least preventing diseases like obesity. Although the root cause of obesity is excessive caloric intake coupled with a sedentary lifestyle (Blaut and Bischoff, 2010), Ley et al. (2005) proposed in their findings that alteration in the populations of mice gut microflora may have caused or may have been an effect of obesity. Because of this, current researches aim in using probiotics in the treatment of diseases such as obesity.

In two related studies (Arroyo and Fabiculana, 2011; Parungao et al., 2012), the effect of a functional food containing *P. acidilactici* 3G3 on microbial community changes in the gut of obese and non-obese mice was determined through PCR-DGGE. Results of these two preliminary studies showed that obese and non-obese mice had different baseline colonic microbiota. There were also indications that treatment with probiotics shifts the microbiota of obese mice towards the normal non-obese type. As these are preliminary studies, more research is warranted to elucidate the nature of the changes in gut microbiota and how it is related to obesity and the anti-obesity effects of probiotics.

Immuno-enhancement

A preliminary *in vitro* study to examine the immune-enhancing properties of viable and heat-killed preparations of two LAB previously isolated from traditional fermented foods (*L. fermentum* 4B1 and *P. acidilactici* 3G3) on murine peritoneal macrophage cells and spleenic T-cells showed that isolate 4B1 was able to induce NO production in murine macrophages but, like 3G3, was unable to stimulate murine T-cell proliferation (Tan et al., 2008). Furthermore, this study showed that preparations of *L. fermentum* 4B1 have the ability to induce NO production in murine macrophage cells and its effects were more potent when it was alive. The study also showed that isolate 4B1 exhibited better immune-enhancing effect than the probiotic species found in a commercial probiotic drink. T-cell proliferation, however, was not observed in any of the treatments in this study and was attributed to the delayed stimulation in cells responding to a first-time exposure to the different probiotic strain preparations used.

Reduction of blood glucose levels

A study by Ngo et al. (2008) showed that oral administration of kefir, a common fermented food consumed by the elderly, significantly decreased blood glucose levels and body weight of diabetic obese male Sprague Dawley rats. The results of the study showed lower blood glucose levels (from 198.5 to 105.6 mg/dL) and clinically lower body weights (from 342.9 to 311.5 g) of the treated diabetic-obese rats than the untreated diabetic-obese control group.

Prevention of hypercholesterolemia

The effect of *P. acidilactici* 3G3 administration on hypercholesterolemic Swiss Albino mice was determined (Parungao et al., 2009). This strain was able to assimilate cholesterol in the *in vitro* plate assay and decrease HDL, LDL, and total cholesterol in the *in vivo* assay using mice. Strain 3G3 was also shown to adhere well to the duodenum and middle colon. Results suggest the potential of *P. acidilactici* 3G3 in preventing hypercholesterolemia.

Development of Functional Foods

The development of functional foods containing known probiotic strains stems from earlier researches on bacteriocins and isolation of potential probiotics from traditional fermented foods. The beneficial effects of probiotic-supplemented chocolate bars (Arroyo et al., 2010; Arroyo and Fabiculana, 2011), fermented mustard leaves (Calapardo et al., 2006), and coffee wine (Parungao, 2007) have been investigated. Initial studies on mango-milk and carrot juice drinks supplemented with probiotic strains have also been conducted (Bugarin et al., 2010; Elegado et al., 2005). These potential functional foods contain probiotic strains, previously isolated from traditional fermented foods such as *P. acidilactici*AA5a (Elegado et al., 2003), *L. plantarum* BS25 (Banaay et al., 2004), and *P. acidilactici* 3G3 (Lim and Gervacio, 2007). Research on functional foods is still in its infancy but this food category shows promise in disease management as well as in contributing to food security in the country. Commercial interest in probiotic food products is increasing due to the growing understanding of its health benefits. This growing industry can derive benefits from the researches conducted on this emerging food category.

FUTURE PERSPECTIVES

Aside from the research works presented earlier in this paper as well as on-going follow-up studies related to them, future goals may include research on a variety of other possible biomedical applications of LAB with potential probiotic properties. The effect of probiotics on *Helicobacter pylori* infections (that may cause peptic ulcers) may be determined. Their ability to modulate inflammatory and hypersensitivity responses as well as their effect on irritable bowel syndrome and colitis may be investigated. Further research on possible anti-cancer properties of probiotics is warranted as follow-up studies on the work done by Villarante et al. (2011). These studies are very important as these have the potential to address some of the more serious health concerns of our society.

Much is still to be learned about the existing probiotic strains. The molecular biology and genomics of these isolates may be pursued in order to further elucidate their properties and mechanisms of action.

Determination of factors affecting probiotic viability in foods is also important as these will determine if their survival in the food, and therefore their delivery into the host, is maintained. This will constitute a quality control for functional foods.

The potential physiological effects of multiple prebiotic strains, as opposed to a single strain, are also interesting areas of research. The delivery of multiple probiotic strains may help ensure its effectiveness in an environment that contains high diversity of resident microflora. The potential benefits of synbiotics, (combination of probiotic and prebiotic) which have synergistic interaction, may also be investigated. A good combination will greatly enhance the health benefits to humans.

REFERENCE

1.   H Abriouel, A Galvez, R. L Lopez, N. B Omar, 2007Bacteriocin-based strategies for food biopreservationInternational Journal of Food Microbiology1205170

2.   Apaga DLT (2012Detection of bacteriocin structural genes and bacteriocinogenic activity of several pediococcus isolatesUndergraduate thesis. University of the Philippines Manila. 56 p.

3.   Arroyo KZOGarcia JAR, Elegado FB, Calapardo M, Parungao MM (2010Survival of Lactobacillus pentosus 3G3 in home-made chocolate bars. 39th Annual Convention and Scientific Meeting,, PSM, Inc. April 29-30, 2010, City of Naga, Camarines Sur.

4.   Arroyo MJJFabiculana PRS (2011Gut microflora assessment of obese and non-obese mice (Mus musculus L.) orally-fed with lp-3g3 chocolate. Department of Biology, University of the Philippines Manila, Manila, Philippines. Unpublished Undergraduate Thesis.

5.   Banaay CGBElegado FB, Dalmacio IF (2004Identification and characterization of bacteriocinogenic Lactobacillus plantarum BS25 isolated from balao-balao, a locally fermented rice-shrimp mixture from the PhilippinesThe Philippine Agricultural Scientist874427438

6.   Bautista RLSEcarma NCA, Balolong ECJr, Hallare AV, Parungao MM (2008The Effects of Orally-Administered Lactobacillus sp. 4B1 on the Adipose Tissues of Diet-Induced Obese Mice (Mus musculus L.). International Symposium on Probiotic from Asian Traditional Fermented Foods for Healthy Gut Function. August 1920Sari-Pan Pacific Hotel, Jakarta, Indonesia.

7.   J. R Besas, E. I Dizon, 2012Influence of salt concentration on histamine formation in fermented tuna viscera (Dayok).Food and Nutrition

Sciences3201206

8.   M Blaut, and S. C Bischoff, 2010Probiotics and Obesity. Ann Nutr Metab 2010; 57(Suppl.1):20-23.

9.   M. A Bugarin, Sison AAD, Elegado FB, Calapardo M, Parungao MM (2010Survival of Lactobacillus plantarum BS25 on Varying Ratio of Mango-Milk Substrates and Storage Temperature. 39th Annual Convention and Scientific Meeting, PSM, Inc. April 29-30, 2010, City of Naga, Camarines Sur.

10. M. R Calapardo, Bueno MOV, Guillermo MKB, Saguibo JD, Parungao MM, Elegado FB (2006Bile and Acid Tolerance of Bacteriocinogenic Lactobacillus plantarum and its Use as Adjunct Inoculum in Pickled Mustard Leaves. Proceedings of the 5th Asia-Pacific Biotechnology Congress and 35th Annual Convention of the PhilippineSociety for Microbiology Inc. (PSM). "Microbiology and Biotechnology: Roadmaps and Milestones for Enhancing Sustainable Productivity in the Asia-Pacific Region." Bohol Tropics Resort. Tagbilaran City, Bohol, Philippines.

11. H Chen, D. G Hoover, 2003Bacteriocins and their food applications. http://www.ift.org/publications/crfsfsaccessed 22 July 2006)

12. J Cleveland, M. L Chikindas, T. J Montville, I. F Nes, 2001Bacteriocins: safe, natural antimicrobials for food preservationInternational Journal of Food Microbiology7112

13. G. O Consignado, A. C Pena, A. V Jacalne, 1994In vitro study on the bacterial activity of Lactobacillus casei (commercial Yakult drink) against four diarrhea-causing organisms: enterotoxigenic E. coli, Salmonella enteritidis, Shigella dysenteriae, Vibrio cholerae. Philipp J Microbiol Infect Dis. 2325055

14. M. A Daeschel, 1993Applications and interactions of bacteriocins of lactic acid bacteria in food and beverages. In: Hoover, D.G., L.R. Steenson, eds. Bacteriocins of Lactic Acid Bacteria. New York: Academic Press Inc. 6370

15. Dalmacio LMMAngeles AKJ, Larcia LLH, Parungao-Balolong MM, Estacio RC (2011Assessment of microbial diversity in Philippine fermented food products through polymerase chain reaction- denaturing gradient gel electrophoresis (PCR-DGGE). Beneficial Microbes, 24273281

16. W. M De Vos, 1993Future prospects for research and applications of nisin and other bacteriocins. In: Hoover, D.G., L.R. Steenson, eds. Bacteriocins of Lactic Acid Bacteria. New York: Academic Press Inc. 249258

17. G. R Dedeles, Caranza MAE, Elegado FB (2011Bacteriocin-like

inhibitory substances (BLIS) from lactic acid bacteria and their potential as biopreservatives in foods. Proccedings of the 6th Asian Conference on Lactic Acid Bacteria and XIII International Congress of Bacteriology and Applied Microbiology. September 610Sapporo, Japan. 120 p.

18. N. M Delzenne, and P. D Cani, 2010Nutritional modulation of gut microbiota in the context of obesity and insulin resistance: potential interest of prebioticsInt. Dairy J. 20277280

19. F. B Elegado, W. J Kim, D. Y Kwon, 1997Rapid purification, partial characterization and antimicrobial spectrum of the bacteriocin, Pediocin ACM from Pediococcus acidilactici M.International Journal of Food Microbiology371111

20. F. B Elegado, D. Y Kwon, 1998Primary structure and conformational studies of Pediocin AcM, a bacteriocin from Pediococcus acidilactici M. The Philippine Journal of Biotechnology 911926

21. F. B Elegado, K Sonomoto, A Ishizaki, 2001Molecular characterization of a bacteriocin of Pediococcus acidilactici and its production in sago and sugarcane-based substrate. Biotechnology for Sustainable Utilization of Biological Resources in the Tropics. Murooka, Y., T. Yoshida, T. Seki, P. matangkasombut, T.M. Espino, U. Soetisna and M.I.A. Karim. eds. JSPS-NRCT/DOST/LIPI/VCC Joint Seminar. Nov. 7-9, 2001. Bangkok, Thailand. 15173180

22. F. B Elegado, Opina ACL, Banaay CGB, Dalmacio IF (2003Purification and characterization of novel bacteriocins from lactic acid bacteria isolated from Philippine fermented rice-shrimp or rice-fish mixturesThe Philippine Agricultural Scientist8616574

23. F. B Elegado, Guerra MARV, Macayan RA, Estolas MT, Lirazan MB (2004Antimicrobial activity and DNA fingerprinting of bacteriocinogenic Pediococcus acidilactici through RAPD-PCRThe Philippine Agricultural Scientist872229237

24. F. B Elegado, M. R Calapardo, J. F Fabregas, Ona SEN, Parungao MM (2005Antilisterial Efficacy of Bacteriocinogenic Pediococcus acidilactici AA5a in Carrot Juice Fermentation. Proceedings of the 3rd Asian Conference on Lactic Acid bacteria. August 2527Sanur Paradise Hotel, Bali, Indonesia.

25. F. B Elegado, Abuel BJA, Te JT Jr.,Calapardo MR, Parungao MM (2007Antagonism against Listeria spp. and Staphylococuus aureus by bacteriocin-producing lactic acid bacteria screened from the intestine of Philippine carabao using polymerase chain reaction.The Philippine Agricultural Scientist904305314

26. F. B Elegado, and Perez MTM (2011Genetic identification of lactic acid bacteria and bacteriocin structural gene elucidation. National Academy of Science and Technology (NAST) Scientific Meeting. July 1112Manila Hotel, Manila, Philippines.

27. Gervasio ATRLim VMT (2007Probiotic characterization of bacteriocinogenic lactic acid bacteria isolated from fermented foods of selected areas. Undergraduate thesis. University of the Philippines Manila. 105 p.

28. Jack RJTRay B (1995Bacteriocin of gram positive bacteria. Microbiological Reviews. 592171200

29. A. V Jacalne, R. R Jacalne, H Hirano, T Suetomi, C. G Villahermosa, I Castaneda, 1990In-vivo studies on the use of Lactobacillus casei (Yakult Strain) as biological agent for the prevention and control of diarrhea. Acta Med Philipp. 262116122

30. LarciaLLH (2010A study of the bacterial profile of Philippine fermented mustard (burong mustasa) through culture-independent methods / Levi Letlet Larcia, II. Thesis (MS Biochemistry)--University of the Philippines Manila. 158LG995B3 L37

31. Laxamana FLMCarillo MCO, Elegado FB (2011Characterization of a bacteriocin of lactic acid bacteria isolated from fermented rice-fish mixture. Proceedings of the 40th Annual Convention and Scientific Meeting of the Philippine Society for Microbiology, Inc. May 1014Manila, Philippines.

32. C Lee, 1999Cereal fermentations in countries of the Asia-Pacific Region. In: FAO, Fermented Cereals: A Global Perspective, FAO Agricultural Services Bulletin 138Food and Agriculture Organization 144 p.

33. R. E Ley, F Bäckhed, P Turnbaugh, C A Lozupone, R. D Knight, J. I Gordon, 2005Obesity alters gut microbial ecology. Proc Natl Acad Sci USA, 102311107011075

34. R. E Ley, P. J Turnbaugh, S Klein, J. I Gordon, 2006Human gut microbes associated with obesity." Nature44410221023

35. A Licaros, A Bautista, 2009Culture-Dependent and Culture-Independent Analysis of Vinegars from Visayas and Mindanao. Department of Biology, University of the Philippines Manila, Manila, Philippines. Unpublished Undergraduate Thesis.

36. V Lim, T. A Gervacio, 2007Bacteriocin-producing LAB from fermented foods of Central Luzon. Department of Biology, University of the Philippines Manila, Manila, Philippines. Unpublished Undergraduate Thesis.

37.   C. G Marilao, M. R Calapardo, C. E Ciron, F. B Elegado, 2007Antilisterial action of Pediococcus acidilactici AA5a in pork sausage fermentationThe Philippine Agricultural Scientist9014045

38.   R. E Ngo, M. Z Estrada, Balolong ECJr., Parungao MM (2008Orally-administered kefir on lowering blood glucose levels and body weight in diet-induced diabetic obese rats. Project Terminal Report. National Institutes of Health University of the Philippines Manila.

39.   J. W Nielsen, J. S Dickson, J. D Crouse, 1990Use of a bacteriocin produced by Pediococcus acidilactici to inhibit Listeria monocytogenes associated with fresh meat.Applied and Environmental Microbiology 5423492353

40.   M Olympia, H Fukuda, H Ono, Y Kaneko, M Takano, 1995Characterization of starch-hydrolyzing lactic acid bacteria isolated from a fermented fish and rice food, "Burong Isda", and its amylolytic enzymeJournal of Fermentation and Bioengineering802124130

41.   M. M Parungao, 2007Probiotic Properties of Philippine Coffee Wine. The 4th Asian Conference on Lactic Acid Bacteria & the 3rd International Symposium on Lactic Acid Bacteria and Health. October 1719Shanghai, China.

42.   M. M Parungao, R. L Castillo, Mortel MRA (2006Effects of Oral Administration of Lactobacillus paracasei K34C on Blood Glucose Levels and Body Weight in Diabetic Obese Mice (Mus musculus L.). December 15-17, 2006. Proceedings of the 11th Biological Sciences Graduate Congress, Chulalongkorn University, Bangkok, Thailand.

43.   M. M Parungao, C Ong, N Laluces, K De Torres, J Usisa, M Calapardo, L Trinidad, F. B Elegado, 2009The Cholesterol-Reducing Ability and Adhesion Properties of Lactobacillus pentosus 3G3. Proceedings of the 5th Conference of Federation for Societies of Lactic Acid Bacteria. July 14National University of Singapore, Singapore.

44.   M Parungao-balolong, Libed AAO, Loma K, Villena JPDS, Villafuerte AR, Balolong EC Jr., Dalmacio LMM (2012Influence of probiotic drinks on the distal gut bacterial flora of mice (Mus musculus L.) fed with standard diet or high-fat diet. Terminal report, National Institutes of Health University of the Philippines Manila.

45.   M. A Riley, J. E Wertz, 2002Bacteriocins: evolution, ecology andapplications. Annual Review of Microbiology 56117137

46.   Sagpao SMNElegado FB, Zamora A (2007Optimization of production and partial purification of pediocin from Pediococcus acidilactici PNCM 10289. Poster paper presented at the 29th National Academy of science

and Technology (NAST) Philippines Annual Scientific Meeting, Manila Hotel, July 1112

47. Samnang (2010Isolation and identification of lactic acid bacteria from fermenting bignay (Antidesma bunius (LSpreng) and mango (Mangifera indica L.) wines. M.S. thesis. University of the Philippines Los Banos. 242 p.

48. P. C Sanchez, 2008Philippine fermented foods: principles and technology.the University of the Philippines Press. Diliman, Quezon City, Philippines. 516 p.

49. M. R Santiago, C. M Lopez, E. L Tenorio, E. E De Guzman, 2007Detection and partial charcaterization of two bacteriocins produced by Lactobacillus casei and Lactobacillus fermentum isolated from fermented rice-fish mixtures indigenous in the Philippines. Proceedings of the 37th Annual Convention and Scientific Meeting of the Philippine Society for Microbiology, Inc. May 7-9, 2008. Boracay Philippines. 32

50. J. D Tan, Galvez FCF, Asano K, Tomita F (2001Isolation and partial purification of bacteriocin produced by microorganisms from agos-os. Biotechnology for Sustainable Utilization of Biological Resources in the Tropics. Murooka, Y., T. Yoshida, T. Seki, P. matangkasombut, T.M. Espino, U. Soetisna and M.I.A. Karim. eds. JSPS-NRCT/DOST/LIPI/ VCC Joint Seminar. Nov. 7-9, 2001. Bangkok, Thailand. 15104110

51. Tan JASYalung PM, Evangelista KV, Parungao MM (2008In vitro Study on the Effect of Lactobacillus sp. 4B1 and Lactobacillus sp. 3G3 on Murine Macrophage Nitric Oxide Production and Splenic T-cell Proliferation. Proceedings of the International Symposium on Probiotic from Asian Traditional Fermented Foods for Healthy Gut Function. August 1920Sari-Pan Pacific Hotel, Jakarta, Indonesia.

52. K. I Villarante, F. B Elegado, S Iwatani, T Zendo, K Sonomoto, E. E De Guzman, 2010Purification, characterization and in vitro cytotoxicity of the bacteriocin from Pediococcus acidilactici K2a2-3 against human colon adenocarcinoma (HT29) and human cervical carcinoma (HeLa) cellsWorld Journal of Microbiology and Biotechnology27975980

53. R Yang, M. C Johnson, B Ray, 1992Novel method to extract large amounts of bacteriocins from lactic acid bacteria.Appl. Environ. Microbiol. 5833553359

Chapter 3

NON-TOXIC EFFECT OF MONASCUS PURPUREUS EXTRACT ON LACTIC ACID BACTERIA SUGGESTED THEIR APPLICATION IN FERMENTED FOODS

Mohan-Kumari Honganoor Puttananjaiah[1], Mohan Appasaheb Dhale[2], Vijayalakshmi Govindaswamy[1]

[1]Department of Food Microbiology, Central Food Technological Research Institute, Council of Scientific and Industrial Research, Mysore, India;

[2]Biological Oceanography Division, National Institute of Oceanography, Council of Scientific and Industrial Research, Dona Paula, India.

ABSTRACT

The effect of Monascus purpureus extract (MPE) on probiotic lactic acid bacteria (LAB) was investigated to ascertain its application in fermented foods. Viable count of LAB was not affected after 24 hours of incubation in Man Rogosa Sharpe (MRS) broth containing MPE. The agar well-diffusion assay did not show any inhibition zone. The biotransforma-tion of isoflavone glycosides by LAB in culture medium supplemented with MPE increased antioxidant activities. These data suggest that, nutritive and biological functionality of fermented foods can be improved by the use of MPE.

INTRODUCTION

Monascus sp. have been used commercially to produce valuable secondary metabolites viz., pigments [1,2], monacolin K (lovastatin), hypotensive agent, γ-aminobutyric acid, 3-hydoxy-4-methoxy-benzoic acid [3,4]. Feeding of M. purpureus red mould rice demonstrated significant decrease TC, TG, and LDL-C levels in hyperlipidemic hamsters. Monacolin K (lovastatin) is a potent competitive inhibitor of 3-hydroxy-3-methylglutaryl coenzyme A (HMG-CoA) reductase enzyme [5]. Dimerumic acid, dihydromonacolin-MV and dehydromonacolin-MV2 isolated from Monascus sp. have been characterized for their antioxidant action [6-8]. Our earlier study on toxicity of RMR did not show any adverse effect on rats and significantly reduced the cholesterol and triacylglycerol levels in serum and liver [9]. Orange pigments (monascorubrin and rubropunctatin) inhibited the growth of Bacillus subtilis, Escherichia coli, some filamentous fungi and yeasts [10]. Amino acid derivatives of Monascus pigments L-Phe, D-Phe, L-Tyr, and D-Tyr exhibited high activities against Gram + ve and Gram -ve bacteria. Derivatives with L-Asp, D-Asp, L-Tyr and D-Tyr were effective against the filamentous fungi Aspergillus niger, Penicillium citrinum and Candida albicans [11]. The use of Monascus cultures as food additives is not approved either in the EU or in the USA, though it is currently permitted in Japan. It has been traditionally used for manufacturing food colorants (e.g. red rice) and fermented foods and beverages in Southern and Far Eastern Asia [3]. Monascus pigments were used as colouring agents in preparation of sausage, hams, fish paste, surimi and tomato ketchup [12]. However, there are no reports on the M. purpureus isoflavones and their bioconversion. In this study we have identified the non-toxic effect of MPE on lactic acid bacteria (LAB) and biotransformation of isoflavone glycosides by LAB in culture medium supplemented with MPE. Consequently, providing food products with aglycones would be considered as a novel trend for the food industry.

MATERIALS AND METHODS

Microorganisms and Culture Conditions

The microorganisms M. purpureus MTCC 410, Lactobacillus acidophilus B4496 (La), L. bulgaricus CFR2028 (Lb), L. casei B1922 (Lc), L. plantarum B4495 (Lp) and L. helviticus B4526 (Lh) were obtained from Microbial Type Culture Collection (MTCC), Institute of Microbial Technology (IMTECH) Chandigarh, India. The M. purpureus and LAB cultures were maintained on PDA and MRS media (Himedia Laboratories, Mumbai, India) respectively at 4°C - 8°C. Seed culture was prepared according to method of Su et al., by inoculating M. purpureus spores into a 500 ml flask containing 100 ml basal medium and incubated for 48 h at 30°C. The seed culture inoculum was used for solid state fermentation [3]. The medium pH was adjusted to 6.0 prior to sterilization.

Sterile, debraned and cooked rice was inoculated with 5% (v/w) seed culture of M. purpureus in 500 ml conical flasks. The inoculum was incubated at 30°C in a slanting position for 14 days with intermittent shaking of the flask. After incubation the fermented red rice was dried at 45°C - 50°C for 24 h and powdered. The powder was used to prepare MPE using methanol (1:4 w/v). The extraction was carried on rotary shaker at 150 rpm for 60 min at 70°C.

This was centrifuged at 20°C with 18,000 rpm for 30 min and supernatant was collected. The supernatant was vacuum concentrated and lyophilized to completely remove methanol and water (Buchi, Flawil, Switzerland) and stored at 4°C for further analysis. The MPE dissolved in DMSO (5 mg/ml) was used to determine viable counts and inhibitory activity.

Five strains of LAB cultures maintained in MRS agar stabs were activated after two successive transfers in MRS broth at 37°C for 12 h - 15 h. Again these activated cultures were inoculated into MRS broth. After incubation at 37°C for 16 h, approximately 7 - 8 $\log_{10}$ CFU/ml were used for experimental purpose.

Determination of Viable Counts of LAB

Viable cell counts of LAB were determined by pour plate method [13]. Each 1 ml LAB suspension and MPE were added to 8 ml of sterile 0.85% saline and incubated at 37°C for 24 hours. After 24 hours the cultures were serially diluted with 9 ml saline and 1 ml of the appropriate dilution was used for selective enumeration by pour plate technique. After 24 h of incubation plates containing 25 to 250 colonies were counted and recorded as CFU. The control set of experiment for each LAB was maintained without MPE. The experiment was carried out in triplicate.

Agar Well-Diffusion Method

In vitro antibacterial activity was determined by the agar well-diffusion method [14]. The 16 hour LAB cultures were centrifuged at 8000 rpm for 10 min at 4°C. The supernatant was discarded and bacterial cells were resuspended in saline to make a suspension of approximately 7 - 8 $\log_{10}$ CFU/ml. Plating was carried out by transferring the bacterial suspension (120 µl) to a 20 ml sterile soft MRS agar medium and allowed to solidify. About 30, 40 and 50 µl of MPE (5 mg/ml) was placed in the wells and allowed to diffuse for 1 h at room temperature. Plates were incubated at 37°C for 24 h to determine inhibition zones. The assay was carried out in triplicate.

Analysis of Isoflavones

Isoflavones were analysed according to the method of Chiou and Cheng [15]. The stock solutions were prepared by dissolving daidzin, genistin, daidzein and genistein in 80% aqueous methanol (1 mg/10 ml, w/v). Each isoflavone standard solution was injected into the HPLC and the peak areas were determined. One ml of appropriately diluted MPE sample was filtered (0.45 µm) and 20 µl of sample was injected into the HPLC system (Shimadzu,

LC 10A, Japan). A reversed-phase water C_{18}Column (Spherisorb ODS 2, 4.6 × 250 mm) was used. A gradient solvent system started with 20% solvent A (methanol) and 80% solvent B (water) and progressed to 80% of A and 20% of B within 16 min followed by holding for an additional 2 min was followed with a flow rate of 1ml/min. The UV detection was carried out at 265 nm using Shimadzu diode array detector. The isoflavones in MPE were identified using standards injected under similar conditions. The concentrations of isoflavones, in 24 hours LAB culture broth supplemented with and without MPE were estimated as above.

DPPH Radical Scavenging Assay

The DPPH radical scavenging [16] activity was measured according to the method of Moon and Terao [17]. The reaction mixture was containing 1ml DPPH (500 µM in ethanol), 0.9 ml Tris-HCl buffer (100 mM, pH 7.4) and 0.1 ml LAB culture supernatant. The reaction mixture was shaken vigorously and incubated at room temperature for 30 min. The absorbance of the resulting solution was measured at 517 nm. All the experiments were carried out by maintaining appropriate blanks and controls. Antioxidant activity was calculated using following formula.

$$\text{Antioxidant activity}(\%) = \left(1 - \frac{A_{\text{sample}(517nm)}}{A_{\text{control}(517nm)}}\right) \times 100$$

Inhibition of Ascorbate Autoxidation

The inhibition of ascorbate autoxidation was estimated according to the method of Kovachich and Mishra [18]. The reaction mixture containing 0.1 ml LAB culture supernatant, 0.1 ml ascorbate solution (5.0 mM, Sigma Chemicals, USA) and 9.8 ml phosphate buffer (0.2 M, pH 7.0). The reaction mixture was incubated at 37°C for 10 min and the absorbance of this mixture was read at 265 nm. All the tests were carried out by maintaining appropriate blanks and controls. Inhibition of ascorbate autoxidation activity was calculated as follows.

$$\text{Inhibition}(\%) = \left(\frac{A_{\text{sample}}}{A_{\text{control}}} - 1\right) \times 100$$

Determination of Reducing Activity

The reducing power was determined according to the method of Oyaizu [19]. Briefly, the reaction mixture was containing 0.5 ml LAB culture supernatant,

0.5 ml sodium phosphate buffer (200 mM, pH 7.0) and 0.5 ml potassium ferricyanide (1%). This mixture was incubated at 50°C for 20 min and 0.5 ml trichloroacetic acid (10%) was added. The mixture was centrifuged at 780 g for 10 min. The upper layer (1.5 ml) was mixed with 0.2 ml of 0.1% ferric chloride and the absorbance was measured at 700 nm against appropriate blanks. A higher absorbance indicates a higher reducing power.

STATISTICAL ANALYSIS

Data was analyzed using Microsoft Excel Windows Vista (Microsoft Co., Redmond, WA, USA) and statistical analysis was performed with statistical analysis software SPSS-10 (SPSS Inc., Chicago, IL, USA). All determinations were carried out in triplicate. Statistical differences between means were determined by ANOVA and Duncan's test for multiple comparisons. P-values of <0.05 was consi-dered significant.

RESULTS

The comparative growth of the five isolates of LAB and its combination with MPE at 37°C for 24 h is shown in **Table 1**. The five LAB isolates grown in MRS broth, the highest viable count was observed in Lc (9.8 $\log_{10}$ CFU/ml) followed by Lp (9.5 $\log_{10}$ CFU/ml) and Lh (8.6 $\log_{10}$ CFU/ml). The isolates grown in MRS broth supplemented with MPE did not show any significant difference in their viable counts. While increase in the viable count of Lh (8.8 $\log_{10}$ CFU/ml) and Lp (9.7 $\log_{10}$ CFU/ml) was observed. The MPE did not show any inhibitory effects (agar well-diffusion assay) on LAB at different concentrations (**Figure 1**). These results revealed the nontoxic effect of MPE on LAB, even though it inhibited gram (+) ve and gram (−) ve bacteria. This indicates that supplementation of MPE will not affect the fermentation process.

Table 1: Cell viability of LAB grown in combination with MPE at 37°C.

Strian	LAB (CFU)	MPE + LAB (CFU)
La	$7.60 \pm 0.02^b \times 10^9$	$7.59 \pm 0.12^b \times 10^9$
Lb	$5.15 \pm 0.02^a \times 10^9$	$5.16 \pm 0.13^a \times 10^9$
Lc	$9.8 \pm 0.17^d \times 10^9$	$9.78 \pm 0.12^d \times 10^9$
Lh	$8.6 \pm 0.46^c \times 10^9$	$8.8 \pm 0.36^c \times 10^9$
Lp	$9.5 \pm 0.44^c \times 10^9$	$9.7 \pm 0.11^d \times 10^9$

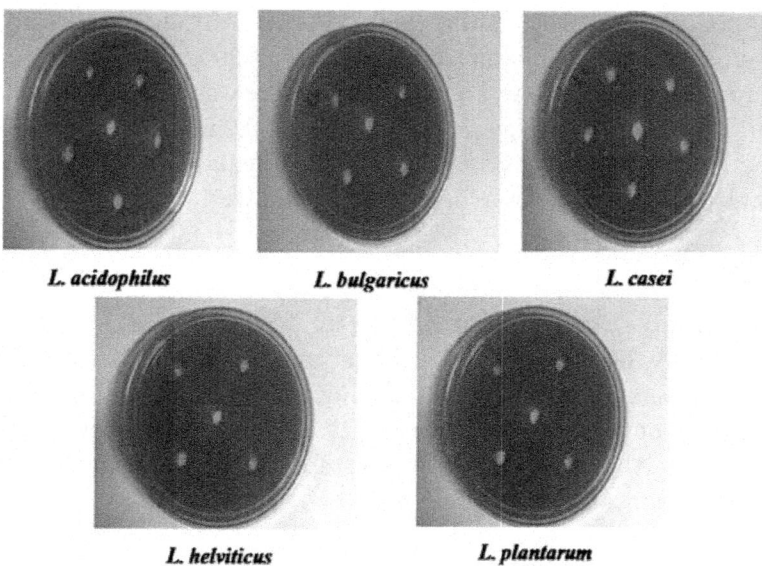

L. acidophilus   *L. bulgaricus*   *L. casei*

L. helviticus   *L. plantarum*

Figure 1: Nontoxic effect of MPE against Lactobacillus acidophilus, L. bulgaricus, L. casei, L. helviticus and L. plantarum was identified by agar well-diffusion assay.

The DPPH radical scavenging activity, inhibition of ascorbate autoxidation, reducing activity of MPE and LAB were shown in Table 2. The culture filtrates of Lc, Lp and Lh have shown higher DPPH scavenging activity (41.68%), inhibition of ascorbate (10.01%) and reducing activity (0.83 µM) respectively.

The concentration of isoflavones in MPE fermented with LAB were identified and estimated by HPLC (Figure 2).

The isoflavone isomers were eluted according to their polarity and hydrophobic interaction with the reversed phase HPLC column. Glucosidic isoflavones eluted first and then the aglycones. Within each glucosidic form of isoflavones, daidzin eluted first followed by genistin and in aglyconic form of isoflavones daidzein eluted first and then the genistein (Figure 3).

Changes occurred in the concentration of glucoside and aglycone isoflavone isomers in MPE fermented by La, Lb, Lc, Lp and Lh for 24 h at 37°C was represented in Table 3. The MPE contained total amount glucosides (20.39 mg/100 ml) and aglycones (0.37 mg/100 ml). After 24 h of incubation, the concentration of total amount aglycones in culture media of LAB was in the range of 27.98 mg/100 ml - 39.56 mg/100 ml and decrease in the concentration of glucosides isoflavones were observed. The highest concentrations of aglycones were measured in the culture medium of La followed by Lh and least concentration was estimated in Lp.

DISCUSSION

Monascus pigments are used as a food colouring agent in

Cultures	Antioxidant activity		
	DPPH scavenging (%)	Inhibition of ascorbate autoxidation (%)	Reducing activity (equivalent cysteine, μM)
MPE	35.81 ± 0.11[a]	7.31 ± 0.24[a]	0.65 ± 0.03[a]
MPE + La	39.08 ± 0.13[d]	9.98 ± 0.15[d]	0.70 ± 0.04[ab]
MPE + Lb	36.19 ± 0.20[a]	8.65 ± 0.10[c]	0.73 ± 0.03[ab]
MPE + Lc	41.68 ± 0.15[e]	8.18 ± 0.17[b]	0.77 ± 0.02[ab]
MPE + Lh	36.90 ± 0.15[b]	9.80 ± 0.02[d]	0.83 ± 0.02[b]
MPE + Lp	38.35 ± 0.28[c]	10.01 ± 0.02[d]	0.75 ± 0.04[ab]

Mean values within each column with different superscripts are significantly different by Duncan's test at $P < 0.05$.

Table 2: Antioxidant activities LAB cultured in MPE.

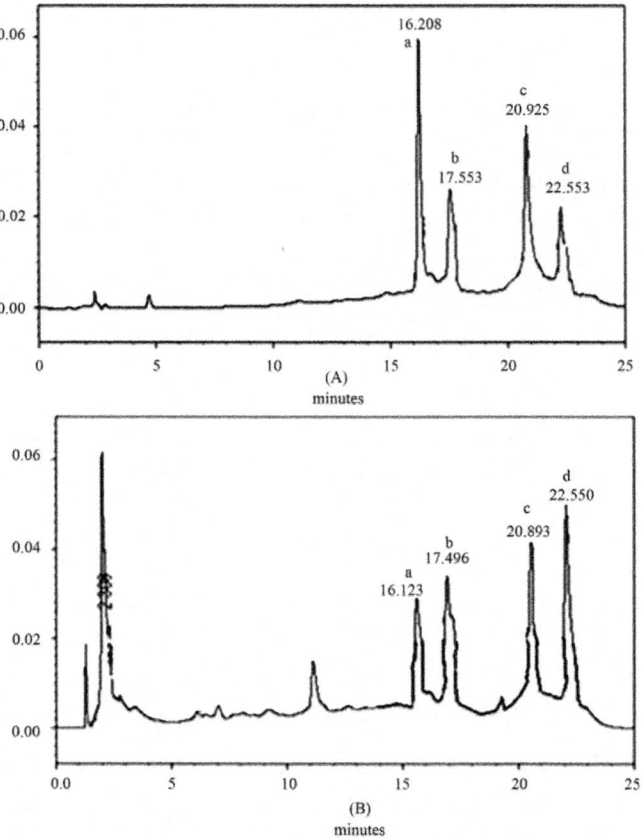

Figure 2: HPLC profile showing elution of standard isoflavones (A) daidzin (a), genistin (b), daidzein (c) and genistein (d) was used to detect the biotransformation of isoflavones by La bacteria in culture media (B) containing MPE.

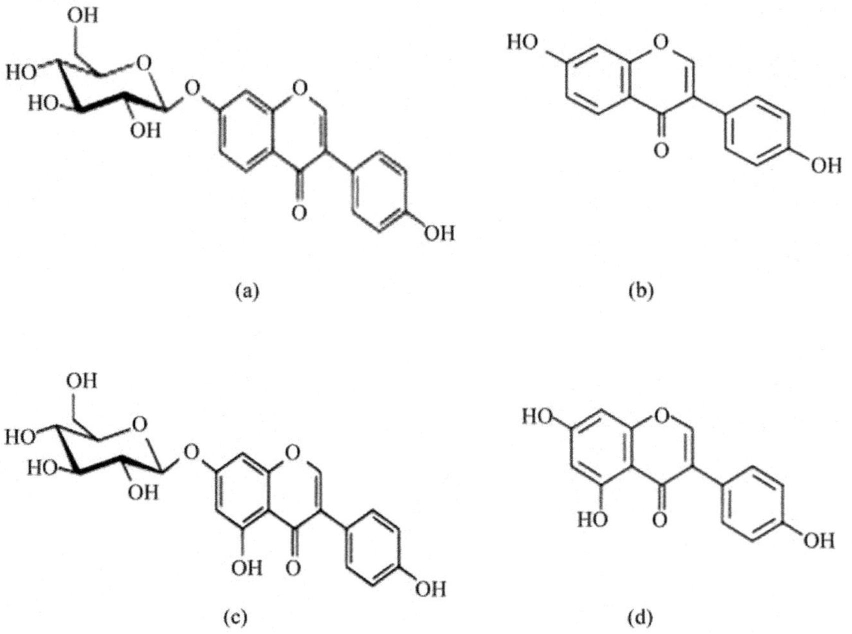

(a)   (b)

(c)   (d)

Figure 3: The isoflavone molecules daidzin (a), daidzein (b), genistin (c) and genistein (d) were identified in MPE by HPLC using standard isofla-vones.

Table 3: Isoflavone contents were estimated in LAB culture media containing MPE before and after 24 hours of incubation.

Cultures	Glucosides (mg/100 ml)			Aglycones (mg/100 ml)			Total (mg/100 ml)
	Daidzin	Genistin	Sub-total	Daidzein	Genistein	Sub-total	
MPE	3.38 ± 0.04^d	17.01 ± 0.07^d	20.39 ± 0.11^e	0.06 ± 0.01^a	0.31 ± 0.03^a	0.37 ± 0.02^a	20.76 ± 0.12^a
MPE + La	0.13 ± 0.03^c	0.79 ± 0.01^c	0.92 ± 0.011^d	7.89 ± 0.03^f	30.75 ± 0.04^f	38.64 ± 0.02^f	39.56 ± 0.02^f
MPE + Lb	0.11 ± 0.03^{bc}	0.59 ± 0.03^b	0.70 ± 0.06^c	5.97 ± 0.04^c	21.98 ± 0.03^c	27.95 ± 0.12^c	28.65 ± 0.08^c
MPE + Lc	0.05 ± 0.03^a	0.30 ± 0.01^a	0.35 ± 0.04^a	4.72 ± 0.06^b	17.16 ± 0.03^b	21.88 ± 0.03^b	22.23 ± 0.01^b
MPE + Lh	0.07 ± 0.02^{ab}	0.52 ± 0.02^b	0.59 ± 0.02^b	$7.04 \pm 0.05e$	26.86 ± 0.07^e	33.90 ± 0.07^e	34.49 ± 0.06^e
MPE + Lp	0.11 ± 0.03^{bc}	0.55 ± 0.06^b	0.66 ± 0.03^{bc}	$6.49 \pm 0.07d$	24.17 ± 0.11^d	30.66 ± 0.04^d	31.32 ± 0.07^d

Mean values within each column with different superscripts are significantly different by Duncan's test at $P < 0.05$.

a limited number of cases such as sausage, hams, fish paste, surimi and tomato ketchup. However, there are no reports on the usage of Monascus pigment in fermented foods. For preparation of fermented foods, lactic acid bacteria including Lactobacilli are the most common bacterial species considered as potential probiotics and important commensal members of the healthy human microbiota. They are useful in promotion of human health and prevention or treatment of several diseases [20]. While, antibacterial property of Monascus sp and effectiveness of monascidin-A against Bacillus, Streptococcus and Pseudomonas has been reported [21]. So, prior to increase the nutritive and nutraceutical values of fermented foods using MPE, it is necessary to evaluate the effect of MPE on LAB.

The agar well diffusion assay and viable count results revealed that, MPE do not have any bacteriostatic or bactericidal effect on LAB. Probiotic microorganisms are increasingly incorporated into food as dietary adjuncts for the purpose to benefit the human health [20] and there are several reports on applications of Monascus metabolites as a colouring [12], and antioxidant agent [6,7]. The aglycone forms of isoflavones are formed by the catalytic action of β-glucosidase [22] on glucosidic form of isoflavones during fermentation. The increase in the antioxidant capacity of LAB can be attributed to the biotransformation of glucosidic isoflavones aglycone isoflavone (genistein and daidzein). Hence, the fermented foods prepared using MPE will be more functional to reduce the oxidative stress related diseases as it contains active aglycone isoflavones. The bioconversion suggested the application of MPE along with probiotic cultures can improves the biological functionality of the fermented food products.

The MPE contains high concentration of glucoside isoflavones that are poorly bioavailable in intestine. The biological effects of M. purpureus isoflavones are not as a result of the glucosides but mainly from their aglycones such as daidzein and genistein. In intestine, the isoflavone aglycones are absorbed in greater amounts than their glycosides and are hydrolysed by the intestinal bacteria [24,25]. The variation in intestinal bacteria may occur because of illnesses, diet or age. The bioconversion and bioavailability of isoflavones depends upon the relative ability of gut microflora. Thus the intestinal bacteria cannot always be relied upon for glucoside deconjugation to release aglycones. Our results confirmed the bioconversion of biologically inactive glucosides to biologically active aglycones and nontoxic effect of MPE on LAB. The fermented foods prepared by using M. purpureus extracts may have more health beneficial effects.

ACKNOWLEDGEMENTS

Mohan-Kumari H. P. acknowledges Council of Scientific and Industrial Research, New Delhi, India and University of Mysore for supporting research through the award of Senior Research Fellowship.

REFERENCES

1. M. A. Dhale, H. P. Mohan-Kumari, S. Umesh-Kumar and G. Vijayalakshmi, "Production of Monascus Purpureus Pigments; Influenced by Amidase and Acid Protease Activity," Journal of Food Biochemistry, Vol. 35, No. 4, 2011, pp. 1231-1241.

2. H. C. Wong and P. E. Koehler, "Production and Isolation of an Antibiotic from Monascus Purpureus and Its Relationship to Pigment Production," Journal of Food Science, Vol. 46, 1981, No. 2, pp. 589-592. doi:10.1111/j.1365-2621.1981.tb04917.x

3. Y. Su, J. J. Wang, T. T. Lin and T. M. Pan, "Production of the Secondary Metabolites c-Aminobutyric Acid and Monacolin K by Monascus," Journal of Industrial Microbiology and Biotechnology, Vol. 30, 2003, pp. 40-46.

4. G. F. Wu and X. C. Wu, "Screening DPPH Radical ScavEngers from Monascus sp," Acta Microbiologica Science, Vol. 40, No. 4, 2000, pp. 394-399.

5. C. L. Lee, T. Y. Tsai, J. J. Wang and T. M. Pan, "In Vivo Hypolipidemic Effects and Safety of Low Dosage Monascus Powder in a Hamster Model of Hyperlipidemia," Applied Microbiology and Biotechnology, Vol. 70, No. 5, 2006, pp. 533-540. doi:10.1007/s00253-005-0137-0

6. Y. Aniya, T. Yokomakura, M. Yonamine, K. Shimada, T. Nagamine, M. Shimabukuro and H. Gibo, "Screening of Antioxidant Action of Various Molds and Protection of Monascus Anka against Experimentally Induced Liver Injuries of rats," General Pharmacology, Vol. 32, No. 2, 1999, pp. 225-231. doi:10.1016/S0306-3623(98)00183-9

7. M. A. Dhale, S. Divakar, S. Umesh-Kumar and G. Vijayalakshmi, "Isolation and Characterization of Dihydromonacolin-MV from Monascus Purpureus for Antioxidant Properties," Applied Microbiology and Biotechnology, Vol. 73, No. 5, 2007, pp. 1197-1202. doi:10.1007/s00253-006-0578-0

8. M. A. Dhale, S. Divakar, S. Umesh-Kumar and G. Vijayalakshmi, "Characterization of Dehydromonacolin-MV2 from Monascus Purpureus Mutant," Journal of Applied Microbiology, Vol. 130, No. 3, 2007, pp. 2168-2173. doi:10.1111/j.1365-2672.2007.03457.x

9. H. P. Mohan-Kumari, K. A. Naidu, S. Vishwanatha, K. Narasimhamurthy and G. Vijayalakshmi, "Safety Evaluation of Monascus Purpureus Red Mould Rice in Albino Rats," Food and Chemical Toxicology, Vol. 47, No. 8, 2009, pp. 1739-1746. doi:10.1016/j.fct.2009.04.038

10. L. Martinkova, P. Juzlova, V. Kren, Z. Kucerouva, V. Havlicek, P. Olsovsky, O. Hovorka, B. Rihova, D. Vesly, D. Vesela, J. Ulrichova and V. Prikrylova, "Biological ActiVities of Oligoketide Pigments of Monascus Purpureus," Food Additive and Contaminants, Vol. 16, No. 1, 1999, pp. 15-24. doi:10.1080/026520399284280

11. C. Kim, H. Jung, Y. O. Kim and C. S. Shin, "Antimicrobial Activities of Amino Acid Derivatives of Monascus Pigments," FEMS Microbiology Letters, Vol. 264, No. 1, 2006, pp. 117-124. doi:10.1111/j.1574-6968.2006.00451.x

12. H. Jung, C. Kim and C. S. Shin, "Enhanced Photostability of Monascus Pigments Derived with Various Amino Acids via Fermentation," Journal of Agriculture and Food Chemistry, Vol. 53, No. 18, 2005, pp. 7108-7114. doi:10.1021/jf0510283

13. J. C. De Man, M. Rogosa and M. E. Sharpe, "A Medium for the Cultivation of Lactobacilli," Journal of Applied Bacteriology, Vol. 23, No. 1, 1960, pp. 130-135. doi:10.1111/j.1365-2672.1960.tb00188.x

14. P. K. Mukherjee, R. Balasubramanian, K. Saha, B. P. Saha and M. Pal, "Antibacterial Efficiency of Nelumbo nucifera (Nymphaeaceae) Rhizomes Extract," Indian Drugs, Vol. 32, No. 6, 1995, pp. 274-276.

15. R. Y. Y. Chiou and S. L. Cheng, "Isoflavone Transformation during Soybean Koji Preparation and Subsequent Miso Fermentation Supplemented with Ethanol and NaCl," Journal of Agriculture and Food Chemistry, Vol. 49, No. 8, 2001, pp. 3656-3660.doi:10.1021/jf0015241

16. M. Blois, "Antioxidant Determinations by the Use of a Stable Free Radical," Nature, Vol. 181, 1958, pp. 1199-1200. doi:10.1038/1811199a0

17. J. H. Moon and J. Terao, "Antioxidant Effect of Caffeeic Acid and Dihydrocaffeeic Acid in Lard and Human Low-Density Lipoprotein," Journal of Agriculture and Food Chemistry, Vol. 46, No. 12, 1998, pp. 5062-5065. doi:10.1021/jf9805799

18. G. B. Kovachich and O. P. Mishra, "Stabilization of AsCorbic Acid and Norepinephrine in Vitro by the Subcellular Fractions of Rat Cerebral Cortex," Neuroscience Letters, Vol. 52, No. 1-2, 1984, pp. 153-158. doi:10.1016/0304-3940(84)90366-5

19. M. Oyaizu, "Studies on Product of Browning Reaction Prepared from Glucose Amine," Japanese Journal of Nutrition, Vol. 44, No. 6, 1986, pp. 307-315.doi:10.5264/eiyogakuzashi.44.307

20. F. Guarner and J. R. Malagelada, "Gut Flora in Health and Disease," Lancet, Vol. 361, No. 9356, 2003, pp. 512-519. doi:10.1016/S0140-6736(03)12489-0

21. H. C. Wong and Y. S. Bau, "Pigmentation and Antibacterial Activity of Fast-Neutron and X-Ray Induced Strains of M. purpureus Went," Plant Physiology, Vol. 60, No. 4, 1977, pp. 578-581. doi:10.1104/pp.60.4.578

22. D. O. Otieno, J. F. Ashton and N. P. Shah, "Stability of β-Glucosidase Activity Processing by Bifidobacterium and Lactobacillus spp in Fermented Soymilk during Processing and Storage," Journal of Food Science, Vol. 70, No. 4, 2005, pp. 236-241.doi:10.1111/j.1365-2621.2005. tb07194.x

23. W. H. Hsu, B. H. Lee and T. M. Pan, "Red Mold Dioscorea-Induced G2/M Arrest and Apoptosis in Human Oral Cancer Cells," Journal Science and Food Agriculture, Vol. 90, No. 15, 2010, pp. 2709-2715. doi:10.1002/ jsfa.4144

24. K. Setchell, N. Brown, N. L. Zimmer, W. Brashear, B. Wolfe, A. Kirschour and J. Heubi, "Evidence for Lack of Absorption of Soy Isoflavone Glycosides in Humans, Supporting the Crucial Role of Intestinal Metabolism for Bioavailability," American Journal of Clinical Nutrition, Vol. 76, No. 2, 2002, pp. 447-453.

25. Y. Kawakami, W. Tsurugasaki, S. Nakamura and K. Osada, "Comparison of Regulative Functions between Dietary Soy Isoflavones Aglycone and Glucoside on Lipid Metabolism in Rats Fed Cholesterol," Journal of Nutritional Biochemistry, Vol. 16, No. 4, 2005, pp. 205-212. doi:10.1016/j. jnutbio.2004.11.005

Chapter 4

PREDOMINANT LACTIC ACID BACTERIA INVOLVED IN THE TRADITIONAL FERMENTATION OF FUFU AND OGI, TWO NIGERIAN FERMENTED FOOD PRODUCTS

Olaoluwa Oyedeji[1], Samuel Temitope Ogunbanwo[2], Anthony Abiodun Onilude[2]

[1] Department of Microbiology, Obafemi Awolowo University, Ile-Ife, Nigeria;

[2] Department of Microbiology, University of Ibadan, Ibadan, Nigeria.

ABSTRACT

Traditional methods of preparation were simulated in the laboratory fermentations of cassava and maize to produce fufu and ogi respectively. Changes in pH, temperature and titratable acidity, as well as the diversity of lactic acid bacteria species were investigated during both fermentations. Lactic acid bacteria strains involved in the fermentation processes were isolated at twelve hourly intervals, characterized and identified using phenotypic and biochemical methods. A rapid decrease in pH, 5.6 to 3.7 in fufu and 5.9 to 3.8 in ogi, were observed with temperature increasing from 26°C to 30°C and 25°C to 31°C in fufu and ogi respectively. Most of the lactic acid bacteria strains isolated were homofermentative and heterofermentative Lactobacillus species and heterofermentative Leuconostoc species. Lactobacillus plantarum and Leuconostoc mesenteroides were the dominant lactic acid bacteria species in fufu while L. cellobiosus, L. plantarum and Lc. lactis were dominant in ogi fermentation. An ecological succession pattern in which Leuconostoc species were mostly isolated during early stages of fermentation with the final stages populated with Lactobacillus species was observed in both cases and is attributable to differential acid tolerance of the two genera. The frequencies of dominance of the strains in fufu were L. plantarum (56.25%), Lc. mesenteroides (18.75%), L. lactis (6.25%), L. coprophillus (6.25%), L.

acidophilus (6.25%) and L. brevis (6.25%). The frequencies of dominance in ogi were L. cellobiosus (26.6%), Lc. lactis (26.6%), L. plantarum (20.0%), L. acidophilus (13.33%) and Lc. paramesenteroides (13.33%). The dominant strains can serve as potential starter cultures for fufu and ogi production.

INTRODUCTION

Fermented foods are consumed throughout the world and traditional fermentation processes such as those involved in the production of fermented dairy products and alcoholic beverages have been performed for thousands of years [1]. These food products result from the activities of microorganisms which modify the flavour and texture and increase long term product stability [2]. Lactic acid bacteria constitute an important group of these organisms and have been associated with production of fermented foods and feeds for many centuries [3]. They are important in the production of many fermented foods such as sauerkraut, silage, sourdough, dry fermented sausages and cheeses [4]. Some of the reasons for their widespread use are the ability to retard spoilage, preserve food as well as improve flavour and texture of foods. They also play fundamental role in microbial ecology of foods by synthesizing a variety of antimicrobial compounds such as organic acids, hydrogen peroxide, diacetyl and bacteriocins [5,6]. They are thus able to inhibit many microorganisms including spoilage and pathogenic organisms. They are increasingly being recognised for their health and nutritional benefits hence some strains are used as probiotics [7-10]. Among the many African, traditionally fermented foodstuffs are fufu and ogi. Fufu, a product of an acidfermented cassava root tuber serves as main course meals in most areas of Nigeria and Africa as a whole [11]. It is prepared by natural fermentation which transforms the cassava root from rapid spoilage after harvest as cassava roots are more perishable than other tuber crops such as yam and sweet potato. Cassava (Manihot esculenta) is a perennial woody shrub with an edible root which grows in tropical and sub-tropical areas of the world [12]. In Africa and Latin America, it is mostly used for human consumption, while in Asia and parts of Latin America, it is also used commercially for the production of animal feeds and starch-based products [13]. Other products obtained from fermented cassava roots are gari and lafun. Ogi, an acid-fermented cereal gruel, is a major staple sour porridge food widely taken in West Africa. It also serves as the traditional infant weaning food [11]. It is generally obtained by fermenting maize grains (Zea mays). In general, a wide spectrum of microorganisms is involved during fermentation processes but a few types usually determine the quality of the end product [14]. Given adequate environmental conditions, a particular microbial community will determine the quality of a specific food. The origin, development and succession of a particular microbial community in any food are thus governed by its ecological factors, which influence the physiological expression of microbial cells [15]. Thus strategies for food processing and preservation can be developed on the basis of ecological factors associated with specific foods and beverages [16]. Yeasts and wide varieties of bacteria strains were found to be associated with the fermentation of cassava roots and cereal grains where they contribute significantly to starch breakdown, acidification, detoxification and flavour enhancement [17]. Yeasts and lactic acid bacteria are therefore the most common microorganisms in a wide variety of traditional food and beverage fermentations [18]. Lactic acid bacteria are found in various stages of fermentation processes where they are useful in flavour and aroma development. They also inhibit spoilage bacteria and pathogens and confer several intestinal health and other health benefits related to blood cholesterol levels, immune competence and antibiotics [19]. The quality of the final product is a factor of the diversity and composition of microorganisms and their dynamics and frequency of occurrence. The objective of this research was therefore to isolate and characterize predominant lactic acid bacteria species during the course of traditional fufu and ogi fermentations.

MATERIALS AND METHODS

Collection of Raw Materials

Unbroken maize grains (Zea mays) and freshly harvested cassava root tubers (Manihot esculenta) were obtained from the Bodija market in Ibadan, Oyo State, Nigeria. They were brought in sterile polythene bags to the laboratory for immediate processing.

Laboratory Preparation of Fufu and Ogi

The laboratory fermentation of cassava tubers and maize grains to produce fufu and ogi respectively, were carried out by simulating the traditional methods of processing (Figures 1 and 2).

Physicochemical Analysis

The hydrogen ion concentration (pH) and temperature changes of the fermenting cassava mash and maize grains were measured at twenty four hourly intervals for 72 h using a Pye-Unicam pH meter and Mercury thermometer respectively [20]. The total titratable acid was determined as percentage lactic acid by titrating 25 ml of the samples used for pH determinations against 0.1 N NaOH.

The volumes of the 0.1 N NaOH used were noted. Duplicate determinations were made for each analysis [21].

Microbiological Isolation

Twenty five millilitre portions were aseptically removed at different stages of fermentation processes: raw water used for steeping cassava roots and maize grains; steep waters (sampled each day for 72 h); water used for mashing steeped cassava and for milling ogi; the steep water used further fermentation of ogi and the final products of fermentation. Each sample was homogenized with 200 ml sterile 0.1% peptone water (Oxoid, UK). This was then serially diluted and 0.1 ml from appropriate dilutions were spread plated on MRS agar plates (Oxoid, UK) in duplicates and incubated in Gas Pak jars (GasPak System, BBL) at 30°C for 72 h. Colonies with distinct morphological differences such as colour, size and shapes were randomly picked from MRS agar plates as presumptive lactic acid bacteria isolates and repeatedly streaked on fresh MRS agar plates to purify the isolates. They were then maintained on appropriate slants at 4°C.

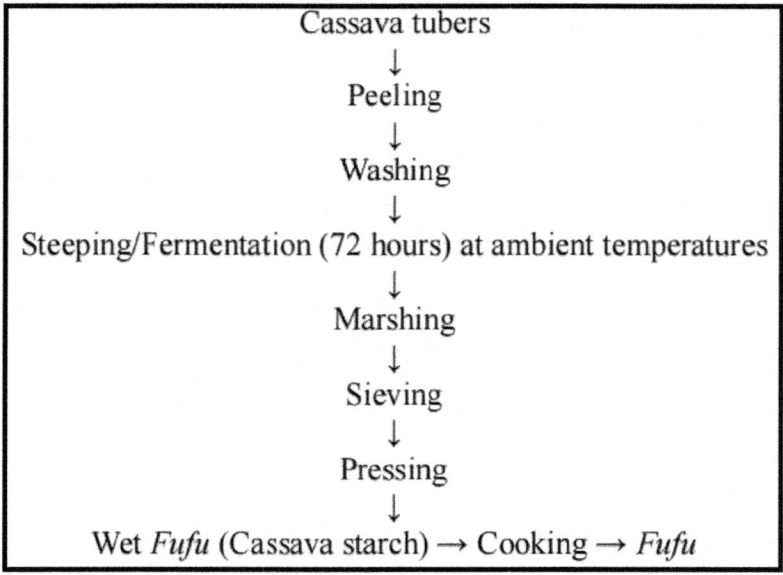

Figure 1: Traditional method of fufu processing

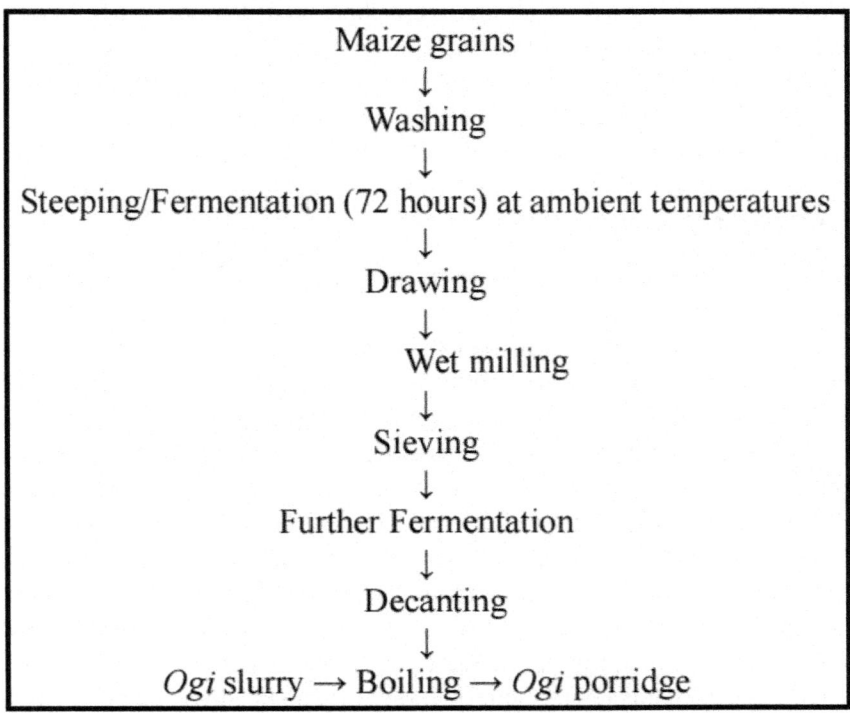

Maize grains
↓
Washing
↓
Steeping/Fermentation (72 hours) at ambient temperatures
↓
Drawing
↓
Wet milling
↓
Sieving
↓
Further Fermentation
↓
Decanting
↓
Ogi slurry → Boiling → *Ogi* porridge

Figure 2: Traditional method of ogi processing.

Characterisation and Identification

Each of the lactic acid bacteria isolate was initially examined for colonial and cell morphologies, cell arrangement, spore formation and motility. Only the Gram positive, catalase negative and non spore forming isolates were then characterized by phenotypic and biochemical tests. An overnight culture (inoculums) of each isolate in MRS broth was used for all tests incubated anaerobically (GasPak System, BBL) at 30°C. The lactic acid bacteria isolates were tested for fermentation of the following carbohydrates (Sigma, Germany): D-Glucose, lactose, sucrose, galactose, maltose, mannitol, sorbitol, mannose, L-arabinose, D-xylose, cellobiose, dulcitol, inositol, raffinose, rhamnose, inulin and salicin. Bromocresol purple broth base was used as basal medium. One percent filtersterilized sugar solution using 0.2 µm Millipore filter (Corning) was added aseptically into sterilized bromocresol purple broth base before inoculation with 18 - 24 h old culture of each lactic acid bacteria strain. The results were assessed with reference to an uninoculated control after anaerobic incubation at 30°C for 5 d. Tubes in which bromocresol purple colour changed to yellow indicated utilisation of sugar or acid production. The various lactic acid bacteria strains were then identified by reference to the Bergey's Manual of Systematic Bacteriology [22] and The Genera of Lactic Acid Bacteria [23] based on the results of the various tests. The identity of the lactic acid bacteria isolates were further confirmed by using the API 50 CHL tests and the Computer Program APILAB Plus (BioMerieux, France).

Results and Discussion

The pH of the fermenting cassava roots decreased from 5.6 to 3.7 during the 72 h fermenting period. Correspondingly, the temperature increased from 26°C to 30°C. The total titratable acidity (in % lactic acid) increased from 0.07 ± 0.01 to 0.21 ± 0.01 and decreased to 0.09 by the end of 72 h fermentation period (Table 1). The pH of fermenting maize grains dropped from 5.9 to 3.8 by the end of the 72 h fermentation period. Correspondingly, the temperature increased from 25°C to 31°C during the same period while the total titratable acidity increased from 0.13 ± 0.01 to 0.28 ± 0.01 after 24 hours and then decreased to 0.14 by the end of fermentation period (Table 1). During both fufu and ogi fermentation processes, the temperature of the fermenting materials increased as fermentation progressed. The two processes are thus exothermic in nature and the heat generated might have resulted from the metabolic activities of the fermenting organisms [24]. The fermenting organisms do contribute to acidity attributable to the production of lactic acid and acetic acid during the processes which exerts a depressive effect on the pH of the fermenting materials [23,25]. Despite the unhygienic wet-milling and wet-sieving processes involved in the traditional preparation of both fufu and ogi, the low pH of the fermented products would make them safe for consumption. Also the acidic fermentation and lactic acid metabolites are responsible for inactivation of enterobacteriaceae including toxin-producing and foodborne infectious pathogens [26-29]. These result in an improvement of the aroma, flavor, texture, safety and shelf life of the food. A total of 32 lactic acid bacteria strains were isolated from the various phases of the fermentation of cassava for fufu production while 14 strains were obtained from fermenting maize steep for ogi production. The isolates were subjected to various morphological, physiological and biochemical characterizations. The characteristics exhibited by the isolates were compared with those of standard strains for their identification [22,30]. All the isolates were Gram positive and catalase negative rods or cocci. All produced no endospore and were non motile. They were facultative anaerobes and were fermentative rather than being oxidative in nature (Table 2). Table 3 shows the periodic distribution of dominant lactic acid bacteria strains during fermentation processes of fufu and ogi. Strains isolated from fufu fermentation included Lactobacillus plantarum, L. lactis, L. coprophillus, L. acidophilus,

L. brevis and Leuconostoc mesenteroides while L. plantarum, L. cellobiosus, L. acidophilus, Lc. lactis and Lc. paramesenteroides were isolated from ogi fermentation. Ecological succession patterns in which the Leuconostoc species occur mostly at initial stages and the Lactobacillus species at later stages were observed in both fermentations. The most frequently isolated LAB species in fufu fermentation were L. plantarum (18 strains, 56.25%) and Lc. mesenteroides (6 strains, 18.75%) while in ogi, both L. cellobiosus and Lc. lactis had the highest percentage occurrence of 26.67%

Table 1: pH, temperature and titratable acid changes during the fermentation of cassava and maize for fufu and ogi production.

	Cassava (*Fufu*)			Maize (*Ogi*)		
Time (h)	Temperature (°C)	pH	Titratable acidity (% lactic acid)	Temperature (°C)	pH	Titratable acidity (% lactic acid)
0	26	5.6	0.07 ± 0.01	25	5.9	0.13 ± 0.01
24	28	5.0	0.21 ± 0.01	27	5.2	0.28 ± 0.01
48	29	4.4	0.13 ± 0.01	29	4.5	0.21
72	30	3.7	0.09	31	3.8	0.14

Table 2: Morphological and biochemical characterisation of isolates.

Sample source	Fufu						Ogi				
No. of strains	18	2	2	2	2	6	2	4	2	4	2
Gram reaction/ Morphology	GPR	GPR	GPR	GPR	GPR	GPC	GPR	GPR	GPR	GPC	GPC
Catalase	-	-	-	-	-	-	-	-	-	-	-
Motility	-	-	-	-	-	-	-	-	-	-	-
Methyl red	+	+	+	+	+	+	+	+	+	+	+
Voges Proskauer	-	-	-	-	-	-	-	-	-	-	-
Spore stain	-	-	-	-	-	-	-	-	-	-	-
Oxidative/ Fermentative	F	F	F	F	F	F	F	F	F	F	F
Dextran from sucrose	-	-	+	-	-	+	-	-	-	-	-
Fermentation pattern	HMF	HMF	HTF	HTF	HMF	HTF	HMF	HTF	HMF	HMF	HTF
O₂ relationship	FA	FA	FA	FA	FA	FA	FA	FA	FA	FA	FA
Growth in 4% NaCl	+	-	-	-	+	-	+	+	-	+	-
Ammonia from arginine	-	+	+	-	-	-	-	-	-	-	-
Fermentation:											
Glucose	+	+	+	+	+	+	+	+	+	+	+
Lactose	+	+	+	+	+	+	+	+	+	+	-
Sucrose	-	+	+	+	+	-	+	-	+	+	-
Galactose	+	+	+	+	+	-	+	+	+	+	+
Maltose	+	+	+	+	+	-	+	+	+	+	+
Mannitol	+	-	-	+	+	-	+	+	+	+	+
Sorbitol	+	-	-	+	+	-	+	+	+	+	+
Mannose	+	-	-	+	+	-	+	+	+	+	-
Arabinose	-	+	+	+	+	-	+	-	+	+	+
Xylose	-	-	-	-	+	-	-	+	-	-	-
Cellobiose	+	-	-	+	-	-	+	-	+	+	-
Dulcitol	+	-	-	-	-	-	+	-	-	-	-
Inositol	+	-	-	-	-	-	-	+	-	-	-
Inulin	+	-	-	+	-	-	+	+	+	+	-
Raffinose	+	+	+	-	-	+	+	-	-	+	+
Rhamnose	+	-	-	+	+	-	-	+	+	-	-
Salicin	+	+	+	+	-	-	+	-	+	+	-
Identification	L. plantarum	L. lactis	L. coprophilus	L. acidophilus	L. brevis	Lc. mesenteroides	L. plantarum	L. cellobiosus	L. acidophilus	Lc. lactis	Lc. paramesenteroides

Table 3: Periodic distribution of dominant LAB strains during the fermentation of cassava and maize for fufu and ogi production.

Period (h)	Fermentation	Dominating strains of LAB
0	Fufu	L. plantarum, L. brevis, L. coprophillus, Lc. mesenteroides
	Ogi	L. plantarum, L. cellobiosus, Lc lactis, Lc. paramesenteroides
24	Fufu	L. plantarum, Lc. mesenteroides, L. acidophilus, L. lactis, L. brevis, L. coprophillus
	Ogi	L. plantarum, Lc. lactis, Lc. paramesenteroides, L. acidophilus, L. cellobiosus,
48	Fufu	L. plantarum, L. acidophilus, L. lactis, L. brevis, Lc. mesenteroides
	Ogi	L. plantarum, L. cellobiosus, L. acidophilus, Lc. lactis
72	Fufu	L. plantarum, L. acidophilus, L. brevis
	Ogi	L. plantarum, L. cellobiosus,

Followed by L. plantarum with occurrence of 20.0% (Figures 3 and 4). The frequencies of dominance of the lactic acid bacteria strains in fufu fermentation were L. plantarum 56.25%, Lc. mesenteroides 18.75%, L. lactis 6.25%, L. coprophillus 6.25%, L. acidophilus 6.25% and L. brevis 6.25%. In ogi fermentation, they were L. cellobiosus 26.67%, Lc. lactis 26.67%, L. plantarum 20.0%, Lc. paramesenteroides 13.33% and L. acidophilus 13.33%. Lc. mesenteroides and L. plantarum have been reported as the dominant microorganisms implicated in several natural and spontaneous lactic acid fermentation of vegetables [31,32]. L. plantarum was also found to dominate the lactic acid bacteria flora in various food fermentation processes for akamu [33], ogi [34], boza [35], cachaca [36], tempoyak [37], ting [38] and togwa [39]. Spontaneous fermentation typically results from the competitive activities of different microorganisms whereby strains best adapted and with the highest growth rate will dominate particular stages of the process [38]. Among the bacteria associated with food fermentation, lactic acid bacteria are of predominant importance. Samples from plant materials were reported to show the greatest diversity of lactic acid bacteria with the Lactobacillus strains being predominant in food-related ecosystems [40]. Yeasts and lactic acid bacteria are implicated in the fermentation of a wide variety of traditional food and beverage fermentations [14,18,33,39]. While yeasts are known to facilitate alcoholic fermentations, lactic acid bacteria produce lactic acid as part or major by product from the fermentation of carbohydrates [25,41]. Fufu and ogi are products of acid fermentations by lactic acid bacteria. This group of bacteria includes several genera that differ considerably in morphological, physiological and functional properties. Lc. mesenteroides mostly occur at early stages of most vegetable fermentations with L. plantarum predominating towards the end of the processes [14]. This ecological succession is attributed to the differences in the acid tolerance of the two genera. It may also be the reason for the disappearance of the entero bacteriaceae which are observed at early periods during fermentation of plant materials [14]. The water used for steeping or surface microflora of raw materials may be sources of such enterobacteriaceae.

The methods of preparation, water used for steeping, surface microflora of raw materials, atmosphere where processing takes place and the handlers all affect the types of microorganisms isolated during fermentation processes for fufu and ogi. The various species of lactic acid bacteria are however better adapted to the substrates and overcome the possible physiological and biochemical hurdles during the phases of each fermentation process. Fufu and ogi, as at present, are traditionally prepared by spontaneous fermentation relying on the indigenous flora of the raw materials. This leads to a poorly controlled process and variations in the products. For consistency in the quality of these fermented food products, the development of specific starter cultures is important.

CONCLUSION

It is concluded that an association of lactic acid bacteria composed of homofermentative and heterofermentative Lactobacillus species and heterofermentative Leuconostoc species are involved from the beginning to the end of both cassava and maize fermentations for the production of fufu and ogi respectively. Factors which influence these food fermentations include temperature, acidity and hydrogen ion concentration (pH). The assessment of the performance of each of the species under controlled fermentation will enable selection of the best strains that can serve as potential starter culture for the production of microbiologically stable and predictable end products.

REFERENCES

1. K. O. H. Steinkraus, "Indigenous Fermented Foods Involving Acid Fermentation/ Acid Fermented Cereal Gruel," In: Handbook of Indigenous Fermented Foods, Marcel Dekker Inc., 1983, pp. 189-218.

2. W. Holzapfel, "Use of Starter Culture in Fermentation on a Household Scale," Food Control, Vol. 8, No. 5-6, 1997, pp. 241-258. http://dx.doi.org/10.1016/S0956-7135(97)00017-0

3. I. Sanni, A. A. Onilude, S. T. Ogunbanwo and S. I. Smith, "Antagonistic Activity of Bacteriocin Produced by Lactobacillus species from Ogi, an Indigenous Fermented Food," Journal of Basic Microbiology, Vol. 39, No. 3, 1999, pp. 189-195. http://dx.doi.org/10.1002/(SICI)1521-4028(199906)39:3< 189::AID-JOBM189>3.0.CO;2-R

4. H. Rose, "Economic Microbiology, Microbiology of Fermented Foods,"

Academic Press, London, 1982.

5.   T. R. Klaenhammer, "Bacteriocins of Lactic Acid Bacteria," Biochemie, Vol. 70, No. 3, 1988, pp. 337-349. http://dx.doi.org/10.1016/0300-9084(88)90206-4

6.   M. V. Leal, M. Baras, J. L. Ruiz-Barba, B. Floriano and R. Jimenez-Diaz, "Bacteriocin Production and Competitiveness of Lactobacillus plantarum LPCO10 in Olive Juice Broth, a Culture Medium Obtained from Olives," International Journal of Food Microbiology, Vol. 43, No. 1-2, 1998, pp. 129-134. http://dx.doi.org/10.1016/S0168-1605(98)00079-8

7.   S. E. Gilliland, "Health and Nutritional Benefits of Lactic Acid Bacteria," FEMS Microbiology Reviews, Vol. 87, No. 1-2, 1990, pp. 175-188.

8.   S. Salminen, M. Deighton and S. Gorbach, "Lactic Acid Bacteria in Health and Disease," In: S. Salminen and A. Van Wright, Eds., Lactic Acid Bacteria, Marcel Dekker Inc., New York, 1993, pp. 234-294.

9.   M. Chapman, G. R. Gibson and I. Rowland, "Health Benefits of Probiotics: Are Mixtures More Effective than Single Strains?" European Journal of Nutrition, Vol. 50, No. 1, 2011, pp. 1-17. http://dx.doi.org/10.1007/S00394-010-0166-Z

10.   M. V. Herias, C. Hessle, E. Telemo, T. Midtvedt, L. A Hanson and A. E. Wold, "Immunomodulatory Effects of Lactobacillus plantarum Colonizing the Intestine of Gnotobiotic Rats," Clinical and Experimental Immunology, Vol. 116, No. 2, 1999, pp. 283-290. http://dx.doi.org/10.1046/j.1365-2249.1999.00891.x

11.   S. A. Odunfa, "African Fermented Foods," In: B. J. B. Wood, Ed., Microbiology of Fermented Foods, Vol. 2, Elsevier Applied Science Publishers, London, 1985, pp. 155-191.

12.   M. M. Burrell, "Starch: The Need for Improved Quality or Quantity and Overview," Journal of Experimental Botany, Vol. 218, No. 382, 2003, pp. 451-456.

13.   FAO, "Production Yearbook," Vol. 44, 1990, FAO, Rome.

14.   K. Abegaz, "Isolation, Characterization and Identification of Lactic Acid Bacteria Involved in Traditional Fermentation of Borde, an Ethiopian Cereal Beverage," African Journal of Biotechnology, Vol. 6, No. 12, 2007, pp. 1469- 1478.

15.   T. Deak and L. R. Beuchat, "Handbook of Spoilage Yeasts," CRC Press Inc., Boca Raton, 1996, pp. 1-36,61- 154.

16.   G. W. Gould, "Ecosystem Approaches to Food Preservation," Journal of Applied Bacteriology, Vol. 73, No. S21, 1992, pp. 58S-68S.

17. O. B. Oyewole, "Fermentation of Cassava for Lafun Production," Food Laboratory News, Vol. 17, No. 2, 1991, pp. 29-31.

18. S. K. Soni, D. K. Sandhu, K. S. Vikhu and N. Kamra, "Microbiological Studies on Dosa Fermentation," Food Microbiology, Vol. 3, No. 1, 1986, pp. 45-53. http://dx.doi.org/10.1016/S0740-0020(86)80025-9

19. W. E. Sandine, "Looking Backward and Forward at the Practical Applications of Genetic Research on Lactic Acid Bacteria," FEMS Microbiology Reviews, Vol. 46, No. 3, 1987, pp. 205-220.

20. AOAC, "Official Methods of Analytical Chemists," AOAC, Arlington, 1984.

21. Pearson, "Laboratory Techniques in Food Analysis," Butterworth, London, Boston, 1973, pp. 50-57.

22. P. H. A. Sneath, N. S. Mair, M. E. Sharpe and J. G. Holt, "Bergey's Manual of Systematic Bacteriology," Williams Wilkins, Baltimore, 1986.

23. B. J. B. Wood and W. H. Holzapfel, "The Genera of Lactic Acid Bacteria," Blackie Academic and Professional, Chapman and Hall, Glasgow, 1995.

24. P. K. Sarkar and J. P. Tamang, "The Influence of Process Variation and Inoculum Composition on the Sensory Qualities of Kinema," Food Microbiology, Vol. 11, No. 4, 1994, pp. 317-325. http://dx.doi.org/10.1006/fmic.1994.1036

25. K. H. Steinkraus, "Fermentations in World Food Processing," Comprehensive Reviews in Food Science and Food Technology, Vol. 1, No. 1, 2002, pp. 23-32.

26. W. S. M. Lorri, "Nutritional and Microbiological Evaluation of Fermented Cereal Weaning Foods," Ph.D. Thesis, Department of Food Sciences, Chammers University of Technology, Gotenborg, 1993.

27. Y. B. Byaruhanga, B. H. Bester and T. G. Watson, "Growth and Survival of Bacillus cereus in Mageu, a Sour Maize Beverage," World Journal of Microbiology and Biotechnology, Vol. 15, No. 3, 1999, pp. 329-333. http://dx.doi.org/10.1023/A:1008967117381

28. N. F. Kunene, J. W. Hastings and A. Von Holy, "Bacterial Populations Associated with a Sorghum-Based Weaning Cereal," International Journal of Food Microbiology, Vol. 49, No. 1-2, 1999, pp. 75-83. http://dx.doi.org/10.1016/S0168-1605(99)00062-8

29. T. C. Ana, A. P. Rosinea, C. M. Hilario and A. M. Celia, "Inhibition of Listeria monocytogenes by Lactic Acid Bacteria Isolated from Italian Salami," Food Microbiology, Vol. 23, No. 3, 2006, pp. 213-219. http://dx.doi.org/10.1016/j.fm.2005.05.009

30. M. E. Sharpe, "Identification Methods for Microbiologists," 2nd Edition, Acad. Press Soc., 1979, pp. 233-259.

31. L. G. McDonald, H. P. Flemming and H. M. Hassan, "Acid Tolerance of Leuconostoc mesenteroides and Lactobacillus plantarum," Applied and Environmental Microbiology, Vol. 57, No. 7, 1990, pp. 2120-2124.

32. Makimattila, M. Kahala and V. Joutsjoki, "Characterization and electrotransformation of Lactobacillus plantarum and Lactobacillus paraplantarum Isolated from Fermented Vegetables," World Journal of Microbiology and Biotechnology, Vol. 27, No. 2, 2011, pp. 371-379. http://dx.doi.org/10.1007/s11274-010-0468-6

33. P. C. Obinna-Echem, V. Kuri and J. Beal, "Evaluation of the Microbial Community, Acidity and Proximate Composition of Akamu, a Fermented Maize Food," Journal of the Science of Food and Agriculture, 2013. http://dx.doi.org/10.1002/.jsfa.6264

34. M. L. Johansson, A. Sanni, C. Lonner and G. Molin, "Phenotypically Based Taxonomy using API 50CH of Lactobacilli from Nigerian Ogi, and the Occurrence of Starch Fermenting Strains," International Journal of Food Microbiology, Vol. 25, No. 2, 1995, pp. 159-168. http://dx.doi.org/10.1016/0168-1605(94)00096-O

35. V. Getcheva, S. S. Pandiella, A. Angelov, Z. G. Roshkova and C. Webb, "Microflora Identification of the Bulgarian Cereal-based Fermented Beverage Boza," Process Biochemistry, Vol. 36, No. 1-2, 2000, pp. 127-130. http://dx.doi.org/10.1016/S0032-9592(00)00192-8

36. C. O. Gomes, C. L. C. Silva, C. R. Vianna, I. C. A. Lacerda, B. M. Borreli, A. C. Nunes, G. R. Franco, M. M. Mourao and C. A. Rosa, "Identification of Lactic Acid Bacteria Associated with Traditional Cachaca Fermentations," Brazilian Journal of Microbiology, Vol. 41, No. 2, 2010, pp. 486-492. http://dx.doi.org/10.1590/S1517-83822010000200031

37. J. I. Leisner, M. Vancanneyt, G. Rusul, B. Pot, K. Lefebvre, A. Fresi and L. K. Tee, "Identification of Lactic Acid Bacteria Constituting Predominating Microflora in an Acid-Fermented Condiment (Tempoyak) Popular in Malaysia," International Journal of Food Microbiology, Vol. 63, No. 1-2, 2001, pp. 149-157. http://dx.doi.org/10.1016/S0168-1605(00)00476-1

38. Madoroba, E. T. Steenkamp, J. Theron, G. Huys, I. Schierlinck and T. E. Cloete, "Polyphasic Taxonomic Characterization of Lactic Acid Bacteria Isolated from Spontaneous Sorghum Fermentations Used to Produce Ting, a Traditional South African Food," African Journal of Biotechnology, Vol. 8, No. 3, 2009, pp. 458-463.

39. J. K. Mugula, S. A. M. Ninko, J. A. Narvhus and T. Sorhaug, "Microbiological and Fermentation Characteristics of Togwa, a Tanzanian Fermented Food," International Journal of Food Microbiology, Vol. 80, No. 3, 2003, pp. 187-199. http://dx.doi.org/10.1016/S0168-1605(02)00141-1

40. L. H. Damelin, G. A. Dykes and A. Von Holy, "Biodiversity of Lactic Acid Bacteria from Food-Related Ecosystem," Microbios, Vol. 83, No. 334, 1995, pp. 13-22.

41. C. F. Williams and C. W. Dennis, "Food Microbiology," 4th Edition, McGraw Hill, 2011, p. 330.

Chapter 5

FUNCTIONAL SCREENING OF ANTIBIOTIC RESISTANCE GENES FROM A REPRESENTATIVE METAGENOMIC LIBRARY OF FOOD FERMENTING MICROBIOTA

Chiara Devirgiliis,[1] Paola Zinno,[1] Mariarita Stirpe,[1,2] Simona Barile,[1] andGiuditta Perozzi[1]

[1]CRA-NUT, Food & Nutrition Research Center, Agricultural Research Council, Via Ardeatina 546, 00178 Rome, Italy
[2]Department of Biology and Biotechnology Charles Darwin, Sapienza University of Rome, Piazzale Aldo Moro 5, 00185 Rome, Italy

ABSTRACT

Lactic acid bacteria (LAB) represent the predominant microbiota in fermented foods. Foodborne LAB have received increasing attention as potential reservoir of antibiotic resistance (AR) determinants, which may be horizontally transferred to opportunistic pathogens. We have previously reported isolation of AR LAB from the raw ingredients of a fermented cheese, while AR genes could be detected in the final, marketed product only by PCR amplification, thus pointing at the need for more sensitive microbial isolation techniques. We turned therefore to construction of a metagenomic library containing microbial DNA extracted directly from the food matrix. To maximize yield and purity and to ensure that genomic complexity of the library was representative of the original bacterial population, we defined a suitable protocol for total DNA extraction from cheese which can also be applied to other lipid-rich foods. Functional library screening on different antibiotics allowed recovery of ampicillin and kanamycin resistant clones originating from Streptococcus salivarius subsp. thermophilus and Lactobacillus helveticus genomes. We report molecular characterization of the cloned inserts, which were fully sequenced and shown to confer AR phenotype to recipient bacteria. We also show that metagenomics can be applied to food microbiota to identify underrepresented species carrying specific genes of interest.

INTRODUCTION

Bacterial fermentation products provide specific sensory properties which characterize a wide variety of foods. Foodborne fermenting microorganisms can either be added to sterilized matrices as commercial starter mixtures composed of specific strains [1] or they can originate from the environment as in the case of the raw ingredients employed for artisanal food production. This latter condition is the most frequent in traditional cheese manufacturing, which does not employ selected industrial starters as it relies on the microflora naturally present in raw material, often represented by complex microbial consortia whose species profile reflects local microenvironments. Lactic acid bacteria (LAB) are prevalent microorganisms within the fermenting food microbiota. Complex environmental bacterial communities have been extremely difficult to characterize, mostly due to the limitations imposed by culture-dependent approaches [2]. The proportion of bacteria from natural environments that are not readily culturable was estimated to about 99% [3]. Therefore, the majority of environmental strains have never been described and cannot be exploited for research and for biotechnological applications. Metagenomics represents, at the moment, the most promising culture-independent, DNA-based molecular method to overcome such difficulties [4, 5]. Food microbiology has taken advantage of the application of such innovative strategies, which were applied to study the composition and the evolution, as well as the spatial distribution of fermenting microbial ecosystems [6, 7].

Metagenomic libraries can be constructed from a variety of sources and through several methods, depending on the objective to be pursued. Taxonomic analysis requires comparison of conserved genome stretches, and therefore total DNA extracted from environmental microbiota is mostly PCR-amplified prior to cloning into the appropriate vectors, resulting in gene-specific metagenomic libraries (most frequently 16S-ribosomal DNA libraries) that are easily analyzed using bacterial genome databases and tools. However, the PCR step introduces a bias in DNA complexity, by altering the relative species proportions with respect to their relative abundance within the original microbiota. On the other hand, direct cloning of total DNA extracted from complex microbial communities, although quantitatively more reliable, requires very high cloning efficiencies to avoid selection against the least represented genomes. The choice of methodological approach is therefore strictly dependent on the purpose of the study, although the majority of metagenomic libraries described in the literature employ PCR-amplified DNA as starting material.

Our laboratory has focused on studying antibiotic resistance (AR) genes from microbial food sources and their corresponding genomic context which represents the main driver of horizontal transfer to human opportunistic pathogens [8, 9]. AR genes are widely distributed in several different environments, including food production systems [10]. Recent findings suggest the possibility of horizontal gene transfer among bacteria within food matrices, since fermented and minimally processed foods contain high titers of live microbial cells [11]. We have chosen a specific water buffalo fermented cheese as a model; that is, Mozzarella di Bufala Campana (MBC), which is produced in restricted geographical regions of Italy, is consumed fresh and therefore supplies high titers of live bacteria [12, 13]. Fermentation in this product is performed by specific thermophilic microbial communities provided by natural whey starter cultures (NWSC) [14], which, together with the microbiota of raw milk, contribute a wide variety of uncharacterized, environmental strains to the final cheese ready for consumption. Although PCR amplification with gene-specific primers of total DNA extracted from MBC had shown the presence of several AR genes, when applying culture-dependent approaches to isolate the corresponding AR strains, we were able to identify AR colonies only from the raw materials employed for cheese production, in which microbial titers are higher than in the final product [13]. Previous studies by other laboratories demonstrated the efficacy of culture-independent approaches in the identification of AR clones from oral metagenome libraries [15]. We turned therefore to metagenomics, with the aim of constructing a representative library of the entire cheese microbiome that could allow detection and analysis of AR genes carried by nonculturable or underrepresented species within the microbiota of fermented food products. Our experimental design involved construction of a fosmid metagenomic library containing large fragments of total DNA extracted from MBC, followed by functional screening of recombinant clones on representative antibiotics belonging to different pharmacological classes and employed in the past in animal farming and/or presently used in human therapy, namely, ampicillin, kanamycin, gentamycin, and tetracycline. To best reflect the complexity of the fermenting microbiota, the metagenomic library needed to be quantitatively representative of the different species present in the starting material, and we thus had to confront with several technical aspects representing crucial steps towards our goal. We describe in this paper the choices deriving from such a challenge, which resulted in the construction and screening of a cheese metagenomic library leading to the identification of fosmidborne, LAB derived genes expressing an AR phenotype in the E. coli host. To the best of our knowledge, this is the first report of direct, nonamplified metagenomic cloning of microbial genes from a complex fermented food matrix.

MATERIALS AND METHODS

Mozzarella Processing and Sampling

Samples of MBC were received on the day of production from four dairy factories located in different provinces of central and southern Italy (Latina-LT; Salerno-SA; Caserta-CE; Foggia-FG). We exclusively selected dairy plants with associated animal farming, which guarantees reproducible sources of milk and associated microbiota profiles for cheese production. Samples were stored at 4°C and processed within 12 h. Pooled or single samples of MBC were homogenized with a BagMixer400 (Interscience, France) in sodium citrate solution (2% w/v) at a concentration of 0.5 g/mL. In order to test the titer of mesophilic cultivable LAB, serial dilutions were made in Quarter Strength Ringer's solution and plated on MRS agar medium (Oxoid Ltd, Basingstoke, Hampshire, England), as previously reported [13]. Plates were incubated at 30°C for 48 h, under aerobic and anaerobic conditions (Anaerocult A, Merck, Germany).

DNA Extraction

Total DNA extraction from MBC was performed by a modified version of a published method [16]. Relevant methodological modifications are described in Section 3. Microscopic observation of sample aliquots during the lysis procedure was carried out to monitor the progressive disappearance of intact microbial cells. The yield of total DNA obtained from MBC samples was about 0.5 μg/g.

Library Construction

Metagenomic library construction was performed using the EpiFos Library Production Kit (Epicentre Technologies, Madison, Wisconsin, USA), following manufacturer's indications with the following modifications: ligation reaction was carried out with ligase enzyme from Stratagene, and incubation time of E. coli host cells with phage particles during the infection process was extended to 40 min. Such modifications resulted in increased packaging efficiency as well as in improved titer of packaged fosmid clones by about 4-fold.

Antibiotics and Reagents

Antibiotics (ampicillin, chloramphenicol, erythromycin, gentamycin, kanamycin, tetracycline, and vancomycin) were purchased from Sigma (Italy). Restriction enzymes were provided by Takara (Italy). PCR reagents were obtained from Invitrogen (Italy).

Bacterial Strains and Growth Conditions

E. coli EPI100- T1R [F− mcrA Δ(mrr-hsdRMS-mcrBC) ϕ80dlacZΔM15 ΔlacX74 recA1 endA1 araD139 Δ (ara, leu)7697 galU galK λ−rpsL nupG tonA] was grown in LB medium (Difco) overnight at 37∘ C with shaking. Recombinant libraries were stored at −80∘ C in LB-Cm (LB medium added with chloramphenicol at a final concentration of 12.5 mg/L) containing glycerol (15% v/v). For screening purposes, libraries were plated on LB-Cm agar plates and a total of 20.000 recombinant E. coli clones were picked and stored in 96-multiwell plates containing 10 clones/well. This plates were then replica-plated on LB-agar added with the appropriate antibiotic, with the aid of a metallic replica plater for 96-multiwell (Sigma, Italy), and grown overnight at 37∘ C

DNA Amplification and Molecular Analysis

Microbial DNA was amplified by PCR as previously described [9]. Fosmid DNA was isolated with FosmidPrep kit (Epicentre Technologies, Madison, Wisconsin, USA) according to manufacturer's instructions. Primers used are listed in Table 1. Restriction analysis and southern hybridization were performed by standard protocols, using probes labelled with digoxigenin-11-dUTP (Roche Diagnostics, Milan, Italy).

Table 1: Primers used for PCR experiments

Primer pair	Sequence	Target gene	Reference
P0	GAGAGTTTGATCCTGGCT	*Bacterial 16S rDNA*	[46]
P6	CTACGGCTACCTTGTTAC		
SINE-F	GGATCCGGCATTGCCGTTAG	*Swine short interspersed nuclear elements*	[47]
SINE-R	GTCTTTTTTTGCCATTTCTTGG		
ITS1	TCCGTAGGTGAACCTGCG	*Yeast 5.8S rDNA*	[48]
ITS4	TCCTCCGCTTATTGATATGC		
BufGH-F	TTGGGCCCCTGCAGTTC	*Buffalo growth hormone*	[49]
BufGH-R	GGTCCGAGGTGCCAAACAC		

The two-step gene walking method consisted of a walking-PCR (step 1) followed by direct sequencing of the PCR product (step 2) [17]. Walking-PCRs were performed as described [18], with the specific primer Epifos-FW (Epicentre). PCR products were purified using a NucleoSpin Extract II kit according to the manufacturer's instructions and sequenced with T7 primer.

Full Sequencing of Recombinant AR Fosmids

Sequencing was performed at the DNA sequencing facility of GenProbio s.r.l., Italy (http://www.genprobio.com/).

RESULTS AND DISCUSSION

Microbial Representativeness for Metagenomic Analysis

The first step towards library construction concerned sampling of MBC from different sources to ensure metagenome representativeness of the entire microbiota that characterizes this specific food product. To this aim, MBC cheeses were collected from four different dairy plants located in Italian provinces where the majority of producers are present (see Section 2). Some of the selected geographical areas are over 300 Km apart from each other and represent different pedoclimatic microenvironments leading to diverse milk microbial profiles [13]. We reasoned that pooling these samples should lead us to obtain genomic DNA representing the great majority of microbial genera/species entering the human GI tract through MBC consumption. Moreover, the titer of the mesophilic LAB component of the MBC microbiota resulted in about 10^6 Cfu/g (data not shown), in accordance with our previous findings [13]. The four MBC samples were therefore pooled in equal proportions and microbial DNA for library construction was extracted from the resulting homogenate.

Food-Derived DNA as a Source of Bacterial Genomes

A strategic aspect that we had to confront with in order to achieve representative, nonamplified metagenomic DNA of good quality was the optimization of qualitative/quantitative steps in the DNA extraction procedure. Fat represents a major component in dairy products, and its presence can impair bacterial recovery and lysis, which in turn greatly affects DNA yields. In order to obtain high molecular weight genomic DNA required for fosmid library construction, we therefore modified a previously published protocol [16], improving fat removal and DNA extraction efficiencies by introducing serial washes of dairy homogenates in Na-citrate buffer, followed by a combination of freeze-thaw cycles and mechanical as well as enzymatic lysis.

The presence of contaminating DNA from eukaryotic cells is another crucial aspect affecting representativeness of microbial genomes within the library, which is usually overcome by PCR amplification. Unlike meat fermentation products, dairy foods should contain almost exclusively microbial DNA, with very low contamination from higher eukaryotic cell DNA [8], but this aspect needed to be assayed before proceeding with our approach of direct cloning unselected high molecular weight DNA extracted from food. To this aim, a PCR approach was carried out including DNA extracted from fermented swine meat sausages for comparison, with primers specific for either microbial or eukaryotic species-specific genes, namely, bacterial 16S rDNA, yeast 5.8S rDNA, buffalo growth hormone gene, and swine SINE (short interspersed nuclear element). Primer sequences are reported in Table 1. The results shown in Figure 1 confirm that total DNA extracted from MBC is almost exclusively of microbial origin. Bacterial DNA represented the major component, while yeast DNA accounted for about 10% of the amplicons (Figure 1(b)). On the other hand, eukaryotic DNA was almost undetectable in MBC samples, while representing a great proportion of the total DNA extracted from fermented sausages. These results unequivocally show that microbial genomes constitute the great majority of unamplified DNA extracted from a dairy food matrix such as MBC, which could then be used directly for metagenomic library construction.

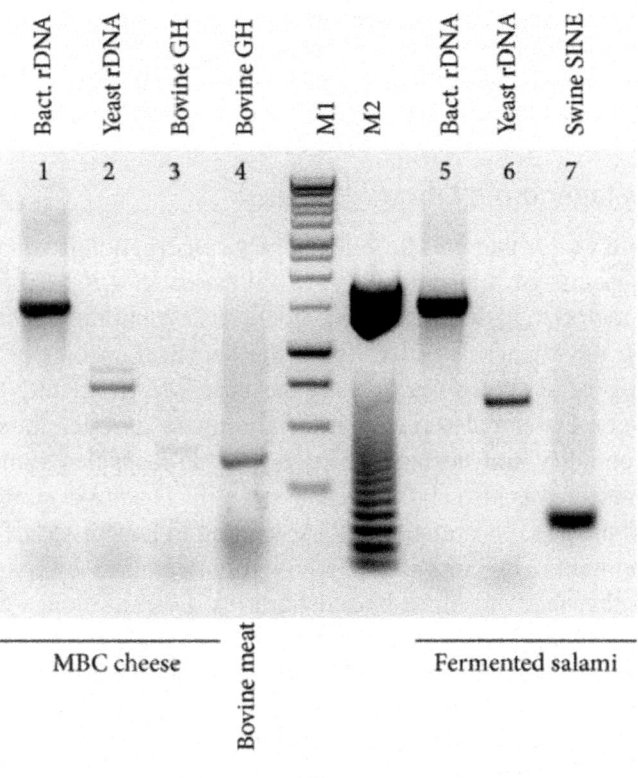

(a)

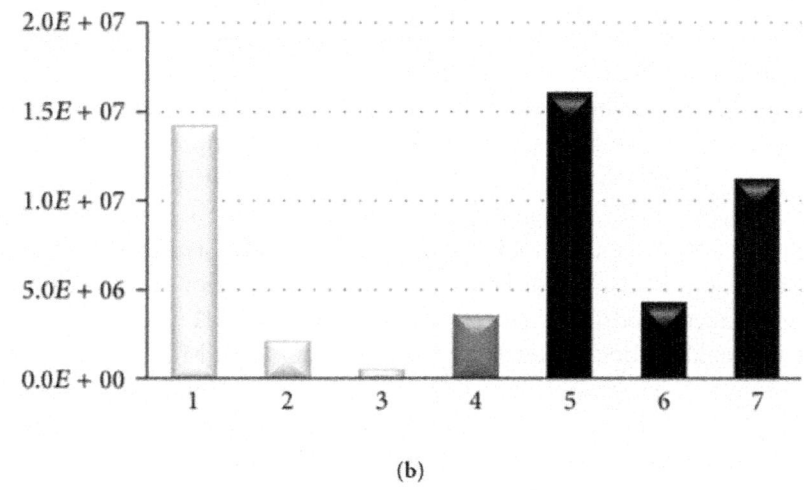

(b)

Figure 1: Total DNA extracted from dairy products contains almost exclusively microbial DNA with undetectable contamination from higher eukaryotic cell DNA. (a) PCR amplifications of DNA extracted from cheese (left) or meat (right) matrices using the species-specific primers listed in Table 1. M1: 1 Kb DNA ladder. M2: 50 bp DNA ladder. (b) Amplicon quantification obtained with the freely available ImageJ densitometry software [45]. Numbers indicate the corresponding PCR amplicons in (a).

MBC Metagenomic Library

The fosmid vector that we chose for library construction is suitable for cloning genomic inserts of approximately 40 kilobases in size. This feature allows us to characterize also the genomic context surrounding specific genes, thus increasing the chances of identifying the bacterial species of origin through sequencing of flanking regions. In the case of AR genes, analysis of the genomic context can also reveal association with mobile elements, indicative of a potentiality for horizontal inter- and intraspecies transfer [19]. The EpiFOS vector was also chosen because it utilizes a novel strategy for cloning randomly sheared, end-repaired DNA, leading to generation of highly random DNA fragments, in contrast to DNA fragmentation by partial restriction digestion that leads to more biased libraries. Fosmid clones containing high molecular weight fragments ranging between 35 and 45 Kb were used to infect the recipient E. coli EPI-100T1R strain, resulting in a 4×10^6 CFU/mL library titre. We estimated the minimum required representativeness of the library using the formula $N = \ln(1 - P)/ \ln(1 - f)$, where P is the desired probability (expressed as a fraction) of a given sequence being present in the library, f is the proportion of the metagenome within a single clone, and N is the number of clones required. Metagenomic samples introduce additional constraints, due to the unpredictable number of different species/strains that constitute the original microbiota; thus, only a rough estimate can be derived on the relative abundance of different populations within the complex bacterial community. For example, assuming an average genome size of 4 Mb, a library with 40 kb average inserts would require at least 100 clones to provide coverage of the entire genome, provided all clone inserts contain distinct sequences. If the genome of this reference organism represents about 10% of the total metagenome, screening 1.000 clones would likely provide a reasonable chance of detecting a specific sequence of interest. Basing on these calculations and considering an average fragment length of 25–30 Kbp, we estimated a total number of 20.000 clones to account for a well-represented MBC fermenting microbiome, as the overall size encompasses 1 Gbp which corresponds to approximately 250 times the size of the E. coli genome (4×10^6 bp).

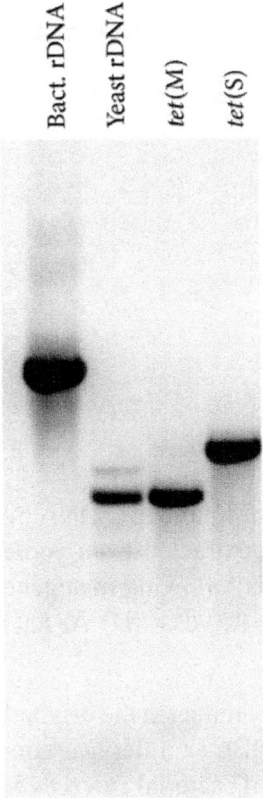

Cheese

(a)

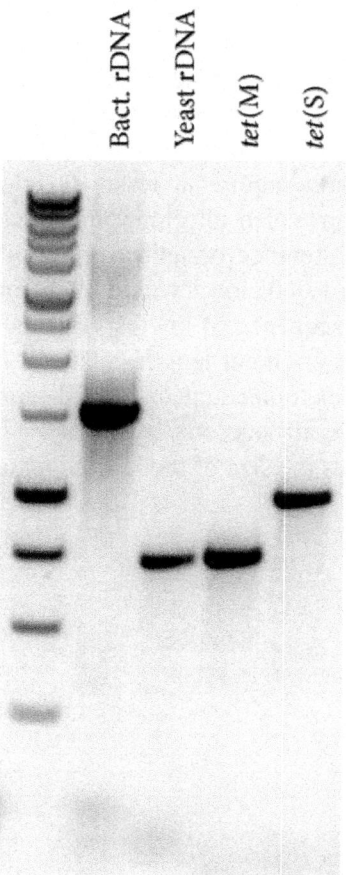

Pool of library fosmids

(b)

Figure 2: Fosmid cloning of total MBC DNA does not alter complexity PCR amplification of total DNA extracted from pooled MBC samples (a) or from pooled recombinant fosmids following metagenomic library construction (b). Primer pairs: bacterial rDNA, yeast rDNA, tet(M), and tet(S) (Table 1). M: 1 Kb DNA ladder.

To ensure that the library reflected the original DNA complexity, total DNA extracted from the pooled MBC samples was compared to pooled library DNA through PCR amplification of bacterial 16S rDNA and yeast 18S-28S intergenic sequences. The results in Figure 2 show that the DNA was qualitatively similar before and after library construction, thus proving that our cloning strategy can preserve the DNA complexity of foodborne microbial genomes.

Moreover, sequence analysis of randomly selected clone inserts followed by sequence similarity searches in public genome databases (Blast, http:// blast.ncbi.nlm.nih.gov/Blast.cgi) confirmed the presence of both bacterial and yeast genomes in the original proportions within the MBC metagenomic library (data not shown). As a control for the presence of specific AR genes, we also confirmed that tet(M) and tet(S), which are among the best characterized tetracycline resistance determinants in LAB, are well represented in both total MBC DNA and library clones.

Functional Screening for Antibiotic Resistance Genes

Functional metagenomics requires heterologous expression of exogenous genes, coupled with activity-based assays that can be easily performed on plates to select specific protein functions. This approach is more efficient than other two-step molecular methods based on detection of specific gene sequences and subsequent demonstration of their functionality, but it can be hampered by potential incompatibility between donor and host expression machineries [20–22]. The emergence and spread of antibiotic resistance determinants in the fermenting microbiota from different foods has been increasingly reported and reviewed by several groups worldwide, including ours [23–25], pointing at the need for deeper understanding of the mechanisms for horizontal transfer of AR genes, which are still partially unknown. AR genes can be easily selected on antibiotic containing media and were therefore chosen in this work to test the efficiency of recovery of LAB genes, which are the most represented species in the MBC microbiome under study. Moreover, AR genes for some of the most common antibiotics are not as well characterized in Gram-positives as they are in Gram-negative pathogens, and functional screening could therefore lead to the possible identification of novel proteins conferring AR in LAB. We therefore sought to test the MBC metagenomic library through functional screening with antibiotics belonging to five different pharmacological classes (tetracycline, aminoglycosides, beta-lactams, macrolides, and glycopeptides), which were chosen on the basis of their relevance in animal and human therapy and/or due to their widespread use in the past as growth promoters. Tetracyclines have been widely used in livestock farming and several tetR determinants were later identified in foodborne LAB from different fermented food sources [8, 26–28]. Along with tetracycline, the macrolide antibiotic erythromycin has also been intensively used in the past as growth promoter, and erythromycin-resistance genes represent, together with the TetR genes, the most widespread resistance determinants in foodborne bacteria [8, 27, 29]. Aminoglycosides and beta-lactams, on the other hand, have never been used as growth promoters, but they represent clinically relevant antibiotics whose corresponding resistance genes have also been described in foodborne LAB strains (lactobacilli and lactococci) [23].

As a first step towards functional screening for AR clones within the MBC library and to avoid interference from AR potentially present in the E. coli Epi100T1^R host, minimum inhibitory concentrations (MIC) were determined for the E. coli Epi100 T1^R strain on each antibiotic to be tested. Streptomycin was not considered as the corresponding resistance gene rpsL is known to be carried by the Epi100 strain. The resulting MIC values are reported in Table 2, showing that the E. coli host strain is phenotypically resistant to erythromycin and vancomycin, while displaying susceptibility to tetracycline, kanamycin, gentamycin, and ampicillin, with MIC values of 5, 25, 12, and 25 mg/L, respectively (Table 2). These latter four antibiotics were therefore chosen for functional screening of the MBC library at concentrations corresponding to their respective MIC values for E. coli. As positive control, clones were replicated on LB agar containing chloramphenicol, whose resistance determinant represents a selective marker (chloramphenicol acetyl transferase) encoded by the fosmid vector. Functional screening by replica plating of 20.000 independent library clones on antibiotic containing plates led to the selection of 4 TetR, 2 KanR, and 6 AmpR colonies. No colonies were rescued on gentamycin containing plates.

Table 2: Summary of screening procedure and resulting AR recombinant library clones.

Pharmacological class	Antibiotic	Target	E. coli MIC (mg/L)	Library clones identified	Verified by secondary screening
Tetracyclines	Tetracycline	Ribosome	5	4	0
Aminoglycosides	Kanamycin, gentamycin	Ribosome	25 12	2 0	2 —
Macrolides	Erythromycin	Ribosome	Resistant	—	—
Beta-lactams	Ampicillin	Cell wall	25	6	6
Glycopeptides	Vancomycin	Cell wall	Resistant	—	—

To confirm that phenotypic resistance in the surviving colonies was conferred by resistance determinants encoded by cloned inserts, fosmid DNA was extracted from each AR clone, packaged into phage particles, and used to infect the E. coli host Epi100T1^R. Secondary screening of the resulting clones was performed on LB agar plates containing the appropriate antibiotic. All kanamycin and ampicillin resistant bacteria confirmed their ability to grow on the corresponding antibiotic-containing medium following this secondary screening (Table 2). Unexpectedly, the tetracycline resistant colonies identified by primary screening resulted in false positives. A possible explanation is that they arose by spontaneous mutations in the E. coli genome induced by the mutagenic effect of chloramphenicol [30]. Unlike kanamycin and ampicillin resistant clones, TetR colonies had indeed been selected on plates containing both antibiotics (tetracycline and chloramphenicol) in the growth medium to increase the selective pressure. Several antibiotics, among which chloramphenicol, are known to induce mutagenesis and recombination within bacterial genomes, which may facilitate bacterial adaptation to different types of stress, including antibiotic pressure [30]. For this reason chloramphenicol was excluded from screening plates used for selection of ampicillin and kanamycin resistant colonies.

Molecular Characterization of AR Recombinant Clones

To further characterize the genomic features of the AR clones, fosmid DNA was extracted from both AmpR (clones Amp1-6) and KanR (clones Kan1 and 10) colonies and subjected to restriction analysis with the HindIII endonuclease, which cuts the pEpiFOS-5 vector at a unique site. The results are reported in Figures3(a) and 3(b) for the AmpR and KanR clones, respectively. With the exception of clone Amp4, which remained undigested by HindIII, the remaining AmpR clones displayed different restriction patterns with almost no overlapping bands, suggesting that the cloned inserts likely originate from distinct genomes within the metagenomic DNA. On the other hand, restriction of the two kanamycin resistant clones yielded fully overlapping restriction bands, indicating identity of the inserts. We therefore considered them as a single resistant clone in our subsequent analysis.

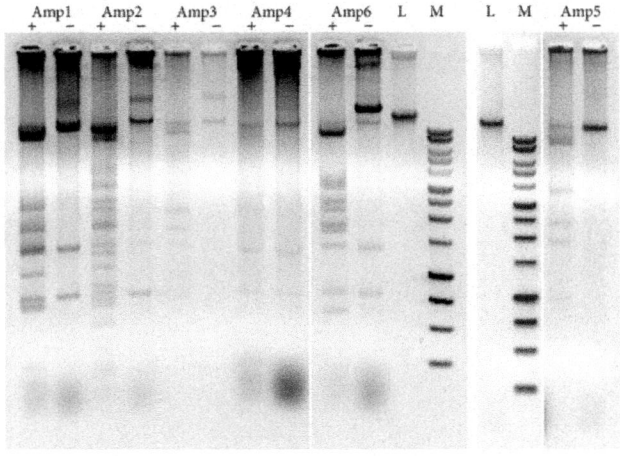

(a)

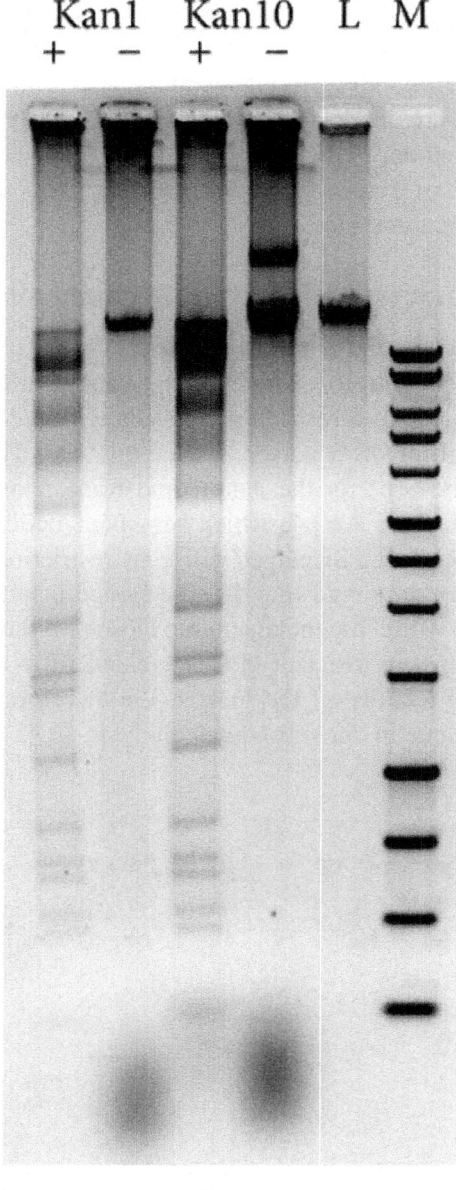

(b)

Figure 3: Restriction analysis of AR recombinant fosmids. Fosmid DNA extracted from each clone was digested (+) or not (−) with Hind III and fractionated by agarose gel electrophoresis. (a) AmpR clones. (b) KanR clones. L: undigested phage lambda DNA. M: 1 Kb DNA ladder.

The presence of the AR gene within a large genomic fragment allows species identification even before the full sequence of the cloned fragment is obtained. Preliminary analysis in this direction, performed by two-step gene walking [17], led to associate Streptococcus salivarius subsp. thermophilus genomic sequences to clones Amp1, 2, 3, and 6 and Lactobacillus helveticus genomic sequences to clones Amp4 and 5 (data not shown).

Full sequencing of the clone inserts, performed for 3 AmpR clones and for the KanR clone, confirmed species identification. S. thermophilus is expected to be a very abundant species in MBC, especially within the first few days of cheese production, as the last processing step for this specific product includes heating at 95°C for a few minutes. The full sequences, whose deposition in public databases is in progress, are provided as supplementary data (See Supplementary Material available online at http://dx.doi.org/10.1155/2014/290967), while the most relevant features for each clone are summarized in Table 3. Fragment size in the four sequenced clones ranged between 14 and 38 Kbp with correspondingly increasing number of predicted ORFs (14–43). Sequence analysis revealed the presence of two genes encoding penicillin-binding proteins (PBP) in clone Amp3 and of RNA methyltransferase genes in clones Amp3 and Kan10. Synthesis of low-affinity PBPs represents an important mechanism of resistance in some Gram-positive bacteria. Several PBPs have been described in resistant strains, including PBP2a from methicillin resistant Staphylococcus aureus (MRSA), PBP2x from penicillin resistant Streptococcus pneumoniae, and PBP5fm from drug resistant Enterococcus faecium [31, 32]. Blast similarity searches revealed that ORFs PBP1A and PBP2A from clone Amp3 were homologous to S. thermophilus penicillin-binding proteins. Two distinct ORFs encode rRNA methyltransferases in clones Amp3 and Kan10, and only the gene present in the Kan10 clone can be specifically identified as 23S rRNA methyltransferase on the basis of sequence similarity searches, while the Amp3 clone cannot be specifically attributed to 16S or 23S. Ribosomal RNA methylation is a frequent mechanism for macrolide and aminoglycoside resistance. RNA methyltransferases were shown to specifically target 16S rRNA in the case of resistance to aminoglycosides such as kanamycin [33]. However, the 23S rRNA methyltransferase encoded by Cfr gene of S. aureus and E. coli, which confers a wide spectrum of resistance to five chemically distinct classes of antimicrobials, was not tested with aminoglycosides [34]. We therefore need to confirm this gene as a possible basis for AR in the Kan10 clone with more detailed genotypic/phenotypic associations. As for the Amp3 clone, it also contains an acetyl transferase sequence belonging to the GCN5-related N-acetyltransferase (GNAT) superfamily of previously characterized gentamicin and kanamycin resistant bacteria [35]. Noteworthy, the remaining two sequenced clones

(Amp1 and Amp6) do not appear to contain ORFs encoding protein functions commonly described in AR bacteria. They do, however, contain at least one ORF with the capacity to mediate bacterial antimicrobial resistance (Table 3).

Table 3: Summary of insert sequencing results of AR fosmids. ORFs with a possible function in AR, as well as transposase genes, are listed

	Amp1	Amp3	Amp6	Kan10
Insert length (bp)	14.380	38.386	21.644	33.491
Predicted ORFs (*n*)	14	43	21	35
Species	S. thermophilus	S. thermophilus	S. thermophilus	S. thermophilus
Relevant ORFs for AR	Serine endopeptidase	Penicillin-binding protein 2A Penicillin-binding protein 1A RNA methyltransferase GNAT family acetyltransferase	Phosphoglucomutase	23S rRNA methyltransferase
Transposase sequences (*n*)	2	5	10	1
MIC of the corresponding antibiotic (mg/L)	50	50	50	25

In particular, clone Amp1 encodes a serine protease whose function includes serine beta-lactamase activities, which deactivate beta-lactam antibiotics by hydrolyzing the beta-lactam ring [36, 37]. The Amp6 clone, on the other hand, contains a phosphoglucomutase (PGM) ORF encoding the key enzyme catalyzing interconversion between glucose-1-phosphate (G1P) and glucose-6-phosphate (G6P) [38]. PGM plays a role in the biosynthesis of several bacterial exoproducts. Increased susceptibility to several antimicrobial agents was observed in pgmdeletion mutants, suggesting a possible role in AR [39, 40]. Noteworthy, almost all clones also contain ORFs annotated as encoding "hypothetical proteins," whose function might be related to AR. Although each clone contains a variable number of genes that could be related to AR, all of them display identical MIC values, suggesting that no additive effects due to the activity of multiple AR genes should be in place in any of the clones. Functional characterization of the putative AR gene sequences within the cloned fragments requires therefore further investigation. Another important feature deserving deeper analysis is the genomic context, as the sequencing output identifies a variable number of transposase genes within all sequenced inserts, usually clustered at a single site that likely represents an insertional hotspot. Transposases are integral parts of IS elements which mediate insertion/excision events known to promote lateral gene transfer events [41] and are especially important in horizontal gene transfer of AR genes.

CONCLUSIONS

We have reported in this work a novel metagenomic approach to identify AR genes within a complex, foodborne microbiome derived from a traditional fermented dairy product and constituted mainly by environmental strains of commensal bacteria. To increase the probability of identifying genes carried by underrepresented species, as well as to enhance representativeness of the library, we adopted a strategy based on direct cloning of total, unamplified DNA extracted from the food matrix, into a fosmid vector that can bear up to 40 Kbp inserts. Functional screening of the resulting metagenomic library, which we have calculated as representative of the entire microbiome, was carried out on antibiotics belonging to different pharmacological classes allowing recovery of ampicillin and kanamycin resistant clones. AmpR and KanR resistance genes are poorly characterized in LAB, although an important role for these bacterial genera as reservoir of transmissible AR genes is increasingly recognized [42]. Molecular characterization of the cloned inserts identified them as distinct regions of the S. thermophilus and L. helveticus genomes, hosting several ORFs which could confer AR phenotypes. The presence of several transposase sequences also emerged from full sequencing of the clone inserts, suggesting potential for lateral gene transfer of the surrounding genomic regions. This aspect is of special relevance, as IS mediated lateral gene transfer events represent the mechanistic basis for AR spreading from the reservoir of nonpathogenic, commensal bacteria to opportunistic pathogens [43]. From the food safety viewpoint, gene transfer events are particularly important as they might also occur through consumption of fermented foods and subsequent gene exchanges, which are known to occur between the food and the gut microbiota of the host [8, 44]. However, the low frequency of recovery of AR clones from our metagenomic library likely reflects a correspondingly low occurrence of AR bacteria in the food product, thus indicating its safe use for human consumption. Our results further support the evidence that metagenomic approaches can overcome the limitations of culture-dependent methods, representing an efficient and sensitive tool to detect genes occurring at low frequencies. Noteworthy, sequence analysis of the cloned inserts, which we had shown to retain the specific AR phenotype following transfer to new E. coli host cells, highlighted a number of genes whose involvement in AR might be novel. This observation points at the power of a screening strategy employing phenotypic selection, as, unlike primer-based methods that require known sequences as starting point, it can uncover novel genes performing similar functions. This work can therefore be considered a pioneer example of the application of metagenomics to food microbiota, and we hope it will pave the way to extend the strategy to other fermented foods, towards a deeper understanding of bacterial metabolic functions which could be beneficial to human health or of technological interest.

CONFLICT OF INTERESTS

The authors declare that there is no conflict of interests regarding the publication of this paper.

ACKNOWLEDGMENTS

The authors wish to thank the Consorzio Mozzarella di Bufala Campana for assistance in sample collection. They also thank Kariklia Pascucci for her kind support in daily laboratory work. This study was supported by Grants NUME (DM 3688/7303/08) and MEDITO (DM12487/7303/11) from the Italian Ministry of Agriculture, Food and Forestry (MiPAAF).

REFERENCES

1. E. B. Hansen, "Commercial bacterial starter cultures for fermented foods of the future," International Journal of Food Microbiology, vol. 78, no. 1-2, pp. 119–131, 2002.

2. W. R. Streit and R. A. Schmitz, "Metagenomic—the key to the uncultured microbes," Current Opinion in Microbiology, vol. 7, no. 5, pp. 492–498, 2004.

3. R. I. Amann, W. Ludwig, and K.-. Schleifer, "Phylogenetic identification and in situ detection of individual microbial cells without cultivation," Microbiological Reviews, vol. 59, no. 1, pp. 143–169, 1995.

4. T. S. Lusk, A. R. Ottesen, J. R. White, M. W. Allard, E. W. Brown, and J. A. Kase, "Characterization of microflora in Latin-style cheeses by next-generation sequencing technology," BMC Microbiology, vol. 12, article 254, 2012.

5. K. M. Tuohy, C. Gougoulias, Q. Shen, G. Walton, F. Fava, and P. Ramnani, "Studying the human gut microbiota in the trans-omics era—focus on metagenomics and metabonomics," Current Pharmaceutical Design, vol. 15, no. 13, pp. 1415–1427, 2009.

6. D. Ercolini, "High-throughput sequencing and metagenomics: Moving forward in the culture-independent analysis of food microbial ecology," Applied and Environmental Microbiology, vol. 79, no. 10, pp. 3148–3155, 2013.

7. S. A. F. T. van Hijum, E. E. Vaughan, and R. F. Vogel, "Application of state-of-art sequencing technologies to indigenous food fermentations," Current

Opinion in Biotechnology, vol. 24, no. 2, pp. 178–186, 2013.

8. C. Devirgiliis, S. Barile, and G. Perozzi, "Antibiotic resistance determinants in the interplay between food and gut microbiota," Genes and Nutrition, vol. 6, no. 3, pp. 275–284, 2011.

9. C. Devirgiliis, D. Coppola, S. Barile, B. Colonna, and G. Perozzi, "Characterization of the Tn916 conjugative transposon in a food-borne strain of Lactobacillus paracasei," Applied and Environmental Microbiology, vol. 75, no. 12, pp. 3866–3871, 2009.

10. L. M. Durso, D. N. Miller, and B. J. Wienhold, "Distribution and quantification of antibiotic resistant genes and bacteria across agricultural and non-agricultural metagenomes," PLoS ONE, vol. 7, no. 11, Article ID e48325, 2012.

11. F. Rossi, L. Rizzotti, G. E. Felis, and S. Torriani, "Horizontal gene transfer among microorganisms in food: current knowledge and future perspectives," Food Microbiology, vol. 42, pp. 232–243, 2014

12. D. Ercolini, G. Moschetti, G. Blaiotta, and S. Coppola, "The potential of a polyphasic PCR-DGGE approach in evaluating microbial diversity of natural whey cultures for water-buffalo Mozzarella cheese production: bias of culture-dependent and culture-independent analyses," Systematic and Applied Microbiology, vol. 24, no. 4, pp. 610–617, 2001.

13. C. Devirgiliis, A. Caravelli, D. Coppola, S. Barile, and G. Perozzi, "Antibiotic resistance and microbial composition along the manufacturing process of Mozzarella di Bufala Campana," International Journal of Food Microbiology, vol. 128, no. 2, pp. 378–384, 2008.

14. D. Ercolini, F. de Filippis, A. La Storia, and M. Iacono, ""Remake" by high-throughput sequencing of the microbiota involved in the production of water Buffalo mozzarella cheese," Applied and Environmental Microbiology, vol. 78, no. 22, pp. 8142–8145, 2012.

15. M. L. Diaz-Torres, A. Villedieu, N. Hunt et al., "Determining the antibiotic resistance potential of the indigenous oral microbiota of humans using ametagenomic approach," FEMS Microbiology Letters, vol. 258, no. 2, pp. 257–262, 2006.

16. J. H. A. Apajalahti, L. K. Särkilahti, B. R. E. Mäki, J. Pekka Heikkinen, P. H. Nurminen, and W. E. Holben, "Effective recovery of bacterial DNA and percent-guanine-plus-cytosine- based analysis of community structure in the gastrointestinal tract of broiler chickens," Applied and Environmental Microbiology, vol. 64, no. 10, pp. 4084–4088, 1998.

17. M. Pilhofer, A. P. Bauer, M. Schrallhammer et al., "Characterization of bacterial operons consisting of two tubulins and a kinesin-like gene by

the novel Two-Step Gene Walking method," Nucleic Acids Research, vol. 35, no. 20, article e135, 2007.

18. S. Barile, C. Devirgiliis, and G. Perozzi, "Molecular characterization of a novel mosaic tet(S/M) gene encoding tetracycline resistance in foodborne strains of Streptococcus bovis," Microbiology, vol. 158, no. 9, pp. 2353–2362, 2012.

19. G. Churchward, "Conjugative transposons and related mobile elements," in Mobile DNA II, N. Craig, R. Craigie, and M. Gellert, Eds., ASM Press, Washington, DC, USA, 2002.

20. G. Dantas, M. O. Sommer, P. H. Degnan, and A. L. Goodman, "Experimental approaches for defining functional roles of microbes in the human gut," Annual Review of Microbiology, vol. 67, pp. 459–475, 2013.

21. T. C. Hazen, A. M. Rocha, and S. M. Techtmann, "Advances in monitoring environmental microbes,"Current Opinion in Biotechnology, vol. 24, no. 3, pp. 526–533, 2013.

22. E. C. Pehrsson, K. J. Forsberg, M. K. Gibson, S. Ahmadi, and G. Dantas, "Novel resistance functions uncovered using functional metagenomic investigations of resistance reservoirs," Frontiers in Microbiology, vol. 4, article 145, 2013.

23. C. Devirgiliis, P. Zinno, and G. Perozzi, "pdate on antibiotic resistance in foodborne and species,"Frontiers in Microbiology, vol. 4, p. 301, 2013.

24. F. Clementi and L. Aquilanti, "Recent investigations and updated criteria for the assessment of antibiotic resistance in food lactic acid bacteria," Anaerobe, vol. 17, no. 6, pp. 394–398, 2011.

25. M. S. Ammor, A. B. Flórez, and B. Mayo, "Antibiotic resistance in non-enterococcal lactic acid bacteria and bifidobacteria," Food Microbiology, vol. 24, no. 6, pp. 559–570, 2007.

26. R. Comunian, E. Daga, I. Dupré et al., "Susceptibility to tetracycline and erythromycin of Lactobacillus paracasei strains isolated from traditional Italian fermented foods," International Journal of Food Microbiology, vol. 138, no. 1-2, pp. 151–156, 2010.

27. M. Nawaz, J. Wang, A. Zhou et al., "Characterization and transfer of antibiotic resistance in lactic acid bacteria from fermented food products," Current Microbiology, vol. 62, no. 3, pp. 1081–1089, 2011.

28. A. B. Flórez, P. Reimundo, S. Delgado et al., "Genome sequence of Lactococcus garvieae IPLA 31405, a bacteriocin-producing, tetracycline-resistant strain isolated from a raw-milk cheese," Journal of

Bacteriology, vol. 194, no. 18, pp. 5118–5119, 2012.

29. L. Feld, E. Bielak, K. Hammer, and A. Wilcks, "Characterization of a small erythromycin resistance plasmid pLFE1 from the food-isolate Lactobacillus plantarum M345," Plasmid, vol. 61, no. 3, pp. 159–170, 2009.

30. E. López and J. Blázquez, "Effect of subinhibitory concentrations of antibiotics on intrachromosomal homologous recombination in Escherichia coli," Antimicrobial Agents and Chemotherapy, vol. 53, no. 8, pp. 3411–3415, 2009.

31. A. Zervosen, E. Sauvage, J. Frère, P. Charlier, and A. Luxen, "Development of new drugs for an old target—the penicillin binding proteins," Molecules, vol. 17, no. 11, pp. 12478–12505, 2012.

32. K. Poole, "Resistance to β-lactam antibiotics," Cellular and Molecular Life Sciences, vol. 61, no. 17, pp. 2200–2223, 2004.

33. I. Moric, M. Savić, T. Ilić-Tomić, et al., "rRNA Methyltransferases and their role in resistance to antibiotics," Journal of Medical Biochemistry, vol. 29, pp. 165–174, 2010.

34. K. S. Long, J. Poehlsgaard, C. Kehrenberg, S. Schwarz, and B. Vester, "The Cfr rRNA methyltransferase confers resistance to phenicols, lincosamides, oxazolidinones, pleuromutilins, and streptogramin A antibiotics," Antimicrobial Agents and Chemotherapy, vol. 50, no. 7, pp. 2500–2505, 2006.

35. M. W. Vetting, L. P. S. de Carvalho, M. Yu et al., "Structure and functions of the GNAT superfamily of acetyltransferases," Archives of Biochemistry and Biophysics, vol. 433, no. 1, pp. 212–226, 2005.

36. L. P. Tripathi and R. Sowdhamini, "Genome-wide survey of prokaryotic serine proteases: Analysis of distribution and domain architectures of five serine protease families in prokaryotes," BMC Genomics, vol. 9, article 549, 2008.

37. B. G. Hall and M. Barlow, "Evolution of the serine β-lactamases: past, present and future," Drug Resistance Updates, vol. 7, no. 2, pp. 111–123, 2004.

38. R. Mehra-Chaudhary, J. Mick, J. J. Tanner, M. T. Henzl, and L. J. Beamer, "Crystal structure of a bacterial phosphoglucomutase, an enzyme involved in the virulence of multiple human pathogens,"Proteins, vol. 79, no. 4, pp. 1215–1229, 2011.

39. G. K. Paterson, D. B. Cone, S. E. Peters, and D. J. Maskell, "The enzyme phosphoglucomutase (Pgm) is required by Salmonella enterica serovar

Typhimurium for O-antigen production, resistance to antimicrobial peptides and in vivo fitness," Microbiology, vol. 155, no. 10, pp. 3403–3410, 2009.

40. G. A. McKay, D. E. Woods, K. L. MacDonald, and K. Poole, "Role of phosphoglucomutase ofStenotrophomonas maltophilia in lipopolysaccharide biosynthesis, virulence, and antibiotic resistance,"Infection and Immunity, vol. 71, no. 6, pp. 3068–3075, 2003.

41. J. Mahillon and M. Chandler, "Insertion sequences," Microbiology and Molecular Biology Reviews, vol. 62, no. 3, pp. 725–774, 1998.

42. B. M. Marshall, D. J. Ochieng, and S. B. Levy, "Commensals: underappreciated reservoir of antibiotic resistance," Microbe, vol. 4, no. 5, pp. 231–238, 2009.

43. L. A. Old and R. R. B. Russell, "Distribution and activity of IS elements in Streptococcus mutans," FEMS Microbiology Letters, vol. 287, no. 2, pp. 199–204, 2008.

44. V. Martins dos Santos, M. Müller, and W. M. de Vos, "Systems biology of the gut: the interplay of food, microbiota and host at the mucosal interface," Current Opinion in Biotechnology, vol. 21, no. 4, pp. 539–550, 2010.

45. C. A. Schneider, W. S. Rasband, and K. W. Eliceiri, "NIH Image to ImageJ: 25 years of image analysis,"Nature Methods, vol. 9, no. 7, pp. 671–675, 2012.

46. F. di Cello and R. Fani, "A molecular strategy for the study of natural bacterial communities by PCR-based techniques," Minerva Biotecnologica, vol. 8, no. 3, pp. 126–134, 1996.

47. J. H. Calvo, R. Osta, and P. Zaragoza, "Quantitative PCR detection of pork in raw and heated ground beef and pâté," Journal of Agricultural and Food Chemistry, vol. 50, no. 19, pp. 5265–5267, 2002.

48. S.I. Fujita, Y. Senda, S. Nakaguchi, and T. Hashimoto, "Multiplex PCR using internal transcribed spacer 1 and 2 regions for rapid detection and identification of yeast strains," Journal of Clinical Microbiology, vol. 39, no. 10, pp. 3617–3622, 2001.

49. R. M. Lopparelli, B. Cardazzo, S. Balzan, V. Giaccone, and E. Novelli, "Real-time TaqMan polymerase chain reaction detection and quantification of cow DNA in pure water buffalo mozzarella cheese: method validation and its application on commercial samples," Journal of Agricultural and Food Chemistry, vol. 55, no. 9, pp. 3429–3434, 2007.

Chapter 6

FERMENTATION PROCESSES USING LACTIC ACID BACTERIA PRODUCING BACTERIOCINS FOR PRESERVATION AND IMPROVING FUNCTIONAL PROPERTIES OF FOOD PRODUCTS

Grazina Juodeikiene[1], Elena Bartkiene[2], Pranas Viskelis[3], Dalia Urbo-naviciene[3], Dalia Eidukonyte[1] and Ceslovas Bobinas[3]

[1]Kaunas University of Technology,, Lithuania

[2]Veterinary Academy, Lithuanian University of Health Sciences,, Lithuania

[3]Institute of Horticulture, Lithuanian Research Centre for Agriculture and Forestry,, Lithuania

INTRODUCTION

During the recent years health-conscious consumers are looking for natural foods without chemical preservatives that will fit in their healthy lifestyle. The increasing consumption of precooked food, prone to temperature abuse, and the import of raw foods from developing countries are among the main causes of this situation. Biopreservation refers to extended shelf life and enhanced safety of foods using microorganisms and/or their metabolites (Ross et al., 2002). LAB is generally employed because they significantly contribute to the flavor, texture and, in many cases, to the nutritional value of the food products (McKay and Baldwin, 1990). LAB are used as natural or selected starters in food fermentations and exert the antimicrobial effect as a result of different metabolic processes (lactose metabolism, proteolytic enzymes, citrate uptake, bacteriophage resistance, bacteriocin production, polysaccharide biosynthesis, metal-ion resistance and antibiotic resistance) (Zotta, T., 2009, Corsetti, 2004). Lactic acid bacteria (LAB) play a key role in food fermentations where they not only contribute to the development of the desired sensory properties in the final product but also to their microbiological safety. LAB has a GRAS status (generally recognized as safe) and it has been estimated that 25% of the European diet and 60% of the diet in many developing countries consists of fermented foods (Stiles, 1996). Fermentation is one of the most ancient and most important food processing technologies. Fermentation is a relatively efficient, low energy preservation process, which increases the shelf life and decreases the need for refrigeration or other forms of food preservation technology. Currently, fermented foods are increasing in popularity (60% of the diet in industrialized countries) and, to assure the homogeneity, quality, and safety of products, they are produced by the intentional application in raw foods in different microbial systems (starter/protective cultures) (Holzapfel et al., 1995).

Examples of vegetable lactic acid fermentations are: sauerkraut, cucumber pickles, and olives in the Western world; Egyptian pickled vegetables in the Middle East; Indian pickled vegetables and Korean kim-chi, Thai pak-sian-don, Chinese hum-choy, Malaysian pickled vegetables and Malaysian tempoyak. Lactic acid fermented cereals and tubers (cassava) include: Mexican pozol, Ghanaian kenkey, Nigerian gari; boiled rice/raw shrimp/raw fish mixtures: Philippine balao-balao, burong dalag; lactic fermented/leavened breads: sourdough breads in the Western world; Indian idli, dhokla, khaman, Sri-lankan hoppers; Ethiopian enjera, Sudanese kisra and Philippine puto; Lactic acid fermented cheeses in the Western world and Chinese sufu/tofu-ru. Lactic acid fermented yogurt/wheat mixtures: Egyptian kishk, Greek trahanas, Turkish tarhanas.

Moreover, because of the improved organoleptic qualities of traditional fermented food, extensive research on its microbial biodiversity has been carried out with the goal of reproducing these qualities, which are attributed to native microbiota, in a controlled environment.

Recent years the interest increased in bacteriocin-like inhibitory substances (BLIS) producing LAB because of their potential use as natural antimicrobial agents to enhance the safety of food products. Bacteriocins from LAB are described as "natural" inhibitors, in regard to LAB having a GRAS status. Bacteriocin-like inhibitory substances (BLIS) from LAB are antimicrobial compounds that possess bacteriocin requisites but that have not yet been characterized for their amino acid sequence (Jack et al., 1995). Bacteriocins from the generally recognized as safe LAB, have received significant attention as a novel approach to the control of pathogens in foods (Klaenhammer, 1993, Settani et al., 2005). Nisin is the first antimicrobial polypeptide found in LAB (Rogers, 1928); at the time of discovery, the producer strains were identified as *Streptococcus lactis* [later classified as *Lactococcus lactis*] (Schleifer et al., 1985). Today nisin is a permitted preservative in at least 48 countries, in which it is used in a variety of products, including cheese, canned food and cured meat (Delves-Broughton, 1990). Another commercially produced bacteriocin is pediocin PA-1 produced by *Pediococcus acidilactici* and marketed as

ATTA ™ 2431 (Kerry Bioscience, Carrigaline, Co, Cork, Ireland). The source of natural or controlled microbiota and/or antimicrobial compounds could be traditional fermented foods. The health benefits attributed to peptides in these traditional products have, so far, not been established, however. Several factors can affect the bacteriocin activity including interaction with other bacteriocins, constituents from the cells as well as the growth medium, purity and concentration of exogenous added enzymes (Moreno et al., 2000). Enzymes present during food making originate from different sources: there are those that already exist in the plant raw material, those associated with the metabolic activity of yeasts or LAB, and those intentionally added to the formulations. For example, amylases are added for the intensification of the saccharification stage (Stauffer, 1994), while microbial lipases induce changes in lipid and short-chain fatty acid compositions (Martınez-Anaya, 1996). Up to date, numerous recent review articles focused on the isolation of novel LAB bacteriocin-like inhibitory substances produced by LAB, evaluation their inhibitory activities, classification, biochemical and genetic characterization, studying the sensitivity of the BLIS antimicrobial activity to different factors, e.g. enzymes in fermentation media and the mode of action of LAB bacteriocins, as well as on some of their food application (mainly animal origin) have been published. Unfortunately literature lacks of concentrated articles dealing with the use of bacteriocins or bacteriocin-producing strains for the biopreservation and improving functional properties of vegetable and fruit products.

CHARACTERIZATION OF BACTERIOCINS PRODUCED BY LACTIC ACID BACTERIA

The antimicrobial ribosomal synthesized peptides produced by bacteria, including members of the LAB, are called bacteriocins. Such peptides are produced by many, if not all, bacterial species and kill closely related microorganisms (Jack et al., 1995). Due to their nature, they are inactivated by proteases in the gastrointestinal tract. Most of the LAB bacteriocins identified so far are thermo stable cationic molecules that have up to 60 amino acid residues and hydrophobic patches. Electrostatic interactions with negatively charged phosphate groups on target cell membranes are thought to contribute to the initial binding, forming pores and killing the cells after causing lethal damage and autolysin activation to digest the cellular wall (Gálvez et al., 1990). Example of damage caused by bacteriocin on *L. monocytogenes* CECT 4032 cells is presented in Figure 1.

The LAB bacteriocins have many attractive characteristics that make them suitable candidates for use as food preservatives, such as:

- Protein nature, inactivation by proteolytic enzymes of gastrointestinal tract
- Non-toxic to laboratory animals tested and generally non-immunogenic
- Inactive against eukaryotic cells
- Generally thermo resistant (can maintain antimicrobial activity after pasteurization and sterilization)
- Broad bactericidal activity affecting most of the Gram-positive bacteria and some, damaged, Gram-negative bacteria including various pathogens such as *L. monocytogenes, Bacillus cereus, S. aureus,* and *Salmonella*
- Genetic determinants generally located in plasmid, which facilitates genetic manipulation to increase the variety of natural peptide analogues with desirable characteristics

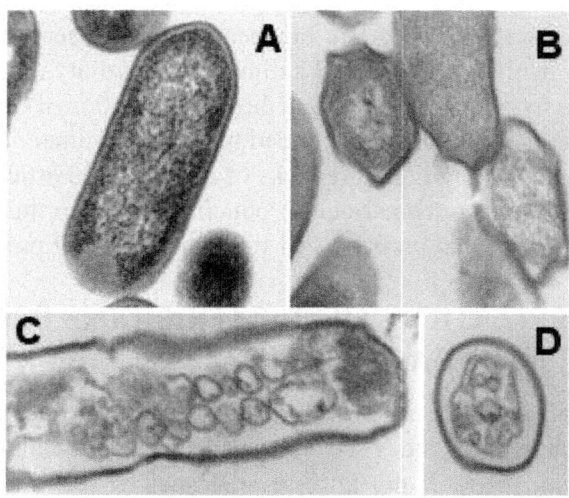

Figure 1: Example of damage caused by bacteriocin on *L. monocytogenes* CECT 4032 cells. (A) cells without enterocin AS-48; (B) cells treated with 0.1 µg/ml of AS-48 for 2 h; (C and D) cells treated with 3 µg/ml of enterocin AS-48 for 10 min (adapted from Mendoza et al., 1999).

For these reasons, the use of bacteriocins has, in recent years, attracted considerable interest for use as biopreservatives in food, which has led to the discovery of an ever-increasing potential of these peptides. Undoubtedly, the most extensively studied bacteriocin is nisin, which has gained widespread applications in the food industry. This FDA approved bacteriocin is produced by the GRAS microorganism *Lactococcus lactis* and is used as a food additive in at least 48 countries, particularly in processed cheese, dairy products and canned foods. Nisin is effective against food-borne pathogens such as *L. monocytogenes* and many other Gram-positive spoilage microorganisms (Thomas et al., 2000, Thomas and Delves-Broughton, 2001). Nisin is listed as E-234, and may also be cited as nisin preservative or natural preservative. In addition to the work on nisin, several authors have outlined issues involved in the approval of new bacteriocins for food use (Harlander, 1993).

The Biosynthetic Pathway of Bacteriocins

The biosynthetic pathways of bacteriocins with the focus on class II bacteriocins (mainly produced by LAB from fermented plant products) will be discussed in this section. All bacteriocins are synthesized as a biologically inactive prepeptide carrying an N-terminal leader peptide attached to the C-terminal propeptide (Hoover and Chen, 2005).

The mode of action of lactic acid bacteria bacteriocins belonging to class I, II and III are presented inFigure 2.

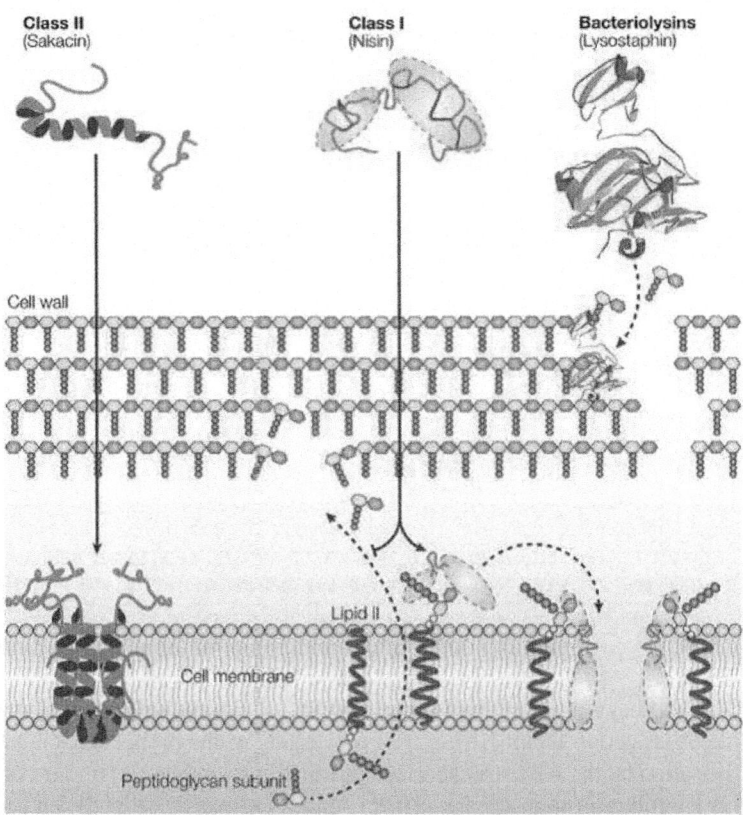

Figure 2: mode of action of bacteriocins by lactic acid bacteria.

Some bacteriocins (or lantibiotics) of class I (e.g. nisin) have a dual mode of action: (1) they prevent correct cell wall synthesis that leads to cell death by binding to lipid II – the main transporter of peptidoglycan subunits from the cytoplasm to the cell wall and (2) they employ lipid II as a docking molecule to initiate a process of membrane insertion and pore formation leading to a rapid cell death (Gillor et al., 2008, Cotter et al., 2005, Wiedemann et al., 2001). The majority of class II bacteriocins kill by inducing membrane permeabilization and the subsequent leakage of molecules from target bacteria (Gillor et al., 2008). Bacteriolysins (bacteriolytic proteins belonging to class III) function directly on the cell wall of Gram-positive targets leading to death and lysis of the target cell (Cotter et al., 2005).

Class II bacteriocins are synthesized as a prepeptide containing a conserved N-terminal leader and a characteristic double-glycine proteolytic processing site, and in contrast to lantibiotics, they do not undergo extensive posttranslational modification (Hoover and Chen, 2005). The examples of the class II bacteriocins are Sakacin P, G, A and Pediocin PA-1/AcH.

A general scheme of the biosynthetic pathway of class II bacteriocins is shown in Figure 3.

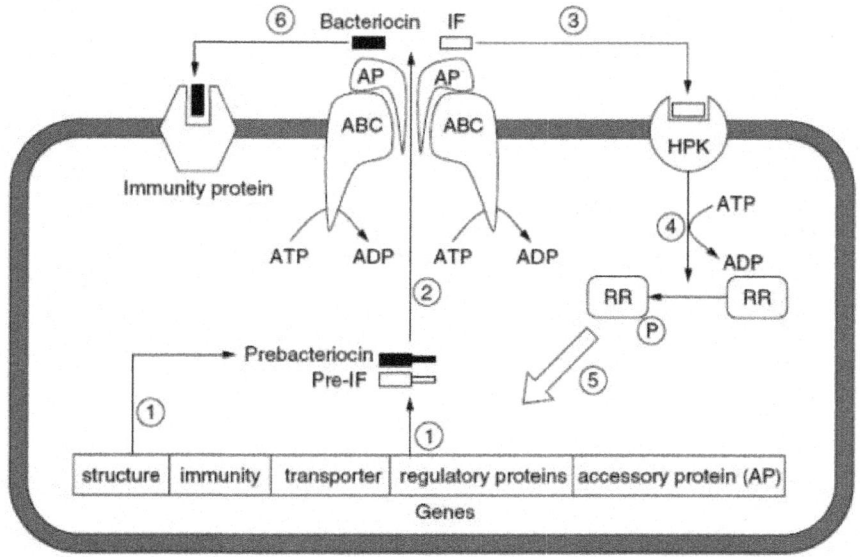

Figure 3: The biosynthesis of class II bacteriocins. 1. Formation of prebacteriocin and prepeptide of induction factor (IF). 2. The processing of the prebacteriocin and pre-IF, and translocation by the ABC-transporter, resulting in the release of mature bacteriocin and IF. 3. Histidine protein kinase (HPK) senses the presence of IF and autophosphorylates. 4. The transfer of the phosphoryl group (P) to the response regulator (RR). 5. The activation of the regulated genes transcription by the RR. 6. Producer immunity.

The production of most class IIa bacteriocins is regulated by a three-component system which includes a histidine protein kinase, a response regulator, and an induction factor. Some class IIa bacteriocins are autoregulated by a two-component signal transduction system, which is a well-known phenomenon in lantibiotics (Sahl et al., 1998). A threshold concentration of the bacteriocin, which functions as a signal molecule accumulating during growth, triggers the transcription of the genes coding for bacteriocin production, suggesting a self-inducing cell density (quorum-sensing)-regulated system (Sahl et al., 1998).

Class IIc bacteriocins are an exceptional case due to their production with a typical N-terminal signal sequence of the *sec*-type, processing and excretion through the general secretory pathway (Leer et al., 1995, Worobo et al., 1995). Once the prepeptide is formed, it is processed to remove the leader peptide concomitant with export from the cell through a dedicated ABC-transporter and its accessory protein (Nes et al., 1996).

Activity Spectra and Biochemical Properties of Lab and Their Produced Bacteriocins

As mentioned before, LAB bacteriocins tend to be active against a wide range of mostly closely related Gram-positive bacteria (Jack et al., 1995). Gänzle (1998) corroborated this while reviewing the inhibitory spectrum of bavaricin A, BLIS C57, and plantaricin ST31, produced by sourdough LAB, indicating no inhibition of Gram-negative bacteria, whereas a variety of Gram-positive bacteria were sensitive. The insensitivity of Gram-negative bacteria to bacteriocins from LAB strains might be explained by their outer membrane providing them with a permeability barrier (Messens et al., 2002). Furthermore, the producer strains are found to be immune towards their own bacteriocin (Gänzle, 1998). Studies on the resistance of *List.monocytogenes* strains towards bavaricin A confirmed that only 3 of the 245 strains examined were resistant to bavaricin A (Larsen and Nørrung, 2003), meanwhileRasch and Knøchel (1998) reported about the correlation between bavaricin A sensitivity and pediocin PA-1 sensitivity of the strains.

The seldom inhibition of commonly encountered enteropathogenic bacteria (*Enterobacter, Klebsiella,*or *Salmonella* was announced as the weakness of the bacteriocins produced by Gram-positive bacteria. Gram-positive bacteriocins are restricted to kill other Gram-positives and the killing range varies significantly (Gillor et al., 2008). Martínez-Cuesta et al. (2006) reported about the relatively narrow range of lactococcins A, B and M able to inhibit only *Lactococcus*, meanwhile some type A lantibiotics (e.g. nisin A, mutacin B-Ny266) had a wide range while killing *Actinomyces, Bacillus,*

Clostridium, Corynebacterium, Enterococcus, Gardnerella, Lactococcus, Listeria, Micrococcus, Mycobacterium, Propionibacterium, Streptococcus, and *Staphylococcus* as shown by Mota-Meira et al. (2005). These particular bacteriocins were found to be active against some medically important Gram-negative strains of *Campylobacter, Haemophilus, Helicobacter,* and *Neisseria* (Morency et al., 2001).

Furthermore, IIa class bacteriocins have relatively narrow killing spectrum as compared to I class bacteriocins and inhibit only closely related Gram-positive bacteria (Heng et al. 2007). However, pediocin was reported to have a fairly broad activity spectrum while inhibiting *Streptococcus aureus*and vegetative cells of *Clostridium* spp., *Listeria* and *Bacillus* spp. (Nes and Holo, 2000; Eijsink et al., 2002). Besides, maximum antimicrobial activity against *Escherichia coli, Staphylococcus aureus* and*Bacillus cereus,* though it was more effective against *E. coli* than others, showed bacteriocin produced by strain CA 44 of *Lactobacillus* genus isolated from carrot fermentation medium. This bacteriocin was stable at up to 100 C but its activity declined compared to that at 68 C and was completely lost at 121 C. The maximum antimicrobial activity was retained within the pH range of 4–5, but it was adversely affected by the addition of papain. Bacteriocin was also effective against *B. cereus* in different fruit products (pulp, juice and wine) indicating its potential application as a biopreservative in fruit products (Joshi et al., 2006).

Quite a few studies on a bacterocin activity possessed by *L. sakei* strains have been performed.Schillinger and Lücke (1989) reported about 221 surveyed lactobacilli strains, among those 19 *L. sakei*strains, 3 *L. plantarum* strains and 1 *L. curvatus* strain were found to inhibit other lactobacilli. Bacteriocins were not identical according to the evaluation of supernatants antimicrobial spectra. Sakacin A produced by *L. sakei* Lb706 was reported to be active against *List. monocytogenes* strains 8732 and 17a, moreover 4 other strains of *L. sakei* and 1 strain of *L. plantarum* also shown antilisterial activity. Mørtvedt and Nes (1990) identified bacteriocin Lactosin S produced by *L. sakei* and described it as moderately heat stable, sensitive to protease and having antimicrobial activity against*Lactobacillus, Pediococcus* and *Leuconostoc* genera members. Furthermore, the instability of bacteriocin production and immunity was revealed by the plasmid biology investigation in *L. sakei* L45. Antagonistic effect of *L. sakei* CTC494 and sakacin K to different extents against *List.innocua*CTC1014 was demonstrated by Hugas and co-workers (1998). While Axelsson with co-workers (1988) developed a system for heterologous expression using a bacteriocin-negative *L. sakei* Lb790 strain, where into Lb790 introduced two plasmids allowed the production of various bacteriocins (sakacin P, pediocin PA-1, and piscicolin 61) at levels equal to or exceeding levels in correspondence with the wild type cultures. Cuozzo et al. (2000) investigated II b class bacteriocin lactocin 705 from *L. casei* CRL 705 and reported about the required of both two peptides presence for the inhibitory activity.

The cloning, expression, and nucleotide sequence of the genes involved in the synthesis of pediocin PA-1 was reported by Marugg et al. (1992). The genes were cloned and expressed in *E. coli* and 5.6-kbp fragment from the plasmid was found to be necessary for the bacteriocin production. Hoover and co-workers (1989) noted about the surveyed 37 pediococci cultures for the antagonistic effect against eight*List. monocytogenes* strains and indicated that a bacteriocin effect of these LABs against *List. monocytogenes* may not be limited to a few industrial starter cultures. 15 strains containing the*Lactobacillus, Pediococcus, Lactococcus,* and *Leuconostoc* genera were examined for the inhibition against eight strains of *List. monocytogenes* (Harris et al., 1989) and only cell-free supernatants from*Lactobacillus* species UAL11, *P. acidilactici* PAC 1.0, and *Leuconostoc* species UAL14 were reported to inhibit all eight strains of *List. monocytogenes*. The addition of proteolytic enzymes caused the prevention of the inhibition. Ennahar et al (1996) found out that *Lactobacillus plantarum* WHE 92 produce a bacteriocin identical to pediocin AcH from *P. acidilactici* H, though pediocin AcH was produced more effectively in *L. plantarum* WHE 92 in the pH range of 5.0 to 6.0 as compared to *P. acidilactici* H. Moreover, *L. plantarum* WHE 92 seems to have more effective means of antagonism against *List. monocytogenes,* since dairy products are normally higher than pH 5.0 (Hoover and Chen, 2005). Miller and co-workers (1998) applied PCR random mutagenesis for the construction of pediocin AcH amino acid substitution mutants. One mutant peptide was found to have a 2.8-fold higher activity against *L. plantarum* NCDO955, while other mutations were inactive and shown a reduced antagonism.Johnsen et al. (2000) increased the stability of pediocin PA-1 with the replaced methionine residue with alanine, isoleucine, or leucine in order to protect from oxidation, since this peptide was found to lose its activity while stored as refrigeration or ambient temperatures.

Lacticin 3147 is a two peptide bacteriocin with a broad-spectrum activity and genetically resides on a self-transmissable plasmid that can be moved to other lactococci strains (Ross et al., 2000). McAuliffe and co-workers (1999) reported about a 3-log10 reduction in CFU/g of *List. monocytogenes* with the used *Lc. lactis* subs. *lactis* culture producing lacticin 3147. Bacteriocin S50 is another antimicrobial compound produced by lactococci, though it has a relatively narrow activity spectrum (Hoover and Chen, 2005). Lacticin FS92, containing 32 amino acids, is a heat-stable bacteriocin and appears to be active against *List. monocytogenes* as noticed by Mao et al. (2001). *List.*

monocytogenes resistant mutants were found to remain sensitive to lacticin FS92, but not to pediocin PA-1, curvaticin FS47 and lacticin FS56. The susceptibility of *List. monocytogenes, List. innocua,* and *List. seeligeri* strains was found to strain-dependent at each pH examined in response to lactocin 705, enterocin CRL35, and nisin (Castellano et al., 2001).

OPTIMIZATION OF MEDIA AND GROWTH CONDITIONS FOR INCREASED BACTERIOCIN PRODUCTION

The incorporation of bacteriocins as a biopreservative ingredient into model food systems has been studied extensively and has been shown to be effective in the control of pathogenic and spoilage microorganisms (Neysen and De Vuyst, 2005). LAB can also be considered as protective cultures because they improve the microbiological quality as well as the safety of the food (Messens and De Vuyst, 2002) and can be a way to prevent product spoilage (Verluyten et al., 2004a). Lactic acid bacteria (LAB) are a group of microorganisms nutritionally exigent. They need a wide range of nutrients to grow and synthesize metabolic products, some nutritional requirements usually being strain specific. De Man-Rogosa-Sharpe (MRS) and yeast autolysate-peptone-tryptone-tween 80-glucose (LAPTg) are standard culture media commonly used to support the growth of lactobacilli in the starting point of their cultivation. These media contain carbon and energy sources (carbohydrates, e.g. glucose), complex nitrogen sources (yeast extract, meat extract, tryptone and peptone) and supplements derived from oleic acid (Tween 80). MRS also includes inorganic and organic salts that have shown a stimulating effect or are essential for the growth of most of the species of this genus. Different components of culture media strongly affect the growth and bacteriocin production of several microorganisms that are mainly considered for food applications and must be included in fermentation processes, which are used on a production scale. Extruded wheat material is one of the candidates to be included as fermentation media for cultivation of bacteriocins producing LAB (*Lactobacillus sakei*MI806, *Pediococcus pentosaceus* MI810 and *Pediococcus acidilactici* MI807), previously isolated from spontaneous Lithuanian sourdoughs, in fermented products preparation formula for wheat bread production to have a higher positive effect compared with the control medium on LAB growth and their antimicrobial activities (Juodeikiene et al., 2011).

Frequently, the conditions that lead to high bacteriocin production are similar to those prevailing during food fermentation processes (Leroy et al., 2002, Delgado et al., 2005; Neysen and De Vuyst, 2005). Bacteriocin production is usually proportional to growth and shows primary metabolite kinetics (Moretro et al., 2000) but often the correlation is weak (Delgado et al., 2005) and this is particularly evident for bacteriocins produced during the stationary phase (Jim´enez-D´ıaz et al., 1993). Food preservation using *in situ* bacteriocin production requires a better understanding of the relationship between growth and bacteriocin production. Different bacteriocin exhibits different inhibition profile on food spoilage and pathogenic microorganisms. Therefore, they could be natural replacements for synthetic food preservatives. In order to increase the productivity of the bacteriocins, a better understanding on the factors affecting their production is essential.

Bacteriocin titres change with environmental factors (Leal-S´anchez et al., 2002; Delgado et al., 2005), such as pH, temperature, and NaCl and ethanol concentrations. These environmental factors may influence growth negatively and thereby the secretion of the induction factor (Leal-S´anchez et al., 2002). Further, it has been suggested that some environmental factors reduce the binding of the induction factor to its receptor (Delgado et al., 2005). Bacteriocin production is strongly dependent on pH, nutrient sources and incubation temperature. Activity levels do not always correlate with cell mass or growth rate of the producer (Kim et al., 1997, Bogovic-Matijasic & Rogelj, 1998). Increased levels of bacteriocin production are often obtained at conditions lower than required for optimal growth (Bogovic-Matijasic & Rogelj, 1998; Todorov et al., 2000, Todorov & Dicks, 2004). Understanding the influence of food-related environmental factors on the induction of bacteriocins is essential for the effective commercial application of bacteriocin-producing LAB in the preservation of foods.

Leal et al. (1998) optimized the production of bacteriocins by *Lactobacillus plantarum* LPCO10 to allow the use of bacteriocins as natural food additives in canned vegetables and other food systems. Results obtained indicated that the best conditions for bacteriocin production were shown with temperatures ranging from 22°C to 27°C, salt concentration from 2.3 to 2.5%, and *L. plantarum*LPCO10 inoculum size ranging from $10^{7.3}$ to $10^{7.4}$ CFU/ml, fixing the initial glucose concentration at 2%, with no aeration of the culture. Under these optimal conditions, about 3.2×10^4 times more bacteriocin per liter of culture medium was obtained than that used to initially purify plantaricin S from*L. plantarum* LPCO10 to homogeneity.

Delgado A. et al. (2007) by modeling studied the effects of some environmental factors on bacteriocin production by *Lactobacillus plantarum* 17.2b. Bacteriocin production by *L. plantarum* 17.2b was very sensitive to environmental conditions and uncoupled from growth. Maximum production required suboptimal growth temperatures, pH values above growth's optimum and no NaCl.

Many studies have focused on optimization of media and growth conditions of LAB for increased bacteriocin production. They have generally focused on the effects of pH, temperature, composition of the culture medium, and general microbial growth conditions (*in vitro* as well as in natural fermentations) on maximal bacteriocin production (FAO-WHO, 2002, ANVISA, 2010, Cruz et al., 2009, Silveira et al., 2009, Minei et al., 2008, Galvez et al., 2008). By supplementing the medium with growth limiting factors, such as carbohydrates, nitrogen, vitamins and potassium phosphate, or by adjusting the medium pH, levels of bacteriocin production is often increased.

Several mechanisms have been proposed for the bacteriocins activity: alteration of enzymatic activity, inactivation of anionic carriers through the formation of selective and non-selective pores and inhibition of spore germination (Parada et al. 2007; Martinez and De Martinis, 2006). Powell et al. (2006),Todorov and Dicks (2005a), (2005b), (2006a), (2006b), Todorov et al. (2000), (2007a), (2007b), (2004) and Todorov (2008) reported higher bacteriocin production levels for *L. plantarum* ST194BZ, *L. plantarum* ST13BR, *L. plantarum* ST414BZ, *L. plantarum* ST664BZ, *L. plantarum* ST23LD, *L.plantarum* ST341LD, *L. plantarum* 423, *L. plantarum* AMAK, *L. plantarum* ST26MS, *L. plantarum*ST28MS, *L. plantarum* ST8KF, *L. plantarum* ST31 in optimized growth media.

In general, the bactericidal/bacteriostatic action of bacteriocins encompasses the increased permeability of the cytoplasmic membrane of the target cells for a broad range of monovalent cations (e.g. Na^+, K^+, Li^+, Cs^+, Rb^+ and choline) leading to the destruction of proton motive force by dissipation of the transmembrane pH gradient and eventually to the cell death (Simova et al., 2009; Oppegård et al., 2007). The bactericidal or bacteriostatic activity possessed by bacteriocins is influenced by the following factors: bacteriocin dose and purification degree, physiological status of the indicator cells (e.g. growth phase) and experimental conditions (e.g. temperature, pH, presence of agents disrupting cell wall integrity and other antimicrobial compounds) (Deraz et al., 2007, Cintas et al., 2001). An increased antibacterial activity of non-lanthionine-containing bacteriocins at low pH can be explained by the following factors: (1) more molecules are available to interact with the sensitive cells due to a lesser probability of the aggregation of hydrophilic peptides; (2) more molecules are available for the bactericidal action, since fewer molecules remain bound to the wall; (3) an enhanced capacity of hydrophilic bacteriocins to pass through hydrophilic regions of the cell wall of the sensitive bacteria;

(4) an inhibited interaction at higher pH values between the non-lanthionine-containing bacteriocins and putative membrane receptors (Parada et al., 2007).

These studies highlight the possibility of increase antimicrobial activity of fermented products with the aim to improve food safety and quality characteristics. Besides the optimization of bacteriocin production and enhancement of its activity are economically important to reduce the production cost.

APPLICATION OF BACTERIOCINS PRODUCING LAB FOR IMPROVING SOME SAFETY CHARACTERISTICS OF PLANT PRODUCTS

Possibilities to Prolong Microbiological Spoilage of Bread Using Novel Blis Producing Lab

Knowledge of fermenting microorganisms plays a defining role in the process of fermentation standardization and it is essential to have an exhaustive view of microbial interactions. The development of starter cultures for food fermentations follows a multidisciplinary approach and requires a thorough ecological study of these ecosystems. LAB are fundamental for the fermented product properties such as lactic fermentation, proteolysis, synthesis of volatile compounds, anti-mould and antiropiness effect. Since LAB are found to be the dominant microorganisms in sourdoughs, the rheology, flavour and nutritional properties of sourdough-based baked products greatly rely on their activity (Corsetti at al., 2003, Hammes and Gänzle, 1998, Gobbetti et al., 2005). Sourdough is used as an essential ingredient for acidification, leavening and production of flavour compounds and biopreservation of bread (Sadeghi, 2008, De Vuyst, 2007, Katina et al., 2005,Hansen, 2004, Clarke et al., 2004). In bakery practice, sourdough is usually sustained by repeated inoculation, whereby a reproducible and controlled composition and activity of the sourdough microflora is paramount to achieve a constant stability of sourdough as well as a constant quality of the end-product. Besides, many researchers have reported about the high resistance of sourdough breads to the microbiological spoilage by moulds and rope-forming bacilli (Valerio et al., 2009, Hassan and Bullerman, 2008, Sadeghi, 2008,Ryan et al., 2008; Mentes et al., 2007). Mould causes mouldiness and bacteria belonging to the genus*Bacillus* (Şimşek et al., 2006) are capable of causing massive economic losses due to the considerable resistance allowing them to survive food processing (Errington, 2003; Driks, 2002). The bacterial spoilage of bread, known as ropiness, occurs as an unpleasant fruity odour (Mentes et al., 2007), and is still of major economic concern in the baking industry.

Rope Production in Bread

Ropiness is mainly caused by *Bacillus subtilis* and *Bacillus licheniformis* (Collins et al., 1991) which reportedly originate from the raw materials, the bakery atmosphere and equipment surfaces (Bailey and von Holy, 1993). These strains are also known to be food-borne pathogens when present at levels of 105 CFU/g in bread crumb (Kramer and Gilbert, 1989). *Bacillus* is a genus of rod-shaped, endospore-forming aerobic or facultatively anaerobic, Gram-positive bacteria (in some species cultures may turn Gram-negative with age) and a member of the division *Firmicutes*. Many species of the genus exhibit a wide range of physiologic abilities that allow them to live in every natural environment. Under stressful environmental conditions, the cells produce oval endospores that can stay dormant for extended periods (Ravel and Fraser, 2005, Kunst et al., 1997). Bacteria belonging to the genus *Bacillus* are capable of causing economic losses to the baking industry due to the food spoilage condition known as rope (Valerio et al., 2008; Thompson et al., 1993). The predominant species involved in bread spoilage are*Bacillus subtilis* and *B. licheniformis*, though *B. pumilus*, *B. megaterium* and *B. cereus* are implicated as well (Şimşek et al., 2006; Rosenkvist and Hansen, 1995, Collins et al., 1991). These strains are also known to be food-borne pathogens when present at levels of 105 CFU/g in bread crumb (Kramer and Gilbert, 1989). Ropiness is noticed as an unpleasant odour, followed by a soft and sticky bread crumb caused by the enzymatic degradation and the production of extracellular slimy polysaccharides (Sadeghi, 2008; Valerio et al., 2008, Pepe et al., 2003).

Figure 4 illustrates an example of —ropy bread. *B. subtilis* spores have been isolated from ropy bread, meanwhile contamination of *Bacillus* have been reported to originate from raw materials, bakery environments and also from additives, including yeast, bread improvers, and gluten (Thompson et al., 1993; Rosenkvist and Hansen, 1995; Collins et al., 1991; Sorokulova et al., 2003; Bailey and von Holy, 1993).

Figure 4: Rope production in bread.

B.subtilis spores being heat resistant can survive the baking process, since the maximum temperature in the loaf centre remains 97°C to 101°C for a few minutes (Östman, 2002, Rosenkvist and Hansen, 1995). During subsequent exposure to the warm (25°C to 30 C) and humid (water activity, ≥0.95) environmental conditions Bacillus spores germinate causing bread spoilage (Volavsek et al., 1992). The spore germination and growth of Bacillus vegetative cells during storage strongly depend on the water activity, pH and temperature (Condón et al., 1996; Quintavalla and Paroli, 1993).

Spore Formation in Bacillus Subtilis

Bacterial spores are very specialized, differentiated cell types and can survive the adverse conditions (e.g. starvation, high temperatures, ionizing radiation, mechanical abrasion, chemical solvents, detergents, hydrolytic enzymes, desiccation, pH extremes and antibiotics). Spores can cause massive problems in the food industry due to the considerable resistance allowing them to survive food processing and conservation methods (Errington, 2003, Driks, 2002).

The mature heat-resistant spore formation takes approx. 8 hours from the initial time of starvation. Numerous alterations in gene expression and a variety of physiological and morphological changes characterize the process of sporulation (Grossman and Losick, 1988). B. subtilis has been used as a model for the sporulation (the process of spore formation) studies (Errington, 2003, Eichenberger et al., 2004, Piggot and Hilbert, 2004; Phillips and Strauch,

2002). Spore formation is a unique and complex process and can be divided into stages 0, II, III, IV, V, and VI (Grossman and Losick, 1988) involving asymmetric cell division, engulfment of the smaller cell and sacrifice of the original bacterial cell for the production of a single spore (Figure 5).

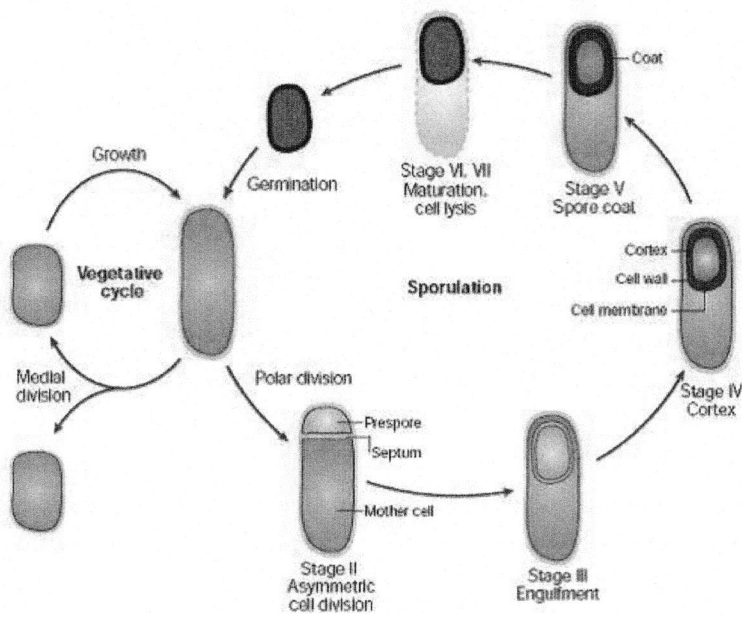

Figure 5: The sporulation cycle of *Bacillus subtilis*.

Stage 0 is characterized as the cell commitment to sporulation that leads to the building of a septum (stage II). As a cell begins the process of forming an endospore, it divides asymmetrically, resulting in the creation of two compartments (the larger mother cell and the smaller prespore). Afterwards, the degradation of thepeptidoglycan in the septum occurs and mother cell engulfs the prespore, leading to the formation of a cell within a cell (stage III). The synthesis of the endospore-specific compounds, formation of the cortex and deposition of the coat (stages IV and V) proceeds due to the activities of the mother cell and prespore. Stages IV and V are followed by the final dehydration and maturation of the prespore (stages VI and VII). Finally, the mother cell is destroyed in a programmed cell death, and the endospore is released into the environment. The endospore remains dormant until it senses the return of more favourable conditions (Errington, 2003, Grossman and Losick, 1988, Phillips and Strauch, 2002).

Strategy for the Control of Bacillus spp. By blis Producing lab in Bread Production

The initial Bacillus spore counts could be reduced by the recommended control procedures such as raw material quality, good sanitation of bakery equipment, stringent temperature control during baking, production cooling and storage environments (Bailey and von Holy, 1993, Viljoen and von Holy, 1993). The use of chemical preservatives (propionic and acetic acids) was reported to be one of the ways for the inhibition of Bacillus germination and growth in bread, although the current trend is to reduce the levels of these substances (Pattison et al., 2004, Marín et al., 2002). The increase in acidity by using traditional sourdough fermentation is an effective way to limit the germination and growth of rope forming bacteria (Sadeghi, 2008). Röcken (1996) reported about the enhanced thermal inactivation of B. subtilis spores by using an increased sourdough contents. Katina and co-workers 2005 examined the ability of LAB to inhibit the growth of rope forming strains in wheat bread and announced about the growth inhibition of B. subtilis and B. licheniformis by Lactobacillus plantarum VTT E-78076 and Pediococcus pentosaceus VTT E-90390. The added heat-treated cultures of L. plantarum E5 and Leuc. mesenteroides A27 were reported to prevent the growth of approximately 10^4 rope-producing B. subtilis G1 spores per cm^2 on bread slices for more than 15 days (Pepe et al., 2003). The inhibition of the rope spoilage of wheat bread was observed with the added 20–30 g of sourdough/100 g of wheat dough (Katina, K. et al., 2002). Kingamkono and co-workers (1994), Svanberg and co-workers (1992) shown that food fermentation by LAB to pH 4.0 or lower inhibited the growth of Bacillus as well as other pathogenic microorganisms. Suomalainen and Mäyrä-Makinen (1999) reported about the inhibitory effect of Lactobacillus rhamnosus LC705 against Bacillus spp. in bakery products. Bogovič-Matijašić and co-workers (1998) found out the antimicrobial activity of Lactobacillus acidophilus LF221 producing two bacteriocins against different pathogens including B. cereus. Røssland and co-workers (2003) demonstrated that a rapid decrease in pH during log phase of fermentation was related with the B. cereus growth inhibition. Meanwhile B. cereus sporulated and existed as endospores with the pH reduced at a slower rate in early log phase.

Several studies have been dedicated for the analysis of the antimicrobial activity of LAB and their produced bacteriocins, and have reported that the antifungal activity of LAB is lost after treatment with proteolytic enzymes. Batish et al. (1989) suggested that the antifungal substance produced by a LAB isolate was of proteinaceous nature since activity disappeared with proteinase treatment. Roy et al. (1996) isolated a Lactococcus lactis subsp. lactis with antagonistic activity against several filamentous fungi, that were lost after enzymatic treatment with chymotrypsin, trypsin and pronase Gourama (1997a) found that the inhibitory effect of a Lactobacillus casei strain against two Penicillium species was slightly reduced by treatment with trypsin and pepsin. Gourama and Bullerman (1995, 1997b) showed that a commercially available silage inoculant with a combination of Lactobacillus species (L. plantarum, L. delbrueckii subsp. bulgaricus and L. acidophilus) had antifungal and antiaflatoxin activity against A. flavus. The inhibitory activity was sensitive to treatments with trypsin and alpha-chymotrypsin, and it was concluded that the activity was due to a small peptide.

The changes in LAB antimicrobial effect upon the interactions with the enzymes is very important in the baking industry where commercial enzyme preparations often are used for fermentation processes intensification and recently became one of the research topics.

In one of the studies (Digaitiene et al., 2005), 270 bacterial strains were isolated from spontaneous sourdoughs and of these, five LAB (Lactobacillus sakei MI806, Pediococcus pentosaceus MI808, MI809 and MI810, Pediococcus acidilactici MI807) isolates were found to produce BLIS (sakacin 806, pediocin 808, 809, 810 and pediocin Ac807 respectively). Isolates of new bacteriocins producing LAB strains depend for subclass II. The results of inhibitory spectra studies and pH sensitivity analysis indicated that the BLIS under investigation were different from each other. These novel BLIS (sakacin 806, pediocin 808, pediocin 809, pediocin 810 and pediocin Ac807) have been tested for their antimicrobial activity against B. subtilis, one of the most important micro-organisms responsible for ropiness in bread; furthermore their sensitivity to various baking enzymes has been examined (Narbutaite et al., 2008). Antimicrobial activity was tested using an overlay assay method; the results showed that the BLIS studied here were effective against B. subtilis. To our knowledge, this is the first report of BLIS-producing LAB isolated from sourdoughs which are active against B. subtilis.

Bacteriocins have gained importance as natural biopreservatives for the control of spoilage and pathogenic organisms in foods. Latest studies highlights the possibility of using LAB exhibiting antimicrobial activity against B. subtilis in sourdough bread making, a desirable characteristic when selecting for more competitive starters. The strains described here can have an impact when used as starter cultures for traditional sourdough fermentation by delaying spore germination and inhibiting the outgrowth of B. subtilis. This opens up the possibility of using such LAB on an industry scale. Future work will also focus on obtaining the amino acid sequences of the BLIS presented here.

Potential of Lactic Acid Bacteria to Degrade Biogenic Amines in Different Fermentation Media

A variety of fermented foods especially protein-rich foods e.g. fermented vegetables, legume products, beers and wines contain biogenic amines (BAs) (Kalač et al., 2002). During the fermentation process protein breakdown products, peptides and amino acids, used by spoilage and also by the fermentation microorganisms represent precursors for BAs formation (Hernandez-Jover et al., 1997, Bodmer et al., 1999). BAs are formed through the decarboxylation of specific free amino acids by exogenous decarboxylases released from the microbial population associated with the raw material.

Some biogenic amines such as histamine (HIS), tyramine (TYR), putrescine (PUT) and cadaverine (CAD) are important for their physiological and toxicological effects on the human body. They may exert either psychoactive or vasoactive effects on sensitive humans. Histamine is physiologically the most important BA. Histamine has been found to cause the most frequent food-borne intoxications associated with BAs; it acts as a mediator and is involved in pathophysiological processes such as allergies and inflammations (Gonzaga et al, 2009). Tyramine can evoke nausea, vomiting, migraine, hypertention and headaches (Shalaby, 1996). Putrescine and cadaverine can increase the negative effect of other amines by interfering with detoxification enzymes that metabolize them (Stratton et al., 1991). The consumption of foods with high concentrations of BAs can induce adverse reactions such as nausea, headaches, rashes and changes in blood pressure (Ladero et al., 2010).

The main BAs associated with such fermented plant product as wine are HIS, TYR and PUT (Ancin-Azpilicueta et al., 2008). Their presence in wine is considered as marker molecules of quality loss, and some EU countries even have recommendations for the amount of histamine acceptable in wine which impacts on the import and export of wines to these countires. Most fermented foods, such as cheese, fermented sausages and beer, which are consumed more frequently than wines, have biogenic amine content (Fernandez et al., 2007). However, the precence of alcohol in wine may enhance the activity of amines because it inhibits monoamine oxidase enzymes. These enzymes depending for the detoxification system in the intestinal tract of mammals convert amines into non-toxic products, which are further excreted out of the organism.

Regarding fruits and vegetables relatively low levels of biogenic amines were found in fruit juice and canned fruit/vegetable samples. The same tendency has been noticed in other publications, but sometimes the results are controversial. Moret et al., (2005) showed that vegetables generally contained low levels in biogenic amines (0.1–9.6 mg kg^{-1}) while Kalač et al., (2002) found relatively high levels of the amines in vegetables (0.8–52.5 mg kg^{-1}). The polyamines PUT and spermidin (SPM) are practically ubiquitous in all vegetables at a few mg/100 g of fresh weight and TYR is less widespread in vegetables (Kalač et al., 2002; Moret et al., 2005). They are implicated in a number of physiological processes, such as cell division regulation, plant growth, flowering, fruit development, response to stress and senescence (Bouchereau et al., 2000). Moreover, although PUT, SPD and other biogenic amines are generated in low quantities in most canned vegetables/fruits, they are not the primary metabolic products produced by the fermenting organisms (Stratton et al., 1991).

With the exception of tempe (Saaid et al., 2009) and taucu, relatively low levels of biogenic amines are found in the soy bean products tested. Studies by Mower and Bhagavan (1989) showed higher level of TYR (450 mg kg^{-1}) in salted black beans. The quantitative analysis of fermented products prepared for wheat bread production revealed that tyramine (32.6–215.8 mg kg^{-1}), histamine (20.8–96.7 mg kg^{-1}), and putrescine (33.7–195.2 mg kg^{-1}) showed as being the major occurring BAs (Bartkiene et al., 2011). Since several varieties of molds, yeasts and lactic acid bacteria are involved in the fermentation processes of such products and the raw material (soy bean) contains considerable amounts of protein, the formation of various amines might be expected during the fermentation (Shalaby, 1996). Studies have shown that biogenic amines in fermented soy bean products are most likely formed by the lactic microflora that is active during fermentation (Kirschbaum et al., 2000). TYR and HIS have been found at various levels in such products (Stratton et al., 1991). The variability of biogenic amines levels in the commercial fermented soy bean products samples had been attributed to the variations in manufacturing processes; variability in the ratio of soy bean in the raw material, microbial composition, conditions and duration of fermentation (Shalaby, 1996).

Knowledge concerning the origin and factors involved in BAs production in fermented products e.g. wine is well documented, and recently several reviews on this topic have been published (Costantini et al., 2009; Moreno-Aribas and Polo, 2010). They are generated either as the result of endogenous decarboxylase-positive microorganisms in raw materials or by the growth of contaminating decarboxylase-positive microorganisms in fermented products. With regards to wine microorganisms, a large amount of literature is available on the production of BAs. Several research group support the view that biogenic amines are formed in winemaking mainly by lactic acid bacteria (LAB) due to the decarboxilation of the free amino acids (Constantini et al., 2006; Lucas et al., 2008). The levels of BAs usually increase during fermentation due to decarboxylase activity of the LAB used as starter culture.

Low acid conditions, such as those occurring during fermentation, favour the decarboxylation of amino acids (De las Rivas et al., 2005). The levels of free amino acids usually increase in fermented products during fermentation due to the action of endogenous and exogenous proteases through proteolysis processes (Hughes et al., 2002). It is thought that proteolysis might provide the nutrient for spoilage microorganisms, leading to a promoted growth of those microorganisms (Riebroy et al., 2004).

In this context, recently published paper (Garcia et al., 2011) reports novel data about the presence of histamine-, tyrosine- and putrescine-degrading enzymatic activities of wine-associated LAB. Of particular interest are the results concerning the degradation of putrescine, since no such degrading ability of any food LAB has previously been reported. The isolates tested (42 strains *Oenococcus oeni*, 7 strains *Pediococcus parvulus*, 4 strains *P. pentosaceus*, 6 strains *Lactobacillus plantarum*, 9 strains *L.hilgardi*, 3 strains *L. zeae*, 7 strains *L. casei*, 5 strains *L. paracasei* and 2 strains Leuconostoc mesenteroides) belong to the principal species of wine LAB and other related ecosystems and were selected because they came from wine cellars that often suffer from the problem of BAs in their wines (Moreno-Aribas and Polo, 2010). In this study the most potent amine-degrading species detected were L. plantarum, P. parvulus and, in particular, P. pentosaceus and L. casei, in spite of the fact that strains of these last species have never be reported to degrade histamine, tyramine and/or putriscine. None of the strains were able to produce these BAs as they did no show the decarboboxylase activity necessary for the production of these compounds in wine. However, this potential for histamine, tyramine and/or putrescine degradation among wine LAB does not appear to be very frequent, since out of the 85 strains examined, only nine displayed noteworthy amine-degrading activity in culture media. Further studies using other LAB species and/or strains may enable more potent amine-degrading enzyme producers to be identified. However, it was observed that positive strains displayed amine-degrading activity against several biogenic amines simultaneously, in accordance with previous works that also reported the presence of either one or two amine oxidases in other food fermenting microorganisms, such as*Micrococcus* varians and *Staphylococcuscarnosus* (Leuschner et al., 1998).

The fact that active bacteria which were able to significantly reduce the concentration of BAs in the conditions used in the study came from different fermentation media such as young wine, wood- aged wines, sherry wines (Table 1), suggest that there are ecological niches for the isolation of potential amine-degrading bacteria.

Table 1: Percentage of degradation of the biogenic amines (histamine, tyramine and putrescine) by wine-associated LAB in culture media. [a] Activity is expressed as a percentage of control without strain and according to HPLC quantitative biogenic amine results.[b] Mean value (n=3); n.e. : no effect was observed.

Strains	Degradation, %[a,b]		
	Histamine	Tyramine	Putrescine
L. casei IFI-CA-52	54	55	65
L. hilgardi IFI-CA-41	n.e.	n.e.	20
L. plantarum IFI-CA-26	33	n.e.	24
L. plantarum IFI-CA-54	23	17	24
O. oeri IFI-CA-32	12	n.e.	16
P. parvulus IFI-CA-30	20	15	53
P. pentosaceous IFI-CA-30	10	12	49
P. pentosaceous IFI-CA-83	19	22	39
P. pentosaceous IFI-CA-86	n.e.	54	69

Recently, homofermentative Pediococcus acidilactici were isolated from spontaneous rye sourdoughs and characterised as producing pediocin Ac807 with antimicrobial activity against Bacillus subtilis (Digaitiene et al., 2005; Narbutaite et al., 2008). Since fermentation by using P. acidilactici could improve or modify flavour, taste, and texture of used plant additives for wheat bread production, the safety characteristics of fermented products rich in proteins are not always predictable. Therefore the BAs investigation in untreated whole lupine flours and fermented products of different lupines species (Lupinus angustifolius and Lupinus luteus) after spontaneous fermentation and fermentation by P. acidilactici has been carried out (Bartkiene et al., 2011). This study showed that the BAs levels were found to be lower after fermentation (by 17%) of L. luteus flour by P. acidilactici. Also the total amount of BAs was significantly reduced (25%) during spontaneous fermentation of L. luteus flour compared to non-treated samples. Opposite to the L. luteus, fermentation of L. angustifolius flour led to an increase of 20.5% and 44% of total amount of BAs in spontaneous and P. acidilactici sourdoughs, respectively. The different BAs in fermented products, which have been prepared using different kinds of lupine can be explained by the quality of the raw material and/or by formation during the fermentation process involving microorganisms and these results are in agreement with previous research (Silla Santos, 1996).

The presented results agree to the findings that the use of decarboxylase negative microorganisms, e.g. LAB as starter cultures could be an important factor to be considered in order to reduce the levels of BAs in fermented foods.

Possible Approaches of Lab and Enzymes to Biodegradation of Mycotoxins

Mycotoxins are secondary metabolites produced by a wide variety of filamentous fungi, including species from the genera *Aspergillus*, *Fusarium* and *Penicillium*. They cause nutritional losses and represent a significant hazard to the food and feed chain. Humans have long been exposed to mycotoxins by several different routes: directly, via foods of plant origin, including cereals from which bread and bakery products are derived; by air (both indoors and outdoors); or indirectly, through foods of animal origin. Many countries, therefore, have established measures to safeguard the health of consumers by establishing regulations for food and feed. The most economically important mycotoxins occurring in food and feed are *aflatoxins*, *ochratoxin* A, *patulin*, and the *Fusarium toxins* (zearalenon, trichothecenes, fumonisins etc.) (Chassy, 2010).

Principally, there are three possibilities to avoid harmful effect of contamination of food and feed caused by mycotoxins: (1) prevention of contamination, (2) decontamination of mycotoxin-containing food and feed, and (3) inhibition of absorption of mycotoxin content of consumed food into the digestive tract (Halász et al., 2009).

The theoretically soundest approach of prevention is doubtless to breed cereals and other food and feed plants for resistance to mould infection and consequently exclude mycotoxin production. Particularly in breeding wheat and corn, significant improvement of resistance has been achieved. The identification of microbial species (and genes coding enzymes degrading mycotoxins) allows transfer of these genes into plants and production of such enzymes by transgenic plants. In this way, the safety problems connected with the use of live microorganisms may be avoided.

Another practical approach to prevention of mycotoxin contamination is the inhibition of the growth of molds and their production of mycotoxins. First, optimal harvesting, storage and processing methods, and conditions may be successful in prevention of mold growth. Although the primary goal is the prevention of mycotoxin contamination, mycotoxin formation appears to be unavoidable under certain adverse conditions.

Treatment of grains by some chemicals to prevent mycotoxin formation is also possible. Most of these compounds work by inhibiting fungal growth. For example, approximately one hundred compounds have been found to inhibit aflatoxin production. Two extensively studied inhibitors of aflatoxin synthesis are dichlorvos (an organophosphate insecticide) and caffeine. As reported by

(Halász et al., 2009) some surfactants have been found to suppress the growth of Aspergillus flavus and aflatoxin synthesis.

When contamination cannot be prevented, physical and chemical decontamination methods have been employed with varying success in the past, principally for feed. Whichever decontamination strategy is used, it must meet some basic criteria:

- The mycotoxin must be inactivated or destroyed by transformation to non-toxic compounds;

- Fungal spores and mycelia should be destroyed, so that new toxins are not formed;

- The food or feed material should retain its nutritive value and remain palatable;

- The physical properties of raw material should not change significantly; and

- It must be economically feasible / the cost of decontamination should be less than the value of contaminated commodity.

Partial removal of mycotoxin may be achieved by dry cleaning of the grain and in the milling process, as well. Milling led to a fractionation, with increased level of mycotoxin in bran and decreased level in flour. The majority of mycotoxins are heat-stable so heat treatment, usually applied in food technology, does not have significant effect on mycotoxin level.

Efforts were made in several countries to find an economically acceptable way of destruction of mycotoxins into non-toxic products using different chemicals such as alkali and oxidative agents. Although such treatment reduces nearly completely the mycotoxin concentration, these chemicals also cause losses of some nutrients and such treatment is too drastic for grain destined for food uses.

Although the different methods used at present have been to some extent successful, most methods have major disadvantages, starting with limited efficacy to losses of important nutrients and generally with high costs.

More recently, biological decontamination and biodegradation of mycotoxins with microorganisms or enzymes, have been used (He et al., 2010; Juodeikiene et al., 2011). In this case no harmful chemicals where used, so no significant losses in nutritive value and palatability of decontaminated food and feed occurred. Today, ruminants appear to be a promising potential source of microbes or enzymes for use in the biotransformation of mycotoxins.

One of the most frequently used strategies for biodegradation of mycotoxins includes isolation of microorganisms able to degrade the given mycotoxin and treatment of food or feed in an appropriate fermentation process. From the food safety point-of-view, fermentation with microorganisms commonly used in food production (fermentation with lactic acid bacteria, alcoholic fermentation, traditional fermentation of vegetable protein used in South Asia, etc.) should be preferred. Knowledge of enzymes that take part in degradation of mycotoxins opens some new approaches: (1) the production of genetically modified species of microorganisms commonly used in food production and their use for production of enzymes mentioned above; or (2) the transfer of genes coding for these enzymes to transgenic plants and use the plants for production of mycotoxin degrading enzymes.

In staple food such as bread and bakery products in the flour sector, yeast and lactobacilli now play an important role. It thus stands to reason that the same microbes and enzymes are the first to have been considered for use as detoxifying or decontaminating agents. This type of biodegradation could therefore prove a useful strategy for partially overcoming the problem of some mycotoxins. Indeed, this already takes place in bread and in sourdough processes (Bartkiene et al., 2008); and OTA in food can also undergo biodegradation (Abrunhosa et al., 2010) and certain antagonistic yeast strains can substantially degrade OTA. This might offer new possibilities for reducing this mycotoxin in bread and bakery products and their raw materials (Patharajan et al., 2011). The use of enzymes or engineered micro-organisms (provided that these are allowed by legislation) as processing aids in the bread and bakery sector would also prove beneficial. Genetic engineering technologies will improve the efficiency with which enzymes can be produced from these organisms, and will allow the production of engineered organisms which have the target genes. They will additionally increase the availability and bioavailability, and will improve the quality of the end product.

Inhibition of Mycotoxins Biosynthesis by Lactic Acid Bacteria

Several papers dealing with the inhibition of mycotoxin biosynthesis by LAB have focused on aflatoxins (Thyagaraja & Hosono, 1994). During cell lysis, it is possible that LAB releases molecules that potentially inhibit mould growth and therefore lead to a lower accumulation of their mycotoxins (Gourama & Bullerman, 1995). These "anti-mycotoxinogenic" metabolites could also be produced during LAB growth. Gourama (1991), using a dialysis assay, demonstrated the occurrence of a metabolite that inhibits aflatoxin accumulation in Lactobacillus cell-free extracts. It was suggested that this inhibition of aflatoxin biosynthesis was not the result of a hydrogen peroxide production or a pH decrease (Karunaratne et al., 1990). These findings were consistent with those of Gourama (1991), who suggested that inhibition of aflatoxin biosynthesis by Lactobacillus cell free supernatants was probably due to specific bacterial metabolites. Coallier-Ascah and Idziak (1985) reported a significant reduction of aflatoxin biosynthesis by Lactobacillus cell free supernatants and suggested that this inhibition was related to a heat stable, low-molecular-weight inhibitory compound. Although Lactobacillus spp. were found to delay aflatoxin biosynthesis, other lactic strains such as Lc. lactis were found to stimulate aflatoxin accumulation (Luchese and Harrigan, 1990).

Decontamination of Mycotoxins Using Microorganisms by Binding or Degradation

Biological detoxification of mycotoxins works mainly via two major processes, sorption and enzymatic degradation, both of which can be achieved by biological systems. Live micro-organisms can absorb either by attaching the mycotoxin to their cell wall components or by active internalization and accumulation. Dead microorganisms too can absorb mycotoxins, and this phenomenon can be exploited in the creation of biofilters for fluid decontamination or probiotics (which have proven binding capacity) to bind and remove the mycotoxin from the intestine.

Another approach to the biological decontamination of mycotoxins involves their degradation by selected micro-organisms. Recently critical review on biological detoxification by (Dalié et al., 2009) summarized different and interesting aspects of the biological detoxification of mycotoxins.

Micro-organism detoxification can be performed in many different ways (Magan and Olsen, 2004):

- The entire organism can be used as a starter culture, as in the fermentation of beer, wine and cider, or in lactic acid fermentation of vegetables, milk and meat.
- The purified enzyme can be used in soluble or immobilized (biofilter) forms.
- The gene encoding the enzymatic activity can be transferred and overexpressed in a heterologous system; interesting candidates for this application include yeasts, probiotics and plants.

Biological methods have been applied for the biodegradation and decontamination of different mycotoxins.

Aflatoxins. As the first mycotoxins to be discovered, were also the first target in screening for microbial degradation. Several examples of the detoxification of the most common and important mycotoxins are reviewed. Almost 40 years ago, several species of micro-organisms – including yeasts, moulds, bacteria, actinomycetes and algae – were screened for detoxification activity; based on this studies only one isolate was found, Flavobacterium auranotiacum, which significantly removed aflatoxin from a liquid medium (Ciegler et al., 1966).

Later aflatoxins decontamination during fermentation was reported in several cases. About 50% reduction in aflatoxins B1 and G1 has been reported during an early stage of miso fermentation. It was attributed to the degradation of the toxin by micro-organisms. Significant losses of aflatoxin B1 and ochratoxin A were observed during beer brewing (Chu et al., 1975). Detoxification of aflatoxin B1 occurred during the fermentation of milk by LAB and in dough fermentation during breadmaking. Digestive tract micro-organisms are able to reduce mycotoxin levels not only by binding and removal but also by detoxification.

Most data dealing with the effects of LAB on the accumulation of mycotoxins are related to aflatoxin-producing moulds. Wiseman and Marth (1981) revealed the existence of an amensalism relationship between Lc. lactis and A. parasiticus. When these authors added the spores of A. parasiticus to a 13-day-old culture of Lc. lactis, they observed the entire repression of aflatoxin production. When the fungal spore suspension and the lactic strain were inoculated simultaneously, an increase in aflatoxin production was observed. In contrast, Coallier-Ascah and Idziak (1985) showed an inhibition of aflatoxin accumulation when both microorganisms were simultaneously cultivated in Lab-Lemco tryptone broth (LTB). Addition of glucose to the cultivation medium during the conidiation phase of the mould did not restore the the production of aflatoxin.

Several LAB have been found to be able to bind aflatoxin B_1 in vitro (Kankaanpää et al., 2000; Gratz et al., 2004), with an efficiency depending on the bacterial strain (Shah & Wu, 1999). El-Nezami with co-workers (1998a) have evaluated the ability of five Lactobacillus to bind aflatoxins in vitro and have shown that probiotic strains such as Lb. rhamnosus GG and Lb. rhamnosus LC-705 were very effective for removing aflatoxin B_1, with more than 80% of the toxin trapped in a 20µg/ml solution (Haskard et al.,1998).

According to Coallier-Ascah and Idziak (1985), the inhibition of aflatoxin accumulation was not related to a pH decrease but rather to the occurrence of a low-molecular-weight metabolite produced by the LAB at the beginning of its exponential phase of growth. Inhibition of aflatoxin production by other LAB belonging to the genus Lactobacillus was also reported (Karunaratne et al., 1990). It was assumed that this inhibition resulted from the production of a metabolite different from hydrogen peroxide or organic acid (Gourama, 1991). Haskard wich co-wokers (2001) demonstrated that Lb. rhamnosus GG (ATCC 53103) and Lb. rhamnosus LC-705 (DSM 7061) were able to eliminate aflatoxin B_1 from the culture medium by a physical process.

Several studies have suggested that the antimutagenic and anti-carcinogenic properties of probiotic bacteria can be attributed to their ability to non-covalently bind hazardous chemical compounds such as aflatoxins in the colon (El-Nezami et al., 1998b; Gratz et al., 2004). Both viable and non-viable forms of the probiotic bacterium Lactobacillus rhamnosus GG effectively removed aflatoxin B1 from an aqueous solution (El-Nezami et al., 1998b). Since metabolic activation is not necessary, binding can be attributed to weak, non-covalent, physical interactions, such as association to hydrophobic pockets on the bacterial surface (Haskard et al., 2000). Coallier-Ascah and Idziak (1985) reported a significant reduction of aflatoxin biosynthesis by Lactobacillus cell free supernatants and suggested that this inhibition was related to a heat stable, low-molecular-weight inhibitory compound.

Ochratoxin-A (OTA) The major OTA producers in food and feed products are considered to be A. alliaceus, A. carbonarius, A. ochraceus, A. steynii, A. westerdijkiae, P. nordicum and P. verrucosum (Frisvad et al., 2006). These are mainly associated with agricultural crops pre-harvest, or in post-harvest storage situations. Biological methods use microorganisms, which can decompose, transform or adsorb OTA to detoxify contaminated products or to avoid the toxic effects when mycotoxins are ingested. These are the technologies of choice for decontamination proposes because they present several advantages from being mediated by enzymatic reactions. For example, they are very specific, efficient, environmentally friendly, and they preserve nutritive quality.

Two pathways may be involved in OTA microbiological degradation (Abrunhosa et al., 2010). First, OTA can be biodegraded through the hydrolysis of the amide bond that links the L-β-phenylalanine molecule to the OTα moiety. Since OTα and L-β-phenylalanine are virtually non-toxic, this mechanism can be considered to be a detoxification pathway. Second, a more hypothetical process involves OTA being degraded via the hydrolysis of the lactone ring. In this case, the final degradation product is an opened lactone form of OTA, which is of similar toxicity to OTA when administered to rats. However, it is less toxic to mice and Bacillus brevis. Although this is hypothetical, it is likely to occur since microbiological lactonohydrolases, which undertake a similar transformation, are common. Several protozoal, bacterial, yeast and filamentous fungal species are able to biodegrade OTA. After success in clarifying the mechanism and degradation products of ochratoxin, three directions in recent research may be observed (1) possibilities of bacterial degradation, study of molds able to degrade this mycotoxin and identification and isolation of enzymes taking part in the degradation process.

Lactobacillus strains were demonstrated to eliminate 0.05 mg OTA/L added to culture medium - in particular, L. bulgaricus, L. helveticus, L. acidophillus, eliminated up to 94%, 72% and 46%, respectively, of OTA (Böhm et al., 2000); L. plantarum, L. brevis and L. sanfrancisco were reported to eliminate 54%, 50% and 37%, respectively, of 0.3 mg OTA/L after 24 h of incubation (Piotrowska and Zakowska, 2000). It is now generally accepted that OTA adsorption to the cells walls is the predominant mechanism involved in this OTA detoxification phenomenon by lactic acid bacteria (LAB). For example, adsorption effects were claimed by Turbic et al. (2002), who found that heat and acid treated cells from two Lactobacillus rhamnosus strains were more effective at removing OTA from phosphate buffer solutions than viable cells. The strains removed 36% to 76% in the buffer solution (pH 7.4) after 2 h at 37 C. Similarly, Piotrowska and Zakowska (2005) verified that L. acidophilus and L. rhamnosus caused OTA reductions of 70% and 87% of 1 mg OTA/L after five days at 37 C, and that significant levels of the OTA were present in the centrifuged bacteria cells. Other LAB (L. brevis, L. plantarum and L. sanfranciscencis) also produced smaller decreases on OTA (approximately 50%). Finally, Del Prete et al. (2007) tested 15 strains of oenological LAB in order to determine the in vitro capacity to remove OTA, and reported Oenococcus oeni as the most effective, with OTA reductions of 28%. The involvement of cell-binding mechanisms was confirmed as (i) up to 57% of the OTA absorbed by the cells was recovered through methanol extraction from the bacteria pellets; (ii) crude cell-free extracts were not able to degrade OTA; and (iii) degradation products were not detected. Nevertheless, some authors consider that metabolism may also be involved. For example, Fuchs et al. (2008) confirmed that viable cells of L. acidophilus removed OTA more efficiently then unviable. A L. acidophilus strain was able to decrease ≥95% the OTA in buffer solutions (pH 5.0) containing 0.5 and 1 mg OTA/L when incubated at 37 C for 4 h. In addition, a detoxification effect was also demonstrated since pre-incubation of OTA with this strain reduced OTA toxicity to human derived liver cells (HepG2) (Fuchs et al., 2008). Other L. acidophilus strains demonstrated only a moderate reduction in OTA contents suggesting that the effect was strain specific. In summary, some LAB adsorbs OTA by a strain specific cell-wall binding mechanism, although some undetected catabolism can also be involved. The detection of this OTA catabolism may only be possible with radiolabeled OTA.

The potential of LAB as mycotoxin decontaminating agents has been studied in different fermentation processes and reviewed (Shetty and Jespersen, 2006). The ochratoxin-A content, its fate during wine-making and possibilities of its degradation have been intensively studied. Overviews concerning presence and fate of this mycotoxin in grapes, wine and beer were published by Mateo et al. (2007) and Varga and Kozakiewicz (2006). Although the decrease of OTA content in liquid phase during vinification process is observed by the majority of researchers, reports are controversial regarding the mechanism of OTA removal. Is it a result of malolactic fermentation due to the action of lactic acid bacteria (Kozakiewicz et al., 2003), or is it adsorption to yeast cell walls (Binder et al., 2000). Reports about the capacity of proteolytic enzymes to hydrolyze OTA can also be found.

Furthermore, although the results of these studies look very promising for reducing OTA contamination, studies on model systems do not guarantee the degradation of OTA in situ, using food or feed. Further studies are needed to characterize the products of degradation and to investigate the activity of these bacteria in food and feedstuffs.

Patulin contamination of apple and other fruit-based foods and beverages is an important food safety issue due to the high consumption of these commodities. Patulin contamination is considered of greatest concern in apples and apple products; however, this mycotoxin has also been found in other fruits, such as pears, peaches, strawberries, blueberries, cherries, apricots and grapes as well as in cheese (Halász et al., 2009). The initial studies concerning degradation of patulin by actively fermenting yeasts were reported in the 1979 by Stinson et al. (1979). However, authors were not able to chemically characterize the products of degradation. More recently, Moss and Long (2002) reported that under fermentative conditions, the commercial yeast Saccharomyces cerevisiae transformed patulin into ascladiol.

In a recent study (Richelli et al., 2007) the ability of Gluconobacter oxydans to degrade patulin was investigated and the degradation products of this mycotoxin determined. More than 96% of patulin was degraded after twelve-hour treatment, due to change of chemical structure (opening of the pyran ring). The degradation product was confirmed to be ascladiol. The genus Gluconobacter, whose taxonomy is at present under worldwide study, is made up of five different species (Tanasupawat et al., 2004;Sievers et al., 1995) which have no health risk, and that are commonly used in food manufacturing. Apple juice inoculated with this bacterium and incubated for 3 days still tasted like juice and was drinkable. However, keeping in mind the toxicity of ascladiol and eventual unsatisfactory organoleptic properties of alcoholic apple (fruit) juice (apple wine), the use of this bacterium at the industrial level needs additional investigation.

In screenings for patulin detoxifying bacteria, has been isolated a bacterium from fermented sausage; it was identified as Lactobacillus plantarum, and it significantly reduced patulin levels via an intracellular enzyme (Halász et al., 2009).

Fusarium toxins Considerable amounts of the Fusarium mycotoxins zearalenone (ZEN) and its derivative α-zearalenol, were bound in vitro to the probiotic bacteria L. rhamnosus GG and L. rhamnosus LC705. Both heat-treated and acid-treated bacteria were capable of removing the toxins, indicating that binding, not metabolism is the mechanism by which the toxins are removed from the media (El-Nezami et al., 2002). Zearalenone was also degraded by a mixed bacterial culture. A few other microbial activities that transform zearalenone have been published but are protected by patents. Several micro-organisms have been found that can degrade DON and T-2. On the basis of morphological and phylogenetic studies, the degrader strain was classified as a bacterium belonging to the Agrobacterium Rhizobium group. Interactions between lactic strains and ZEN and its derivative, β-zearalenol were also investigated.

DON levels did not change during beer malting and the amount of trichothecene did not change during wine alcoholic fermentation. In contrast, trichothecene and iso-trichothecin were decomposed during alcoholic fermentation of grape juice. It was suggested that the yeast epihydroxylase might be involved.

A significant proportion (38–48%) of both toxins was trapped in the bacterial pellet and no degradation product of zearalenone or α-zearalenol was detected (El-Nezami et al., 2002), leading to the conclusion that binding and not metabolism was the mechanism by which the toxins were removed from the media. Similar results were obtained with other mycotoxins including ochratoxin A (Del Prete et al., 2007; Fuchs et al., 2008) and fumonisin B_1 and B_2 (Niderkorn et al., 2006).

Therefore, two specific processes such as binding and inhibition of biosynthesis may be involved in the interaction between LAB and the accumulation of some mycotoxins.

Concerning the mechanisms of action involved in the removal of fumonisins by LAB, Niderkorn (2007) suggested that peptidoglycans were the most plausible fumonisin binding sites. The quenching ability of LAB was increased when bacteria were killed using different physical and chemical treatments, while lysozyme and mutanolysin enzymes that target peptidoglycans partially inhibited it. It was also reported that tricarballylic acid chains found in fumonisin molecules played an important role in the binding process since hydrolysed fumonisin had less affinity for LAB, and free amine group inactivation had no effect on the binding process (Niderkorn, 2007). The same article attempted to explain the low affinity of fumonisin B_1 using a molecular modelling approach. In fact, an additional hydroxyl group in fumonisin B_1 could form a hydrogen bond with one of the tricarballylic acid chains, resulting in a spatial configuration where the tricarballylic acid chain is less available to interact with bacterial peptodoglycans.

Removal of fumonisins by LAB was ascribed to adhesion to cell wall components rather than covalent binding or metabolism, since the dead cells fully retained their binding ability. Peptidoglycans probably play a key rule in this binding process. Therefore, elucidating the differences between bacterial cell wall components of LAB strains might make it possible to select LAB species with the potential to act as biopreservative agents capable of reducing exposure from fumonisins that occur in food and feed.

CONCLUSIONS

The use of bacteriocins and/or bacteriocin-producing strains of LAB are of great interest as they are generally recognized as safe organisms and their antimicrobial products as biopreservatives. Several studies confirmed that microorganisms and enzymes could be a practical way to reduce the concentrations of some contaminants and to avoid the toxic effects via bioremediation. This opens up wide possibilities of using such biotechnological means on an food industry scale. Further experiments like utilization of the strains in fermentation processes or using the enzyme preprations can exhibit great outcomes. Great source of BLIS producing LAB strains could be traditional fermentation processes which should be more widely distributed as novel microorganisms crossing the borders. Further development of these strains for the biopreservation of food products requires an understanding of the mechanisms of action of the antimicrobial activity and of the decontamination of certain contaminants e.g. biogenic amines and mycotoxins. It seems that, according to results of the experiments to date, microorganisms are the main living organisms applicable for the biodegradation of these contaminants. Progress in this field of molecular biology techniques, antimicrobial LAB strains with multi-functional properties, including the degradation of mycotoxins, can be engineered to significantly improve the quality, safety and acceptability of plant foods. Further studies and knowledge of enzymes taking part in mycotoxin degradation allows production of these enzymes and their use for detoxification instead of microorganisms or additionally with LAB strains. Despite the intensive research in this field and of the numerous publications that confirm the ability of various microorganisms to degrade mycotoxins, lack of results achieved until now in the development of practical commercial technologies by using BLIS producing strains hampered progress. The majority of experiments were carried out in model systems and in laboratory conditions. The control of degradation products and the effects of detoxification on nutritive and sensory properties is in every case a decisive part of research and potential application. The use of antifungal LAB instead of chemical preservatives would enable the food industry to meet the request of consumers for natural products. Finally a practical technology should be developed and controlled from an economical point of view. Several studies confirmed that different environmental factors have strongly affected on the growth and bacteriocin production of several LAB strains that are mainly considered for food applications. However, it is desirable to continue the studies on this subject to select the most efficient factors and their combinations, which could be used on a production scale. Future work on LAB bacteriocin production, purification, obtaining the amino acid sequences of the BLIS to increase their activity is required

REFERENCES

1. L. Abrunhosa, R. R. M. Paterson, A. Venâncio, 2010 Biodegradation of ochratoxin A for food and Feed Decontamination. Toxins 2 1078 1099 .

2. C. Ancin-Azpilicueta, A. Gonzalez-Marco, N. Jimenez-Mareno, 2008 Current knowledge about the precense of biogenic amines in wine. Critical Reviews in Food Science and Nutrition 48 257 275 .

3. ANVISA- Brazilian Agency of Sanitary Surveillance. Food with health claims, new foods/ ingredients, bioactive compounds and probiotics. 2010http://www.anvisa.gov.br/alimentos/comissoes/tecno_lista_alega. htm. Accessed May 28, 2010.

4. L. Axelsson, T. Katla, M. Bjornslett, Eijsink. G. H. Vincent, A. Holck, 1988 A system for heterologous expression of bacteriocins in Lactobacillus sakei. FEMS Microbiology Letters 168 (1), 137-143.

5. C. P. Bailey, A. Von Holy., 1993 Bacillus spore contamination associated with commercial bread manufacture. Food Microbiology 10 (4), 287 294 .

6. V. K. Batish, S. Grover, R. Lal, 1989 Screening lactic starter cultures for antifungal activity. Cultured Dairy Products Journal 24 21 25 .

7. P. Castellano, M. E. Farías, W. Holzapfel, G. Vignolo, 2001 Sensitivity variations of Listeria strains to the bacteriocins, lactocin 705, enterocin CRL35 and nisin. Biotechnology Letters 23 (8), 605 608 .

8. B. M. Chassy, 2010 Food safety risk and consumer health. New Biotechnology 27 (5), 534 544 .

9. L. M. Cintas, M. P. Casaus, C. Herranz, I. F. Nes, P. E. Hernández, 2001 Review: Bacteriocins of lactic acid bacteria. Food Science and Techology International 7 (4), 281 305 .

10. C. I. Clarke, T. J. Schober, P. Dockery, K. O'Sullivan, E. K. Arendt, 2004 Wheat sourdough fermentation: effects of time and acidification on fundamental rheological properties. Cereal Chemistry 81 (3), 409 -417.

11. J. Coallier-Ascah, E. Idziak, 1985 Interaction between Streptococcus lactis and Aspergillus flavus on production of aflatoxin. Applied and Environmental Microbiology 49 163 167 .

12. N. E. Collins, L. M. Kirschner, A. Von Holy., 1991 Characterization of Bacillus isolates from ropey bread, bakery equipment and raw materials. South African Journal of Science 87 62 66 .

13. S. Condón, A. Palop, J. Raso, F. J. Sala, 1996 Influence of the incubation temperature after heat treatment upon the estimated heat resistance values of spores of Bacillus subtilis. Letters in Applied Microbiology 22 (2), 149 -152.

14. A. Constantini, M. Cersosimo, V. Del Prete, E. Garcia-Moruno, 2006 Production biogenic amines by lactic acid bacteria: screening by PSR, thin chromatography and high-performance liquid chromatography of strains isolated from wine and must. Journal of Food Protection 69 391 396 .

15. A. Corsetti, M. De Angelis, F. Dellaglio, A. Paparella, P. F. Fox, L. Settanni, M. Gobbetti, 2003 Characterization of sourdough lactic acid bacteria based on genotypic and cell-wall protein analyses. Journal of Applied Microbiology 94 (4), 641 654 .

16. A. Corsetti, L. Settanni, D. Van Sinderen, 2004 Characterization of bacteriocin-like inhibitory substances (BLIS) from sourdough lactic acid bacteria and evaluation of their in vitro and in situ activity. Journal of Applied Microbiology 96 (3), 521 534 .

17. A. Costantini, E. Vaudano, W. D. Prete, M. Danei, E. Garcia-Maruno, 2009 Biogenic amine production by contaminating bacteria found in starter preparations used in winemaking. Journal of Agricultura and Food Chemistry 57 10664 10669 .

18. P. D. Cotter, C. Hill, R. P. Ross, 2005 Bacteriocins: developing innate immunity for food. Nature Reviews Microbiology 3 (10), 777 88 .

19. A. G. Cruz, A. E. C. Antunes, A. L. O. P. Sousa, J. A. F. Faria, S. M. I. Saad, 2009 Ice-cream as a probiotic food carrier. Food Research International 42 1233 1239 .

20. S. A. Cuozzo, F. Sesma, J. M. Palacios, Holgado. A. P. de Ruíz, R. R. Raya, 2000 Identification and nucleotide sequence of genes involved in the synthesis of lactocin 705, a two-peptide bacteriocin from Lactobacillus casei CRL 705. FEMS Microbiology Letters 185 (2), 157 61 .

21. E. Bartkiene, G. Juodeikiene, D. Vidmantiene, P. Viskelis, D. Urbonaviciene, 2011 Nutritional and quality aspects of wheat sourdough bread using L. luteus and L. angustifolius flours fermented by Pedioccocus

acidilactici. International Journal of Food Science and Technology 46 1724 1733 .

22. E. Bartkiene, G. Juodeikiene, D. Vidmantiene, 2008 Evaluation of deoxynivalenol in wheat by acoustic method and impact of starter on its concentration during wheat bread baking process. Food Chemistry and Technology 42 (1), 5-12.

23. E. M. Binder, D. Heidler, G. Schatzmayr, N. Thimm, E. Fuchs, M. Schuh, R. Krska, J. Binder, 2000 Microbial detoxification of mycotoxins in animal feed. In Mycotoxins and Phytotoxins in Perspective at the Turn of the Millenium. Proceedings of the 10 -th International IUPAC Symposium on Mycotoxins and Phytotoxins; De Koe, W.J., Samson, R.A., Van Egmond, H.P., Gilbert, J., Sabino, M., Eds; Garuja, Brazil, May 20-25, IUPAC, 271-277.

24. A. Bouchereau, P. Guenot, F. Larher, 2000 Analysis of amines in plant materials. Journal of Chromatography B 747 49 67 .

25. S. Bodmer, C. Imark, M. Kneubühl, 1999 Biogenic amines in foods: histamine and food processing. Inflammation Research 48 296 300 .

26. B. Bogovic-Matijasic, I. Rogelj, 1998 Bacteriocin complex of Lactobacillus acidophilus LF221- production studies in MRS-media at different pH-values and effect against Lactobacillus helveticus ATCC 15009. Process Biochemistry 33 345 352 .

27. B. Bogovič-Matijašić, I. Rogelj, I. F. Nes, H. Holo, 1998 Isolation and characterization of two bacteriocins of Lactobacillus acidophilus LF221. Applied Microbiology and Biotechnology 49 (5), 606-612.

28. J. Böhm, J. Grajewski, H. Asperger, B. Rabus, E. Razzazi, 2000 Study on biodegradation of some trichothecenes (NIV, DON, DAS, T-2) and ochratoxin A by use of probiotic microorganisms. Mycological Research 16 70 74 .

29. A. Ciegler, B. Lillehoj, H. H. Peterson, 1966 Microbial detoxification of aflatoxin. Journal of Applied Microbiology 14 934 939 .

30. D. K. D. Dalié, A. M. Deschamps, F. Richard-Forget, 2009 A review: Lactic acid bacteria- Potential for control of mould growth and mycotoxins. Food Control 21 (4), 370-380.

31. Rivas. B. De Las, A. Marcobal, R. Muñoz, 2005 Improvedmultiplex-PCR method for the simultaneous detection of foodbacteria producing biogenic amines. FEMS Microbiology Letters 244 367 372 .

32. A. Delgado, D. Brito, C. Peres, F. N. Arroyo-L´opez, A. Garrido-Fern´andez, 2005 Bacteriocin production by Lactobacillus pentosus B96

can be expressed as a function of temperature and NaCl concentration. Food Microbiology 22 521 528 .

33. A. Delgado, N. F. Arroyo-L´opez, D. Brito, C. Peres, P. Fevereiro, A. Garrido-Fern´andez, 2007 Optimum bacteriocin production by Lactobacillus plantarum 17.2b requires absence of NaCl and apparently follows a mixed metabolite kinetics. Journal of Biotechnology 130 193 201 .

34. J. Delves-Broughton, 1990 Nisin and its uses as a food preservative. Food Technology 44 100 117 .

35. S. F. Deraz, E. N. Karlsson, A. A. Khalil, B. Mattiasson, 2007 Mode of action of acidocin D20079, a bacteriocin produced by the potential probiotic strain, Lactobacillus acidophilus DSM 20079. Journal of Industrial Microbiology and Biotechnology 34 (5), 373 9 .

36. L. De Vuyst, M. Vancanneyt, 2007 Biodiversity and identification of sourdough lactic acid bacteria. Food Microbiology 24 (2), 120 127 .

37. V. Del Prete, H. Rodriguez, A. V. Carrascosa, B. D. L. Rivas, E. Garcia-Moruno, R. Munoz, 2007 In vitro removal of ochratoxin A by wine lactic acid bacteria. Journal of Food Protection 70 2155 2160 .

38. A. Digaitiene, Å. Hansen, G. Juodeikiene, J. Josephsen, 2005 Microbial population in Lithuanian spontaneous rye sourdoughs. Ecology and Technology 5 (77), 193-198.

39. A. Driks, 2002 Overview: Development in bacteria: spore formation in Bacillus subtilis. Cellular and Molecular Life Sciences 59 (3), 389 91 .

40. H. S. El -Nezami, P. Kankaanpää, S. Salminen, J. Ahokas, 1998a Ability of dairy strains of lactic acid bacteria to bind a common food carcinogen, aflatoxin B1. Food and Chemical Toxicology 36 321 326 .

41. H. S. El -Nezami, P. Kankaanpää, S. Salminen, J. Ahokas, 1998b Physiochemical alterations enhance the ability of dairy strains of lactic acid bacteria to remove aflatoxin from contaminated media. Journal of Food Protection 61 446 448 .

42. H. S. El -Nezami, N. Polychronaki, S. Salminen, H. Mykkänen, 2002 Binding rather metabolism may explain the interaction of two food-grade Lactobacillus strains with zearalenone and its derivative α-zearalenol. Applied and Environmental Microbiology 68 3545 3549 .

43. P. Eichenberger, M. Fujita, S. T. Jensen, E. M. Conlon, D. Z. Rudner, S. T. Wang, C. Ferguson, K. Haga, T. Sato, J. S. Liu, R. Losick, 2004 The program of gene transcription for a single differentiating cell type during sporulation in Bacillus subtilis. Public Library of Science 2 (10), 328.

44. V. G. Eijsink, L. Axelsson, D. B. Diep, L. S. Håvarstein, H. Holo, I. F. Nes, 2002 Production of class II bacteriocins by lactic acid bacteria; an example of biological warfare and communication. Antonie Van Leeuwenhoek 81 (1-4), 639 654 .

45. S. Ennahar, D. Aoude-Werner, O. Sorokine, A. Van Dorsselaer, F. Bringel, J. C. Hubert, C. Hasselmann, 1996 Production of pediocin AcH by Lactobacillus plantarum WHE 92 isolated from cheese. Applied and Environmental Microbiology 62 (12), 4381 7 .

46. J. Errington, 2003 Regulation of endospore formation in Bacillus subtilis. Nature Reviews Microbiology 1 (2), 117 26 .

47. FAO-WHO 2002 Food and agriculture organization of the United Nations, World Health Organization. Report of a joint FAOWHO working group on drafting guidelines for the evaluation of probiotics in food. Ontario. ftp://ftp.fao.org/docrep/fao/009/a0512e/a0512e00.pdf

48. M. Fernandez, D. M. Linares, A. Rodriguez, M. A. Alvarez, 2007 Factors affecting tyramine production in Enteroccus durans IPLA 655. Applied Microbiology and Biotechnology 73 1400 1406 .

49. J. C. Frisvad, U. Thrane, R. A. Samson, J. I. Pitt, 2006 Important mycotoxins and the fungi which produce them. Advances in Food Mycology 571 3 31 .

50. S. Fuchs, G. Sontag, R. Stidl, V. Ehrlich, M. Kundi, S. Knasmuller, 2008 Detoxification of patulin and ochratoxin A, two abundant mycotoxins, by lactic acid bacteria. Food and Chemical Toxicology 46 1398 1407 .

51. A. Gálvez, E. Valdivia, M. Martínez-Bueno, M. Maqueda, 1990 Induction of autolysis in Enterococcus faecalis S-47 by peptide AS-48. Journal of Applied Bacteriology 69 406 413 .

52. A. Galvez, R. L. Lopez, H. Abriouel, 2008 Application of bacteriocins in the control of food-borne pathogenic and spoilage bacteria. Critical Reviews in Biotechnology 28 125 152 .

53. M. G. Gänzle, 1998 Useful Properties of Lactobacilli for application as protective cultures in food. PhD thesis. University of Hohenheim, Germany.

54. A. Garcia-Ruiz, E. M. Gonzalez-Rompinelli, B. Bartolome, M. V. Moreno-Arribas, 2011 Potential of wine-associated lactic acid bacteria to degrade biogenic amines. International Journal of Microbiology 148 115 120 .

55. O. Gillor, A. Etzion, M. A. Riley, 2008 The dual role of bacteriocins as anti- and probiotics. Applied Microbiology and Biotechnology 81 (4),

591 606 .

56. M. Gobbetti, et al. 2005 Biochemistry and physiology of sourdough lactic acid bacteria. Trends in Food Science and Technology 16 (1-3), 57 69 .

57. V. E. Gonzaga, A. G. Lescano, A. A. Huaman, G. Salmon-Mulanovich, D. I. Blazes, 2009 Histamine levels in fish from markets in Lima, Peru. Journal of Food Protection 72 1112 1115 .

58. H. Gourama, 1997a Inhibition of growth and mycotoxin production of Penicillium by Lactobacillus species. Lebensmittel- Wissenschaft und-Technologie 30 279 283 .

59. H. Gourama, L. B. Bullerman, 1997b Anti-aflatoxigenic activity of Lactobacillus casei pseudoplantarum. International Journal of Food Microbiology 34 131 143 .

60. H. Gourama, L. B. Bullerman, 1995 Inhibition of growth and aflatoxin production of Aspergillus flavus by Lactobacillus species. Journal of Food Protection 58 1249 1256 .

61. H. Gourama, 1991 Growth and aflatoxin production of Aspergillus flavus in the presence Lactobacillus species. P h.D. thesis, University of Nebraska-Lincoln.

62. S. Gratz, H. Mykkänen, A. C. Ouwehand, R. Juvonen, S. Salminen, H. S. El -Nezami, 2004 Intestinal mucus alters the ability of probiotic bacteria to bind aflatoxin B1 in vitro. Applied and Environmental Microbiology 70 6306 6308 .

63. A. D. Grossman, R. Losick, 1988 Extracellular control of spore formation in Bacillus subtilis. Proceedings of the National Academy of Sciences 85 (12), 4369 4373 .

64. S. K. Harlander, 1993 Bacteriocins of Lactic Acid Bacteria, Academic Press, San Diego, CA, 233 247 .

65. A. Halász, R. Lásztity, T. Abonyi, A. Bata, 2009 Decontamination of mycotoxin-containing food and feed by biodegradation. Food Reviews International 25 284 298 .

66. W. P. Hammes, M. G. Gänzle, 1998 Sourdough breads and related products. In Wood, B.J.B. Eds. Microbiology of Fermented Foods (1 . London: Chapman & Hall, UK., 199 216 .

67. Å. S. Hansen, 2004 Sourdough bread. In Hui et al. Eds., Handbook of Food and Beverage Fermentation Technology, Marcel Dekker Inc., Florida, USA, 729 755 .

68. L. J. Harris, M. A. Daeschel, M. E. Stiles, T. R. Klaenhammer, 1989 Antimicrobial activity of lactic acid bacteria against Listeria

monocytogenes. Journal of Food Protection 52 (6), 384-387.

69. C. A. Haskard, H. S. El -Nezami, K. D. Peltonen, S. Salminen, J. T. Ahokas, 1998 Sequestration of aflatoxin B1 by probiotic strains: Binding capacity and localization. Revue de Medecine Veterinaire 149, 571.

70. C. A. Haskard, C. Binnion, J. Ahokas, 2000 Factors affecting the sequestration of aflatoxin by Lactobacillus rhamnosus strain GG. Chemico-Biological Interactions 128 39 49 .

71. C. A. Haskard, H. S. El -Nezami, P. E. Kankaanpää, S. Salminen, J. T. Ahokas, 2001 Surface binding of aflatoxin B1 by lactic acid bacteria. Applied and Environmental Microbiology 67 3086 3091 .

72. Y. I. Hassan, L. B. Bullerman, 2008 Antifungal activity of Lactobacillus paracasei ssp. tolerans isolated from a sourdough bread culture. International Journal of Food Microbiology 121 (1), 112 115 .

73. J. He, T. Zhou, J. C. Young, G. J. Boland, P. M. Scott, 2010 Chemical and biological transformations for detoxification of trichothecene mycotoxins in human and animal food chains: a review. Trends in Food Science & Technology 21 (2), 67 76 .

74. N. C. K. Heng, P. A. Wescombe, J. P. Burton, R. W. Jack, J. R. Tagg, 2007 The Diversity of bacteriocins in Gram-positive bacteria. In Riley, M.A., Chavan, M., Eds., Bacteriocins: ecology and evolution. Springer Berlin Heidelberg, New York, USA, 45 93 .

75. T. Hernandez-Jover, M. Izquierdo-Pulido, M. T. Veciana-Nogues, A. Marine-Font, M. C. Vidal-Carou, 1997 Biogenic amines and polyamine contents in meat and meat products. Journal of Agricultural Food Chemistry 45 2098 2102 .

76. W. H. Holzapfel, R. Geisen, U. Schillinger, 1995 Biological preservation of foods with reference to protective cultures, bacteriocins and food-grade enzymes. International Journal of Food Microbiology 24 343 362 .

77. D. G. Hoover, H. Chen, 2005 Bacteriocins with potential for use in foods. In Davidson, P.M., Sofos, J.N., Branen, A.L. Eds., Antimicrobials in Food (3rd edition). Taylor & Francis Group, LLC, FL, USA, 389 428 .

78. D. G. Hoover, K. J. Dishart, M. A. Hermes, 1989 Antagonistic effect of Pediococcus spp. against Listeria monocytogenes. Food Biotechnology 3 (2), 183-196.

79. M. Hugas, F. Pagés, M. Garriga, J. M. Monfort, 1998 Application of the bacteriogenic Lactobacillus sakei CTC494 to prevent growth of Listeria in fresh and cooked meat products packed with different atmospheres. Food Microbiology 15 (6), 639-650.

80. M. C. Hughes, J. P. Kerry, E. K. Arendt, P. M. Kenneally, P. L. H. Mc Sweeney, E. E. O'Neill, 2002 Characterization of proteolysis during the ripening of semi-dry fermented sausages. Meat Science 62 205 216 .

81. R. W. Jack, J. R. Tagg, B. Ray, 1995 Bacteriocins of Gram-positive bacteria. Microbiological Reviews 59 171 200 .

82. R. Jiménez-Díaz, R. M. Rios-Sanchez, M. Desmazeaud, J. L. Ruiz-Barba, J. C. Piard, 1993 Plantaricin S and T, two new bacteriocins produced by Lactobacillus plantarum LPCO10 isolated from a green olive fermentation. Applied and Environmental Microbiology 59 1416 1424 .

83. L. Johnsen, G. Fimland, V. Eijsink, J. Nissen-Meyer, 2000 Engineering increased stability in the antimicrobial peptide pediocin PA-1. Applied and Environmental Microbiology 66 (11), 4798 802 .

84. V. K. Joshi, S. Sharma, N. S. Ranaet, 2006 Bacteriocin from lactic acid fermented vegetables. Food Technology and Biotechnology 44 (3), 435-439.

85. G. Juodeikiene, J. Salomskiene, D. Eidukonyte, D. Vidmantiene, V. Narbutaite, L. Vaiciulyte-Funk, 2011 Impact of novel fermented products on the base of extruded wheat material on the quality of wheat bread, Food Technology and Biotechnology, 2011 (Article in Press).

86. G. Juodeikiene, L. Basinskiene, D. Vidmantiene, T. Makaravicius, E. Bartkiene, 2011 Benefits of β-xylanase for wheat biomass conversion to bioethanol. Journal of the Science of Food and Agriculture, 2011 Jul 11. (Article in Press).

87. V. Ladero, M. Calles-Enríquez, M. Fernández, M. A. Alvarez, 2010 Toxicological effects of dietary biogenic amines. Current Nutrition and Food Science 6 145 156 .

88. A. G. Larsen, B. Nørrung, 1993 Inhibition of Listeria monocytogenes by bavaricin A, a bacteriocin produced by Lactobacillus bavaricus MI401. Letters in Applied Microbiology 17 (3), 132 134 .

89. M. V. Leal-Sánchez, R. Jiménez-Díaz, A. Maldonado-Barragán, A. Garrido-Fernández, J. L. Ruiz-Barba, 2002 Optimization of bacteriocin production by batch fermentation of Lactobacillus plantarum LPCO10. Applied and Environmental Microbiology 68 4465 4471 .

90. M. V. Leal, M. Baras, J. L. Ruiz-Barba, B. Floriano, R. Jiménez-Díaz, 1998 Bacteriocin production and competitiveness of Lactobacillus plantarum LPCO10 in olive juice broth, a culture medium obtained from olives. International Journal of Food Microbiology 43 129 134 .

91. R. J. Leer, J. M. B. M. Van der Vossen, M. Van Giezen, Johannes. M. Van Noort, P. H. Pouwels, 1995 Genetic analysis of acidocin B, a novel bacteriocin produced by Lactobacillus acidophilus. Microbiology 141 (7), 1629 1635 .

92. F. Leroy, J. Verluyten, W. Messens, L. De Vuyst, 2002 Modeling contributes to the understanding of the different behaviour of bacteriocin-producing strains in a meat environment. International Dairy Journal 12 247 253 .

93. R. S. Leuschner, M. Heidel, W. P. Hammes, 1998 Histamine and tyramine degradation by food fermenting microorganisms. International Journal of Food Microbiology 39 1 10 .

94. R. H. Luchese, W. F. Harrigan, 1990 Growth of and aflatoxin production by Aspergillus parasiticus when in the presence of either Lactococcus lactis or lactic acid and at different initial pH values. Journal of Applied Bacteriology 69 512 519 .

95. P. M. Lucas, O. Gaisse, A. Lonvaud-Funel, 2008 High frequency of histamine producing bacteria in the enological environment and instability of the histidine decarboxylase production phenotype. Applied and Environmental Microbiology 74 811 817 .

96. P. Kalač, J. Šavel, M. Križek, T. Pelikánová, M. Prokopová, 2002 Biogenic amine formation in bottled beer. Food Chemistry 79 431 434 .

97. P. Kalač, S. Švecova, T. Pelikánová, 2002 Levels of biogenic amines in typical vegetable products. Food Chemistry 77 349 351 .

98. P. Kankaanpää, E. Tuomola, H. El -Nezami, J. Ahokas, S. J. Salminen, 2000 Binding of aflatoxin B1 alters the adhesion properties of Lactobacillus rhamnosus strain GG in Caco-2 model. Journal of Food Protection 63 412 414 .

99. K. Katina, E. Arendt, K. H. Liukkonen, K. Autio, L. Flander, K. Poutanen, 2005 Potential of sourdough for healthier cereal products. Trends in Food Science and Technology 16 104 112 .

100. K. Katina, M. Sauri, H. L. Alakomi, T. Mattila-Sandholm, 2002 Potential of lactic acid bacteria to inhibit rope spoilage in wheat sourdough bread. Lebensmittel-Wissenschaft und-Technologie 35 (1), 38 45 .

101. A. Karunaratne, E. Wezenberg, L. B. Bullerman, 1990 Inhibition of mold growth and aflatoxin production by Lactobacillus spp. Journal of Food Protection 53 230 236 .

102. W. S. Kim, R. J. Hall, N. W. Dunn, 1997 The effect of nisin concentration and nutrient depletion on nisin production of Lactococcus lactis. Applied

Microbiology and Biotechnology 50 429 433 .

103. R. Kingamkono, E. Sjögren, U. Svanberg, B. Kaijser, 1994 pH and acidity in lactic-fermenting cereal gruels- effects on viability of enteropathogenic microorganisms. World Journal of Microbiology and Biotechnology 10, (6), 664-669.

104. J. Kirschbaum, K. Rebscher, H. Bruckner, 2000 Liquid chromatographic determination of biogenic amines in fermented foods after derivatization with 3 5 -dinitrobenzoyl chloride. Journal of Chromatography A 881, 517-530.

105. T. R. Klaenhammer, 1993 Genetics of bacteriocins produced by lactic acid bacteria. FEMS Microbiology Reviews 12 39 86 .

106. Z. Kozakiewicz, P. Battilani, I. Cabanes, A. Venancio, G. Mule, E. Tjamos, 2003 Making wine safer. In Meeting the Mycotoxin Menace, van Egmond, H., van Osenbruggen, T., Lopez Garcia, R., Visconti, A., Eds., Wageningen Academic Publisher: Wageningen, 131 140 .

107. J. Kramer, R. Gilbert, 1989 Bacillus cereus and other Bacillus sp. In Doyle, M.P., Eds., Foodborne Bacterial Pathogens. Marcel Dekker Inc., New York, USA,, 22 70 .

108. F. Kunst, et al. 1997 The complete genome sequence of the Gram-positive bacterium Bacillus subtilis. Nature 390 (6657), 249 56 .

109. Y. Mao, P. M. Muriana, M. A. Cousin, 2001 Purification and transpositional inactivation of lacticin FS92, a broad-spectrum bacteriocin produced by Lactococcus lactis FS92. Food Microbiology 18 (2), 165 175 .

110. S. Marín, M. E. Guynot, P. Neira, M. Bernadó, V. Sanchis, A. J. Ramos, 2002 Risk assessment of the use of sub-optimal levels of weak-acid preservatives in the control of mould growth on bakery products. International Journal of Food Microbiology 79 (3), 203 11 .

111. J. D. Marugg, C. F. Gonzalez, B. S. Kunka, A. M. Ledeboer, M. J. Pucci, M. Y. Toonen, S. A. Walker, L. C. Zoetmulder, P. A. Vandenbergh, 1992 Cloning, expression, and nucleotide sequence of genes involved in production of pediocin PA-1, and bacteriocin from Pediococcus acidilactici PAC1.0. Applied and Environmental Microbiology 58 (8), 2360 7 .

112. M. A. Martınez-Anaya, 1996 Enzymes and bread flavor. Journal of Agricultural and Food Chemistry 44 2469 2480 .

113. M. C. Martínez-Cuesta, T. Requena, C. Peláez, 2006 Cell membrane damage induced by lacticin 3147 enhances aldehyde formation in Lactococcus lactis IFPL730. International Journal of Food Microbiology

109 (3), 198 204 .

114. R. C. R. Martinez, E. C. P. De Martinis, 2006 Effect of Leuconostoc mesenteroides 11 bacteriocin in the multiplication control of Listeria monocytogenes. Ciźnia e Tecnologia de Alimentos 26 (1), 52-55.

115. R. Mateo, A. Medina, E. M. Mateo, F. Mateo, M. Jimenez, 2007 An overview of ochratoxin A in beer and wine. International Journal of Food Microbiology 119 (1-2), 79 83 .

116. M. V. Moreno-Aribas, M. C. Polo, 2010 Wine Chemistry and Biochemistry. In Mycotoxins in food Detection and control: Biological decontamination of mycotoxins, Magan, N. and Olsen, M., Eds., Springer New York., Woodhead Publishing Ltd and CRC Press LLC, 2006 2211 .

117. L. L. Mc Kay, K. A. Baldwin, 1990 Application for biotechnology: present and future improvements in lactic acid bacteria. FEMS Microbiology reviews 7 3 14 .

118. O. Mc Auliffe, C. Hill, R. P. Ross, 1999 Inhibition of Listeria monocytogenes in cottage cheese manufactured with a lacticin 3147 -producing starter culture. Journal of Applied Microbiology 86 (2), 251-256.

119. F. Mendoza, M. Maqueda, A. Gálvez, M. Martínez-Bueno, E. Valdivia, 1999 Antilisterial activity of peptide AS-48 and study of changes induced in the cell envelope properties of an AS-48-adapted strain of Listeria monocytogenes. Applied and Environmental Microbiology 65 618 625 .

120. Ö. Mentes, R. Ercan, M. Akçelik, 2007 Inhibitor activities of two Lactobacillus strains, isolated from sourdough, against rope-forming Bacillus strains. Food Control 18 (4), 359 363 .

121. W. Messens, L. De Vuyst, 2002 Inhibitory substances produced by lactobacilli isolated from sourdoughs- a review. International Journal of Food Microbiology 72 31 43 .

122. W. Messens, P. Neysens, W. Vansieleghem, J. Vanderhoeven, L. De Vuyst, 2002 Modeling growth and bacteriocin production by Lactobacillus amylovorus DCE 471 in response to temperature and pH values used for sourdough fermentations. Applied and Environmental Microbiology 68 (3), 1431 5 .

123. K. W. Miller, R. Schamber, O. Osmanagaoglu, B. Ray, 1998 Isolation and characterization of pediocin AcH chimeric protein mutants with altered bactericidal activity. Applied and Environmental Microbiology 64 (6), 1997 2005 .

124. C. C. Minei, B. C. Gomes, R. P. Ratti, C. E. M. D'Angelis, E. C. P. De

Martinis, 2008 Influence of peroxyacetic acid and nisin and coculture with Enterococcus faecium on Listeria monocytogenes biofilm formation. Journal of Food Protection 71 634 638 .

125. H. Morency, M. Mota-Meira, G. La Pointe, C. Lacroix, M. C. Lavoie, 2001 Comparison of the activity spectra against pathogens of bacterial strains producing a mutacin or a lantibiotic. Canadian Journal of Microbiology 47 (4), 322 31 .

126. I. Moreno, A. L. S. Lerayer, V. L. S. Baldini, M. F. F. Leitao, 2000 Characterization of bacteriocins produced by Lactococcus lactis strains. Brazilian Journal of Microbiology 31 184 192 .

127. S. Moret, D. Smela, T. Populin, L. S. Conte, 2005 A survey on free biogenic amine content of fresh and preserved vegetables. Food Chemistry 89 355 361 .

128. T. Moretro, I. M. Aassen, I. Storro, L. Axelsson, 2000 Production of sakacin P by Lactobacillus sakei in a completely defined medium. Journal of Applied Microbiology 88 536 545 .

129. C. I. Mørtvedt, I. F. Nes, 1990 Plasmid-associated bacteriocin production by a Lactobacillus sake strain. Journal of General Microbiology 136 (8), 1601 1607 .

130. M. O. Moss, M. T. Long, 2002 Fate of patulin in the presence of the yeast Saccharomyces cerevisiae. Food Additives & Contaminants 19 387 399 .

131. M. Mota-Meira, H. Morency, M. C. Lavoie, 2005 In vivo activity of mutacin B-Ny266. Journal of Antimicrobial Chemotherapy 56 (5), 869 871 .

132. H. F. Mower, N. V. Bhagavan, 1989 Tyramine content of Asian and Pacific foods determined by high performance liquid chromatography. Food Chemistry 31 251 257 .

133. V. Narbutaite, A. Fernandez, N. Horn, G. Juodeikiene, A. Narbad, 2008 Influence of baking enzymes on antimicrobial activity of five bacteriocin-like inhibitory substances produced by lactic acid bacteria isolated from Lithuanian sourdoughs. Letters in Applied Microbiology 47 (6), 555 560 .

134. I. F. Nes, H. Holo, 2000 Class II antimicrobial peptides from lactic acid bacteria. Biopolymers 55 (1), 50 61 .

135. I. F. Nes, D. B. Diep, L. S. Håvarstein, M. B. Brurberg, V. Eijsink, H. Holo, 1996 Biosynthesis of bacteriocins in lactic acid bacteria. Antonie Leeuwenhoek 70 113 128 .

136. P. Neysen, L. De Vuyst, 2005 Kinetic and modeling of sourdough lactic

bacteria. Trends in Food Science & Technology 16 95 103 .

137. V. Niderkorn, H. Boudra, D. P. Morgavi, 2006 Binding of Fusarium mycotoxins by fermentative bacteria in vitro. Applied and Environmental Microbiology 101 849 856 .

138. V. Niderkorn, 2007 Activites de biotransformation et de séquestration des fusariotoxines chez les bactéries fermentaires pour la détoxification des ensilages de maïs. PhD thesis, Blaise Pascal University, France.

139. C. Oppegård, G. Fimland, L. Thorbæk, J. Nissen-Meyer, 2007 Analysis of the two-peptide bacteriocins Lactococcin G and Enterocin 1071 by site-directed mutagenesis. Applied and Environmental Microbiology 73 (9), 2931 8 .

140. E. M. Östman, M. Nilsson, H. Elmstahl, G. Molin, I. Bjorck, 2002 On the effect of lactic acid on blood glucose and insulin responses to cereal products: mechanistic studies in healthy subjects and in vitro. Journal of Cereal Science 36 (3), 339 346 .

141. J. L. Parada, C. R. Caron, A. Bianchi, P. Medeiros, C. R. Soccol, 2007 Bacteriocins from lactic acid bacteria: purification, properties and use as biopreservatives. Brazilian Archives of Biology and Technology 50 521 542 .

142. S. Patharajan, K. R. N. Reddy, V. Karthikeyan, D. Spadaro, A. Lore, M. L. Gullino, A. Garibaldi, 2011 Potential of yeast antagonists on invitro biodegradation of ochratoxin A. Food Control 22 290 296 .

143. T. L. Pattison, D. Lindsay, Holy. A. von, 2004 Natural antimicrobials as potential replacements for calcium propionate in bread. South African Journal of Science 100 (7-8), 342-348.

144. O. Pepe, G. Blaiotta, G. Moschetti, T. Greco, F. Villani, 2003 Rope-producing strains of Bacillus spp. from wheat bread and strategy for their control by lactic acid bacteria. Applied and Environmental Microbiology 69 (4), 2321 9 .

145. P. J. Piggot, D. W. Hilbert, 2004 Sporulation of Bacillus subtilis. Current Opinion in Microbiology 7 (6), 579 86 .

146. M. Piotrowska, Z. Zakowska, 2000 The biodegradation of ochratoxin A in food products by lactic acid bacteria and baker's yeast. In Progress in Biotechnology (Food Biotechnology); Bielecki, S., Tramper, J., Polak, J., Eds., Elsevier, Amsterdam, The Netherlands, 17 307 310 .

147. M. Piotrowska, Z. Zakowska, 2005 The elimination of ochratoxin A by lactic acid bacteria strains. Polish Journal of Microbiology 54 279 286 .

148. Z. E. Phillips, M. A. Strauch, 2002 Bacillus subtilis sporulation and

stationary phase gene expression. Cellular and Molecular Life Sciences 59 (3), 392 402 .

149. J. E. Powell, S. D. Todorov, C. A. van Reenen, L. M. T. Dicks, R. C. Witthuhn, 2006 Growth inhibition of Enterococcus mundtii in Kefir by in situ production of bacteriocin ST8KF. Le Lait 86 401 405 .

150. M. Rasch, S. Knøchel, 1998 Variations in tolerance of Listeria monocytogenes to nisin, pediocin PA-1 and bavaricin A. Letters in Applied Microbiology 27 (5), 275 8 .

151. J. Ravel, C. M. Fraser, 2005 Genomics at the genus scale. Trends in Microbiology 13 (3), 95 7 .

152. A. Richelli, F. Baruzzi, M. Solfrizzo, M. Morea, F. P. Fanizzi, 2007 Biotransformation of patulin by Gluconobacter oxydans. Applied and Environmental Microbiology 73 785 792 .

153. S. Riebroy, S. Benjakul, W. Visessanguan, K. Kijrongrojana, M. Tanaka, 2004 Some characteristics of commercial Som-fug produced in Thailand. Food Chemistry 88 527 535 .

154. W. Röcken, 1996 Applied aspects of sourdough fermentation. Advances in Food Sciences 18 (5-6), 212-216.

155. L. A. Rogers, 1928 The inhibitory effect of Sreptococcus lactis on Lactobacillus bulgaricus. Journal of Bacteriology 16 321 325 .

156. H. Rosenkvist, Å. Hansen, 1995 Contamination profiles and characterization of Bacillus species in wheat bread and raw materials for bread production. International Journal of Food Microbiology 26 (3), 353-363

157. E. Røssland, Borge. G. I. Andersen, T. Langsrud, T. Sørhaug, 2003 Inhibition of Bacillus cereus by strains of Lactobacillus and Lactococcus in milk. International Journal of Food Microbiology 89 (2-3), 205 12 .

158. R. P. Ross, S. Morgan, C. Hill, 2002 Preservation and fermentation: past, present and future. International Journal of Food Microbiology 79 3 16 .

159. R. P. Ross, C. Stanton, C. Hill, G. F. Fitzgerald, A. Coffey, 2000 Novel cultures for cheese improvement. Trends in Food Science and Technology 11 (3), 96 104 .

160. U. Roy, V. K. Batish, S. Grover, S. Neelakantan, 1996 Production of antifungal substance by Lactococcus lactis subsp. lactis CHD-28.3. International Journal of Food Microbiology 32 27 34 .

161. L. A. M. Ryan, Bello. F. Dal, E. K. Arendt, 2008 The use of sourdough fermented by antifungal LAB to reduce the amount of calcium propionate in bread. International Journal of Food Microbiology 125 (3), 274 278 .

162. M. Saaid, B. Saad, N. H. Hashim, M. A. S. Ali, M. I. Saleh, 2009 Determination of biogenic amines in selected Malaysian food. Food Chemistry 113 1356 1362 .

163. H. G. Sahl, G. Bierbaum, 1998 Lantibiotics: biosynthesis and biological activities of uniquely modified peptides from Gram-positive bacteria. Revista de Microbiologia 52 41 79 .

164. A. Sadeghi, 2008 The secrets of sourdough: A review of miraculous potentials of sourdough in bread shelf life. Biotechnology 7 (3), 413 417 .

165. L. Settani, O. Massitti, D. Van Sinderen, A. Corsetti, 2005 In situ activity of a bacteriocin- producing Lactococcus lactis strain. Influence on the interactions between lactic acid bacteria during sourdough fermentation. Journal of Applied Microbiology 99 670 681 .

166. U. Schillinger, F. K. Lücke, 1989 Antibacterial activity of Lactobacillus sake isolated from meat. Applied and Environmental Microbiology 55 (8), 1901 6 .

167. K. H. Schleifer, J. Kraus, C. Dvorak, R. Kilpper-Bälz, M. D. Collins, W. Fischer, 1985 Transfer of Streptoccus lactis and related streptoccus to the genus of Lactococcus gen nov. Systematic and Applied Microbiology 6 183 195 .

168. M. Sievers, I. C. Garerth, C. Becsh, W. Ludwig, M. Teuber, 1995 Phylogenetic position of Gluconobacterspecies as a coherent cluster from all Acetobacter species on the basis of 16S ribosomal RNA sequences. FEMS Microbiology Letters 126 123 126 .

169. Santos. M. H. Silla, 1996 Biogenic amines: their importance in foods. International Journal of Food Microbiology 29 213 231 .

170. T. F. V. Silveira, C. M. M. Vianna, G. B. G. Mosegui, 2009 Brazilian legislation for functional foods and the interface with the legislation for other food and medicine classes: contradictions and omissions. Physis Revista de Saúde Coletiva 19 1189 1202 .

171. E. D. Simova, D. B. Beshkova, Z. P. Dimitrov, 2009 Characterization and antimicrobial spectrum of bacteriocins produced by lactic acid bacteria isolated from traditional Bulgarian dairy products. Journal of Applied Microbiology 106 (2), 692 701 .

172. Ö. Şimşek, A. H. Çon, Ş. Tulumoğlu, 2006 Isolating lactic starter cultures with antimicrobial activity for sourdough processes. Food Control 17 (4), 263-270.

173. N. Shah, X. Wu, 1999 Aflatoxin B1 binding abilities of probiotic bacteria. Bioscience and Microflora 18 43 48 .

174. A. R. Shalaby, 1996 Significance of biogenic amines to food safety and human health. Food Research International 29 675 690 .

175. P. H. Shetty, L. Jespersen, 2006 Saccharomyces cerevisiae and lactic acid bacteria as potential mycotoxin decontaminating agents. Trends in Food Science & Technology 17 48 55 .

176. I. B. Sorokulova, O. N. Reva, V. V. Smirnov, I. V. Pinchuk, S. V. Lapa, M. C. Urdaci, 2003 Genetic diversity and involvement in bread spoilage of Bacillus strains isolated from flour and ropy bread. Letters in Applied Microbiology 37 (2), 169 173 .

177. E. E. Stinson, S. F. Osman, D. D. Bills, 2006 Water soluble products from patulin during alcoholic fermentation of apple juice. Journal of Food Science 44 (3), 788 789 .

178. J. E. Stratton, R. W. Hutkins, S. I. Taylor, 1991 Biogenic amines in cheese and other fermented foods: a review. Journal of Food Protection 54 460 470 .

179. C. E. Stauffer, 1994 Enzymes used in bakery products. Fundamentals of enzymes. AIB Tech Bull XVI, 1 6 .

180. M. E. Stiles, 1996 Biopreservation by lactic acid bacteria. Antonie van Leuwenhoek 70 331 345 .

181. U. Svanberg, E. Sjögren, W. Lorri, A. M. Svennerholm, B. Kaijser, 1992 Inhibited growth of common enteropathogenic bacteria in lactic-fermented cereal gruels. World Journal of Microbiology and Biotechnology 8 (6), 601 -606.

182. T. H. Suomalainen, A. M. Mäyrä-Makinen, 1999 Propionic acid bacteria as protective cultures in fermented milks and breads. Lait 79 (1), 165 174 .

183. S. Tanasupawat, C. Thawai, P. Yukphan, D. Moonmangmee, T. Itoh, O. Adachi, Y. Yamada, 2004 Gluconobacter thailandicus sp. nov., an acetic acid bacterium in the alpfa-Proteobacteria. The Journal of General and Applied Microbiology 50 159 167 .

184. L. V. Thomas, M. R. Clarkson, J. Delves-Broughton, 2000 In Natural food antimicrobial systems, Thomas, L.V., Clarkson, M.R. and Delves-Broughton, J., Eds., CRC Press, A.S. Naidu, USA,, 463 524 .

185. L. V. Thomas, J. Delves-Broughton, 2001 New advances in the application of food preservative nisin. Re cent Advances in Food Science 2 11 22 .

186. J. M. Thompson, C. E. R. Dodd, W. M. Waites, 1993 Spoilage of bread by Bacillus. International Biodeterioration & Biodegradation 32 (1-3), 55 66 .

187. S. Todorov, B. Gotcheva, X. Dousset, B. Onno, I. Ivanova, 2000 Influence of growth medium on bacteriocin production in Lactobacillus plantarum ST31. Biotechnology & Biotechnological Equipment 14 50 55 .

188. S. D. Todorov, L. M. T. Dicks, 2004 Effect of medium components on bacteriocin production by Lactobacillus pentosus ST151BR, a strain isolated from beer produced by the fermentation of maize, barley and soy flour. World Journal of Microbiology and Biotechnology 20 643 650 .

189. S. D. Todorov, C. A. Van Reenen, L. M. T. Dicks, 2004 Optimization of bacteriocin production by Lactobacillus plantarum ST13BR, a strain isolated from barley beer. Journal of General and Applied Microbiology 50 149 157 .

190. S. D. Todorov, L. M. T. Dicks, 2004 Influence of growth conditions on the production of a bacteriocin by Lactococcus lactis subsp. lactis ST34BR, a strain isolated from barley beer. Journal of Basic Microbiology 44 305 316 .

191. S. D. Todorov, L. M. T. Dicks, 2005a Effect of growth medium on bacteriocin production by Lactobacillus plantarum ST194BZ, a strain isolated from boza. Food Technology and Biotechnology 43 165 173 .

192. S. D. Todorov, L. M. T. Dicks, 2005b Lactobacillus plantarum isolated from molasses produces bacteriocins active against Gram-negative bacteria. Enzyme and Microbial Technology 36 318 326 .

193. S. D. Todorov, L. M. T. Dicks, 2006a Effect of medium components onbacteriocin production by Lactobacillus plantarum strains ST23LD and ST341LD, isolated from spoiled olive brine. Research in Microbiology 161 102 108 .

194. S. D. Todorov, L. M. T. Dicks, 2006b Medium components effecting bacteriocin production by two strains of Lactobacillus plantarum ST414BZ and ST664BZ isolated from boza. Biologia 61 269 274 .

195. S. D. Todorov, H. Nyati, M. Meincken, L. M. T. Dicks, 2007a Partialcharacterization of bacteriocin AMA-K, produced by Lactobacillus plantarum AMA-K isolated from naturally fermented milk from Zimbabwe. Food Control 18 656 664 .

196. S. D. Todorov, J. E. Powell, M. Meincken, R. C. Witthuhn, L. M. T. Dicks, 2007b Factors affecting the adsorption of Lactobacillus plantarum bacteriocin bacST8KF to Enterococcus faecalis and Listeria innocua. International Journal of Dairy Technology 60 221 227 .

197. S. D. Todorov, 2008 Bacteriocin production by Lactobacillus plantarum AMA-K isolated from Amasi, a Zimbabwean fermented milk product and study of adsorption of bacteriocin AMA-K to Listeria spp. Brazilian

Journal of Microbiology 38 178 187 .

198. A. Turbic, J. T. Ahokas, C. A. Haskard, 2002 Selective in vitro binding of dietary mutagens, individually or in combination, by lactic acid bacteria. Food Additives & Contaminants 19 144 152 .

199. N. Thyagaraja, A. Hosono, 1994 Binding properties of lactic acid bacteria from 'Idly' towards food-borne mutagens. Food and Chemical Toxicology 32 805 809 .

200. T. Zotta, E. Parente, A. Ricciardi, 2009 Viability staining and detection of metabolic activity of sourdough lactic acid bacteria under stress conditions. World Journal of Microbiology and Biotechnology 25 (6), 1119 1124 .

201. F. Valerio, M. Favilla, P. De Bellis, A. Sisto, S. de Candia, P. Lavermicocca, 2009 Antifungal activity of strains of lactic acid bacteria isolated from a semolina ecosystem against Penicillium roqueforti, Aspergillus niger and Endomyces fibuliger contaminating bakery products. Systematic and Applied Microbiology 32 (6), 438 448 .

202. F. Valerio, P. De Bellis, S. L. Lonigro, A. Visconti, P. Lavermicocca, 2008 Use of Lactobacillus plantarum fermentation products in bread-making to prevent Bacillus subtilis ropy spoilage. International Journal of Food Microbiology 122 (3), 328 332 .

203. J. Varga, Z. Kozakiewicz, 2006 Ochratoxin-A in grapes and grape-derived products. Trends in Food Science & Technology 1 72 81 .

204. T. L. J. Verellen, G. Bruggeman, C. A. Van Reenen, L. M. T. Dicks, E. J. Vandamme, 1998 Fermentation optimization of plantaricin 423, a bacteriocin produced by Lactobacillus plantarum 423. Journal of Fermentation and Bioengineering 86 174 179 .

205. J. Verluyten, F. Leroy, L. De Vuyst, 2004a Influence of complex nitrogen source on growth of and curvacin A production by sausage isolate Lactobacillus curvatus LTH1174. Applied and Environmental Microbiology 70 5081 5088 .

206. C. R. Viljoen, Holy. A. von, 1997 Microbial populations associated with commercial bread production. Journal of Basic Microbiology 37 (6), 439 44 4.

207. P. J. A. Volavsek, L. A. M. Kirshner, A. von Holy., 1992 Accelerated methods to predict the rope-inducing potential of bread raw materials. South African Journal of Science 87 99 102 .

208. I. Wiedemann, E. Breukink, C. van Kraaij, O. P. Kuipers, G. Bierbaum, B. de Kruijff, H. G. Sahl, 2001 Specific binding of nisin to the peptidoglycan

precursor lipid II combines pore formation and inhibition of cell wall biosynthesis for potent antibiotic activity. The Journal of Biological Chemistry 276 (3), 1772 9 .

209. D. W. Wiseman, E. H. Marth, 1981 Growth and aflatoxin production by Aspergillus parasiticus when in the presence of Streptococcus lactis, Mycopathologia 73 49 56 .

210. R. W. Worobo, M. J. Van Belkum, M. Sailer, K. L. Roy, J. C. Vederas, M. E. Stiles, 1995 A signal peptide secretion-dependent bacteriocin from Carnobacterium divergens. Journal of Bacteriology 177 (11), 3143 9 .

211. S. Quintavalla, G. Parolari, 1993 Effects of temperature, aw and pH on the growth of Bacillus cells and spores: a response surface methodology study. International Journal of Food Microbiology 19 (3), 207 16 .

Chapter 7

FERMENTATION OF VEGETABLE JUICES BY LACTOBACILLUS ACIDOPHILUSLA-5

Lavinia Claudia Buruleanu[1], Magda Gabriela Bratu[1], Iuliana Manea[1], Daniela Avram[1] and Carmen Leane Nicolescu[1]
[1]Department of Food Engineering, Faculty of Environmental Engineering and Biotechnology, Valahia University of Targoviste, Romania

INTRODUCTION

Probiotics foods represent one of the largest sectors in functional food markets. Most of the available probiotic products are some form of dairy, despite the continuous growth of the non-dairy probiotic sector, with products like soy-based drinks, fruit-based foods, and other cereal-based products. Both non-dairy (in general) and soy-based probiotic products represent a huge growth potential for the food industry, and may be widely explored through the development of new ingredients, processes, and products. For this purpose, new studies must be carried out to: test ingredients, explore more options of media that have not yet been industrially utilized, reengineer products and processes, towards potentially meet the demands of lactose-intolerant and vegetarian consumers for new nourishing and palatable probiotic products [1].

Lactic acid bacteria are among the most important probiotic microorganisms typically associated with the human gastrointestinal tract. Traditionally, lactic acid bacteria have been classified on the basis of phenotypic properties, e.g. morphology, mode of glucose fermentation, growth at different temperatures, lactic acid configuration, and fermentation of various carbohydrates. However some species, like the so-called *Lactobacillus acidophilus* group and some bifidobacteria, are not readily distinguishable by phenotypic characteristics [2]. From the physiological point of view, *Lactobacillus acidophilus* strains were characterized as lactic acid bacteria with strictly homofermentative metabolism (> 85% lactic acid). The hexoses are preferential fermented via Embden Meyerhof Parnas (EMP), (as the strains produce aldolase and phosphoketolase), and only then the pentoses and gluconate are fermented. LAB of the *Lactobacillus acidophilus* group *as* well as of the *Bifidobacterium* group isolated from the human faeces or intestine are thought to have beneficial effects on health being thus considered to be probiotic bacteria [3].

For use in food, important criteria for probiotics must be met, in particular that they should not only be capable of surviving passage through the digestive tract, by exhibiting acid and bile tolerance, but also have the capability to proliferate in the gut.

Probiotics must be able to exert their benefits on the host through growth and/or activity in the human body. Although generally recognised as safe a probiotic strains must be characterized by a set of tests that assure its safety to consumer (1, 2, 3, 5, 6).

Inclusion of probiotic bacteria in fermented dairy products enhances their value as better therapeutic functional foods. However, insufficient viability and survival of these bacteria remain a problem in commercial food products. By selecting better functional probiotic strains and adopting improved methods to enhance survival, including the use of appropriate prebiotics and the optimal combination of probiotics and prebiotics (synbiotics), an increased delivery of viable bacteria in fermented products to the consumers can be achieved [5].

The fermentation of vegetable products, applied as a preservation method for the production of finished and half-finished food products, is considered as an important technology, though requiring more research, as a growing number of raw materials are being processed in this way by the food industry. The main reasons for this interest are nutritional, physiological and hygienic aspects of the process [6]. Thus, according to Kelwicka, (2010) [7], the fermentation of beetroot juice requires selected starter cultures made of LAB, naturally present in this vegetable although their number is usually very small. This makes them un-appropriate to, alone, conducting a fermentation that ensures satisfying sensory properties of the fermented juice, with improved health promoting activity.

Thus, probiotic juices represent an alternative to dairy products that suits consumers who don't want to eat dairy foods or are lactose intolerant. Adding probiotics to juices is more complex than formulating in dairy products where the bacteria can be easily added to other cultures.

Despite its potential for healthy products development, there is very little research activity addressing the fermentation of vegetable juices using probiotic bacteria.

MATERIALS AND METHODS

Vegetables Treatments

Fresh vegetables (carrots, cucumbers, beetroot, white cabbage, red cabbage) were purchased from a retail market and specifically processed by removing the non-edible pieces. The raw material processing was made faster, because the possibility of contamination and proliferation of microorganisms in the products is very high in comparison with their intact counterparts (Lee, 2011). Using a domestic extractor the vegetables were turned into juice. The heating treatment of the juice, applied at 80°C with a view to destroy the undesirable microorganisms under the limit of detection, was followed by cooling at 40°C.

Microorganisms and Fermentation Conditions

The strain *Lactobacillus acidophilus* LA-5 from Christian Hansen (Romania) was used in this study.

The lyophilized culture was aseptically inoculated into the vegetable juices and vigorously homogenized for 15 min, according to the producer's specification. The fermentation experiments were carried out using Erlenmeyer flasks containing 50ml of juice, without pH adjustment. The flasks were incubated statically in an incubator chamber at 37±0.2°C. Sampling was taken at regular interval of times for physico-chemical and microbiological analysis.

The tested supplements were: L-cysteine hydrochloride monohydrate (Merck, Darmstadt, Germany), L-lysine hydrochloride (Merck), L-valine (Merck), L-leucine (Calbiochem, San Diego, CA, USA) and yeast extract (Merck). Cysteine, lysine, valine and leucine were separately added in quantity by 0.1% (w/v) into carrot juice, while amounts by 0.2% (w/v) were tested, also individual, in the case of the yeast extract and cysteine. A control sample without supplements was carried out for each experiment.

Physico – Chemical Analysis

Metabolic activity of the strain LA-5 in the conditions mentioned above was evaluated based on the dynamics of pH, respectively end products of fermentation. The pH values were measured with a HACH pH-meter. Lactic acid was determined using commercial kits (K-DLATE from Megazyme International). The calculations were made with Megazyme Mega-Calc™ and expressed as g lactic acid/l. Reducing sugars were analyzed applying the spectrophotometric method with 3.5-dinitrosalicilic acid (DNS) after the removing of other substances with reducing character using basic lead acetate and expressed as g glucose/l. Ascorbic acid was determined applying the 2,6-dichloroindophenol titrimetic method, based on the reduction of the sodium salt of the dye by ascorbic acid (AOAC method). It was expressed as mg/100ml. The amino acids content, expressed as g glycine/100ml, was determined through the Sörensen method.

Microbiological Analysis

The amount of viable cells of *Lactobacillus* sp. was determined by serial tenfold dilution with sterile peptone water. Aliquots of 1ml were plated, in duplicate, in plates with Man-Rogosa-Sharpe agar, enriched with L-cysteine HCl. The Petri plates were incubated for 48-72h at 37°C and the results were expressed as log colony forming units (CFU)/ml juice.

The optical density of biomass was measured with the UV-Visible spectrophotometer at 610nm. In the preparation of the calibration curve for optical density vs. dry cell weight several dilutions of the juices were made. According Altiok [8], for each dilution 2 ml of sample was used to obtain optical densities at 610 nm wavelength and 15 ml of sample was filtered with a pre-weighed cellulose acetate membrane filter having a pore size of 0.45 μm using a vacuum pump. The biomass collected on the filters was washed with 15 ml of water and the filters were dried at 100°C for approximately 24 h until constant weight was observed. The results were expressed as g.

Statistical Analysis

Statistical analysis was carried out using the software SPSS (Statistical Package for the Social Science 17.0 trial version).

RESULTS AND DISCUSSIONS

Effect of Inoculum Size on the Lactic Acid Accumulation and Biomass Growth

A comparative study of the dynamics of lactic acid fermentation of carrot juice using three different concentrations of lyophilized pure culture was realized (Figure 1).

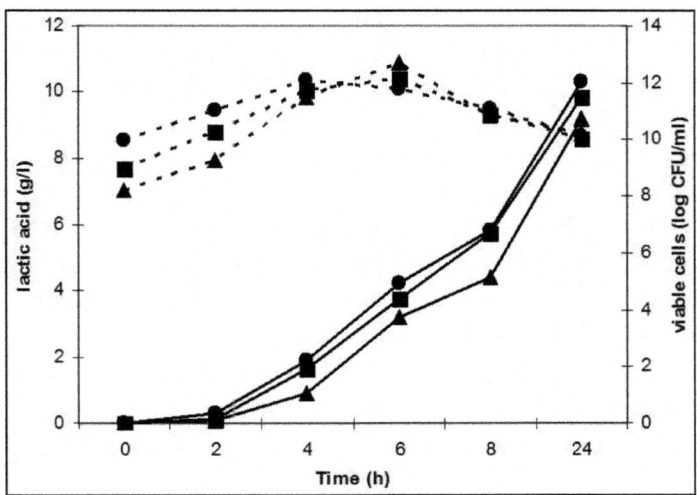

Figure 1: Correlation between lactic acid production by *Lactobacillus acidophilus* LA-5 and number of viable cells during fermentation of carrot juice with different inoculum size ▲ 0.2g/l; ■ 0.3g/l; ● 0.4 g/l (smooth lines - lactate, dashed lines - viable cells count)

Relative higher differences concerning the lactate increasing were observed between the variant with 0.2g/l pure culture initial added and the other two within 24 hours of fermentation. Thus, at the end of this interval, the excess was by 7.06% in the juice with 0.3g/l inoculum and 12.06% in the juice with 0.4g/l inoculum respectively. However, in all the batches the lactic acid accumulation, higher than 9g/l, could be considered satisfactory for the shelf life of the final products. From the other part, the number of viable cells is decisive for the probiotic feature of these ones. A direct proportionality between the amount of the lyophilised culture initial added and the viable cells was observed only in the first 4h of the fermentation. As a general characteristic, in the interval 6 - 24h pH values less than 4.5 have become inhibitory for the useful microbiota in all the experimental samples.

The initial concentration of reducing sugars of the carrot juices, by 25.2g/l, was favourable for the growth of *Lactobacillus acidophilus* LA-5. Testing two strains of Lactobacillus (one genetically selected Mont4+ and the other genetically altered, Mont4+pxyAB-mod). Kiouss [9] established that the Mont4+ had the highest yield of lactic acid fermenting with six percent concentration of glucose, whereas the L strain utilized the sugar best at the four percent concentration. In the same time temperature and pH seemed to play the largest role in the organism's ability to grow and thus affecting its production of lactic acid.

Concluding, higher inoculum densities of *Lactobacillus acidophilus* LA-5 were not significantly influenced the survival yield of the useful microbiota in the lactic acid fermented juices after 24h. In the same time, no parallel relationships between lactic acid concentration and the inoculum size were determined. The result agrees to those obtained by Agarwal, Dutt, Meghwanshi and Saxena [10] using*Enterococcus flavescens* for production of lactic acid. In their opinion, beyond a certain concentration lactic acid yield dropped due to high cell density resulting in fast depletion of essential nutrients, limiting further growth and reducing the yield. Referring to bifidobacteria, Dave and Shah [11] reported also that a higher inoculum did not always improve their viability to a satisfactory level. No data referring to *Lactobacillus acidophilus* were found in the literature.

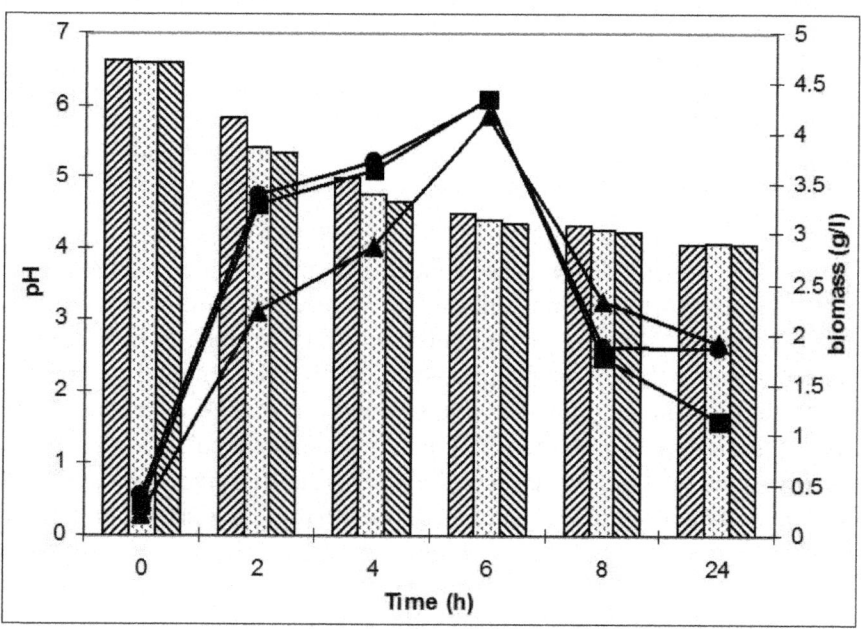

Figure 2: pH and biomass evolution during lactic acid fermentation of carrot juice with different inoculum of*Lactobacillus acidophilus* LA-5: 0.2g/l (▨▨ and ▲); 0.3g/l (▨ and ■); 0.4 g/l (▨ and ●); columns - pH values, lines - biomass

Although the pH dynamics was quite different in the first 6h of the process, the initial amount of the pure culture did not affect the subsequent evolution or the final value of this parameter (Figure 2).

The sharp decrease in biomass from 6 to 8h has been correlated with the viable cells tendency, as result of reaching pH values by 4.34 to 4.47. Being known that *Lb. acidophilus* is more sensitive in acidic environment, this result underlines the necessity to manage the size of inoculum in order to obtain a balance between the lactic acid accumulation and the survival of the probiotic microorganisms.

The maximum rate of acidification v_{max} was calculated as the time variation of pH (dpH/dt) and expressed as pH units/min (Table 1). Other kinetic parameters were also calculated: time to reach v_{max} (t_{max}, hours), time to reach pH 5.0 ($t_{pH\,5.0}$, hours), time to complete the fermentation ($t_{pH\,4.2}$, hours).

Table 1: Acidification kinetic parameters of fermentation of carrot juices by *Lactobacillus acidophilus* LA-5

Inoculum, g/l	$v_{max} \cdot 10^{-3}$(units/min.)	t_{max}(h)	$t_{pH\,5.0}$(h)	$t_{pH\,4.2}$(h)
0.2	7.08	4	2.95	8.4
0.3	9.83	2	2.88	8.2
0.4	10.41	2	2.67	8.05

A double amount of inoculum had an insignificant influence on the time to reach pH 5.0, important parameter from the shelf life of the fermented juices. Thus, $t_{pH\,5.0}$ (h) was 1.1-fold higher in the case of the batch with 0.2g/l lyophilized pure culture initial added to juice than that one with 0.4g/l. A different situation was registered concerning the maximum rate of acidification (v_{max}) and the time to reach this rate (t_{max}). Thus, a polynomial equation of the form $y = -108.5x^2 + 81.75x - 4.93$ correlated the size of inoculum with the corresponding values of v_{max} at R squared $= 1$. Although at the initial moment of fermentation seems to be advantageous to use a higher amount of pure culture, this aspect lessen in time, from the economic point of view being important to obtain a balance between the quantity of inoculum and the targeted parameters which ensure the preservation of the final product.

The values of the biomass content became close after about 6h of fermentation. No parallel relationship between lactic acid concentration and biomass was observed, result that agrees to those obtained by Amrane [12] and Kotzamanidis [13].

However, taking into account the lactic acid accumulation and the dynamics of the number of viable cells, it was obvious that the utilization of higher amount of inoculum is not justified.

Effect of Temperature on the Dynamics of Fermentation

According to the information provided by the producer of the lactic culture, respectively to the data found in literature, two different incubation temperatures were tested: 37^{0}C and 41^{0}C respectively.

The dynamics of both pH and lactic acid (Figure 3) emphasizes the influence of the higher temperature on the rate of acidification. After 24h no significant differences between the pH values were determined, while the lactic acid content of the samples fermented at 41°C was 1.24-fold higher comparatively with those fermented at 37°C. This situation may be due to the higher amino acids content in the samples fermented at 41°C, that act as buffer. Thus, expressed as glycin, the total amount was by 0.165g/100ml at the end of the analyzed interval, which represented an increase by 10% comparatively with the batch fermented at lower temperature.

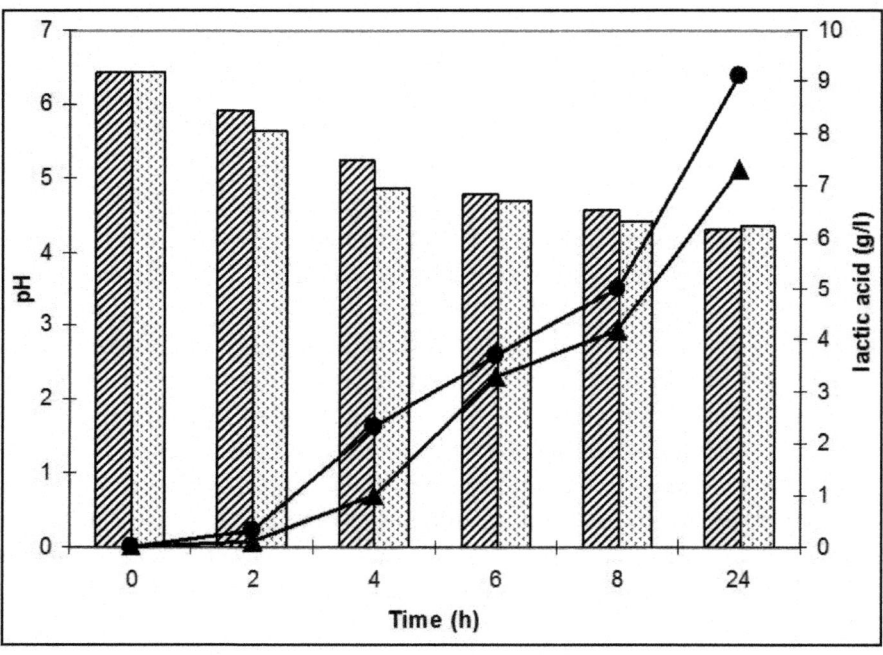

Figure 3: pH and lactic acid dynamics during the lactic acid fermentation of carrot juice at different temperatures: 37°C (▨ and ▲) and 41°C (▨ and ●); columns - pH values, lines - lactic acid content

The rate of acidification has been correlated with the glucose consumption: 38.9% in the case of the juice fermented at 37°C, respectively 53.89% in the case of the juice fermented at 41°C. The different tendency of this parameter became obviously after 4h of fermentation (Figure 4), being the consequence of the different rate of growth of *Lactobacillus acidophilus*, expressed as optical density at 610nm.

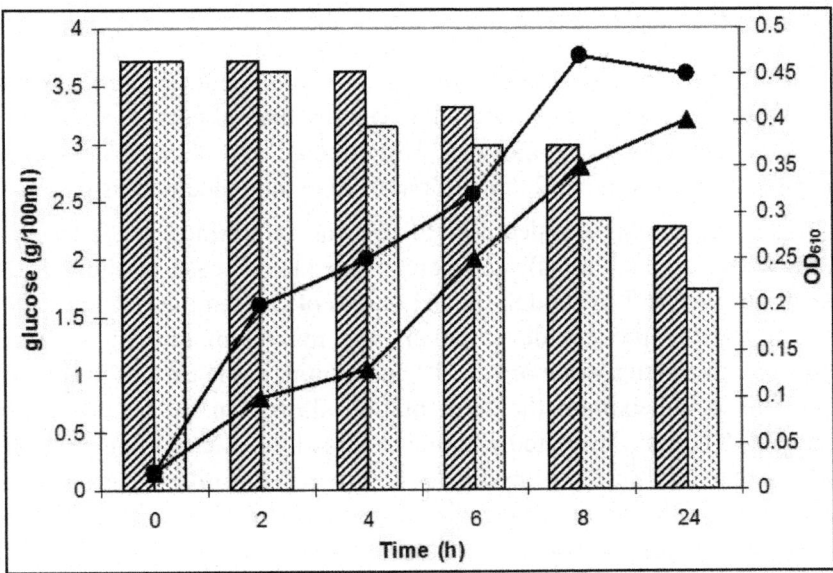

Figure 4: Glucose consumption and microbial evolution during the lactic acid fermentation of the carrot juices at different temperatures: 37°C (▨ and ▲) and 41°C (▨ and ●); columns - glucose, lines - optical density at wavelength by 610nm

Although close, the yields of glucose conversion to lactic acid have inclined the balance in favour of the juices fermented at 37°C, the corresponded value being by 0.5, unlike 0.45 in the case of the juices incubated at 41°C.

The faster consumption of the carbon source, correlated with the growth of the useful microbiota at higher temperature, respectively with the increase of the lactic acid content until the value by 9.1g/l, was followed at 24h by the decline of the viability of *Lactobacillus acidophilus*. Taking into account the dynamics of all the above mentioned parameters, the incubation temperature applied in the further studies was by 37±0.1°C.

The Behaviour of Different Raw Materials during the Lactic Acid Fermentation by Lactobacillus Acidophilus La-5

Fresh white cabbage (*Brassica oleracea* L.), red cabbage (*Brassica oleracea* var. *capitata* f. *rubra*), red beet (*Beta vulgaris* var. *vulgaris*), cucumbers (*Cucumis sativus*) and red onion (*Allium cepa* var.*ascalonicum*) were chosen in order to perform different experimental batches, as follows: Cb - cabbage juice, RCb - red cabbage juice, Rb - red beet juice, Cc - cucumber juice, CcO - cucumber juice with 0.1% (v/v) onion juice added after the heating and cooling of the batches.

pH and lactic acid dynamics during the lactic acid fermentation of vegetable juices with *Lb. acidophilus*are shown in Figure 5 and Figure 6 respectively. The pH values ranged from 6.29 to 3.74, no significant differences between the analyzed batches being observed, excepting the red beet juice. Thus, after one day a higher value by 4.28 was determined, the prolongation of the time of fermentation with other 24h hadn't a positive influence on this parameter.

After 24h, the highest decrease of pH was determined in the case of the cucumber juice (2.51 units), correlated with the increase of the lactic acid amount until 9.36g/l. Although the pH values of the samples Cc and Cb were close during the process development, the maximum rate of acidification v_{max} registered a better value of 9.33 10^{-3} units/min. in the case of the cucumber juice. This could explain the fermentation slowdown in the batch Cb the interval 6 - 8 hours. Correlated with the results of the microbiological analysis, it seems that this time the process was directed towards the growth of the useful microbiota. A minimum value of the maximum rate of acidification, by 6.66 10^{-3} units/min., was determined in the case of CcO, while the time to reach pH 5.0 ($t_{pH\ 5.0}$, hours) ranged between 1.9 (Cb) to 3.5 (CcO).

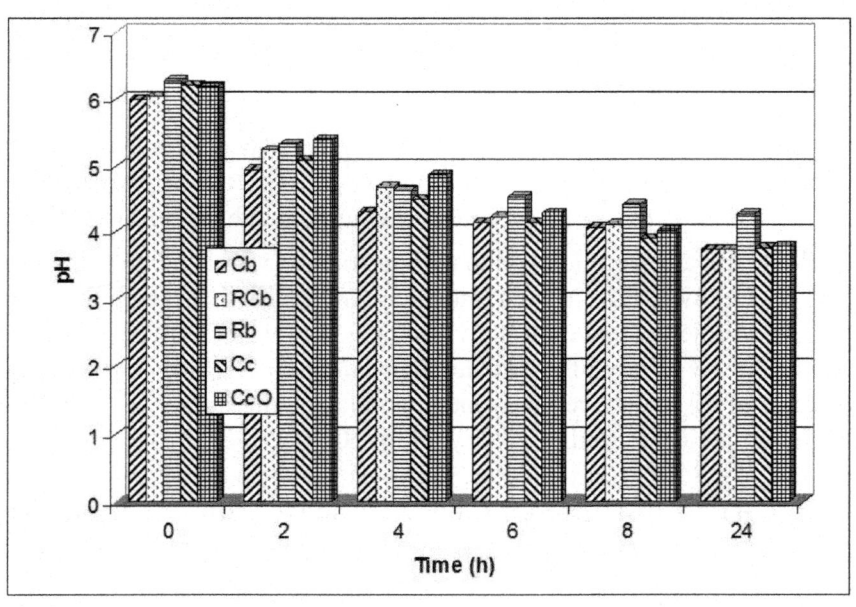

Figure 5: pH dynamics in vegetable juices obtained from different raw materials, during fermentation with*Lactobacillus acidophilus* LA-5

A relative distinct behaviour was observed in the case of red cabbage juice, red beet juice and cucumber juice with onion juice added, in the sense of the slowdown of the metabolism objectified in the dynamics of the parameters that describe the process unfolding. The differences could be explained through the presence of some chemical constituents which can act as inhibitors on useful bacteria, like anthocyanins in the red cabbage, betacyanins in red beet, respectively constituent sulfides in the onion juice. According [14], sulfides, especially those with three or more sulfur atoms, apparently possess potent antimicrobial activity. However, concerning the batch with onion juice added the initial trend was attenuated after 6 hours of fermentation, the oils and their sulfides constituent showing weak antimicrobial activity ([15]).

Referring to the red cabbage juice, although after 24 hours of fermentation the pH values were similar, the lactic acid content was lesser with about 1.5g/l compared with the white cabbage juice. This can be due to the amphoteric nature of the anthocyanins.

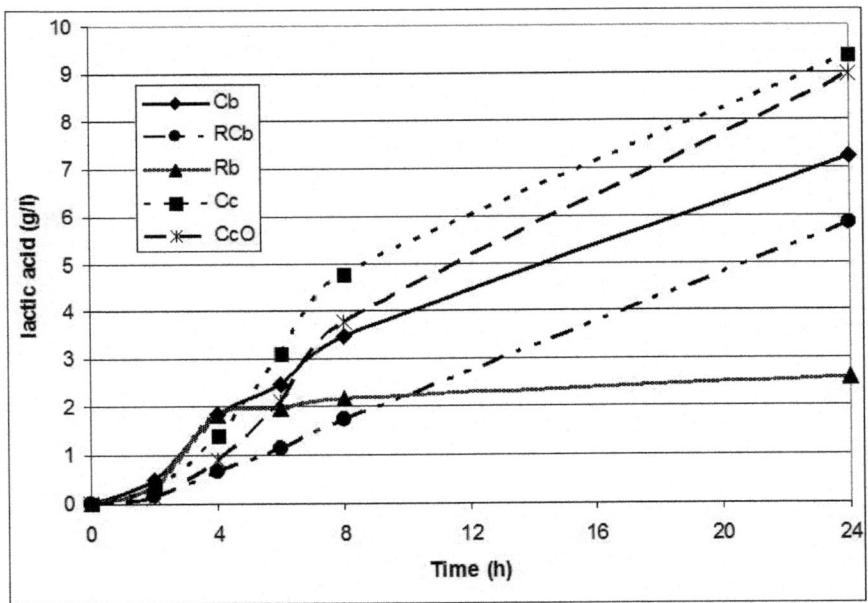

Figure 6: Lactic acid accumulation in vegetable juices obtained from different raw materials, during fermentation with *Lactobacillus acidophilus* LA-5

[16] Studied the fermentation of cucumber juices with a 0.5%, 1% and 2% additions of the onion juices by *Lb. plantarum* CCM 7039. It was found that in the initial stages of fermentation, the presence of onion in the juices positively influenced lactic and acetic acid production. However, in further course of fermentation, slight inhibition effects of onion in the fermented juices were observed, especially at elevated onion/cucumber ratio.

The correlation between the biomass amount and the production of lactic acid (Figure 7) in the case of lactic acid fermentation of red beet juices with *Lactobacillus acidophilus* in the first 24 hours, was described using the Luedeking & Piret model [17]. According to this model, the instantaneous rate of lactic acid formation (dP/dt) can be related to the instantaneous rate of bacterial growth (dN/dt), and to the bacterial density (N), throughout fermentation at a given pH, by the expression:

$$dP/dt = \alpha \, dN/dt + \beta \, N$$

(1)

where the constants α and β are determined by the pH of the fermentation.

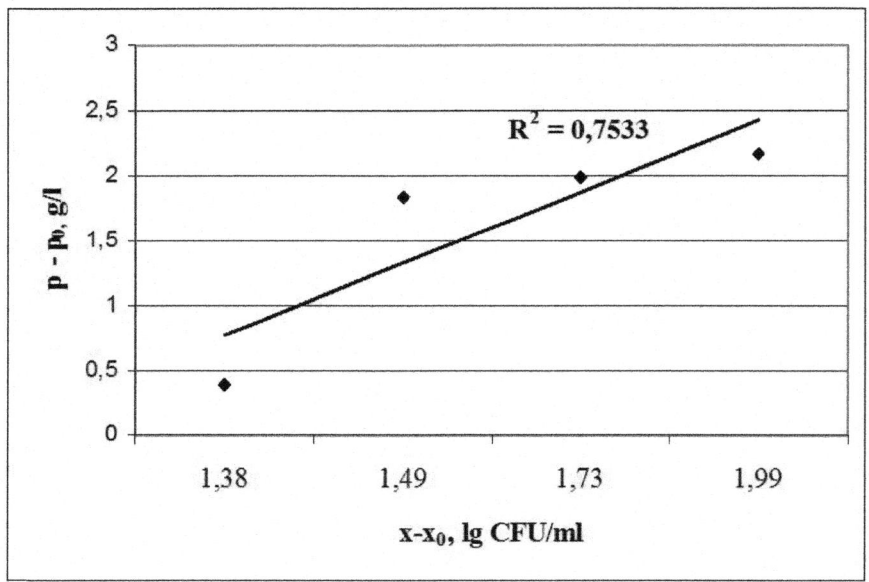

FIGURE 7: The correlation between the lactic acid production and viable cells count of *Lactobacillus acidophilus*LA-5 growing on red beet juices

A simplified presentation of the above model relates to the linear part of the equation which is presented as:

$$(p - p_0) = \alpha \, (x - x_0)$$

(2)

where *p0* and *p* are the concentrations of lactic acid (g/l) initially and at time *t*, respectively, and *x0* and *x*are the increases of the biomass (log CFU/mL) initially and at time *t*, respectively.

The R squared coefficient closed by the ideal value "1" ($R^2 = 0.9989$) in the case of the carrot juices fermented with *Lactobacillus acidophilus* LA-5 (data not shown) highlights a better linear correlation, respectively a strong connection between the lactic acid production and the lactic acid bacteria growth. Not the same situation has registered in the lactic acid fermentation of the red beet juices with the same strain. The highest value of the coefficient (1 $- R^2$) it is caused by the increase of the lactic acid amount in the first 4 hours, followed by a steady interval of evolution of this parameter. From the other hand, according [18], the deviations from the linear dependence are mostly caused by nutritive limitations of the substrates, and are related to the specific bacterial species. Not at least, the initial content of reducing sugars of the red beet, by 21.2g/l, could be limiting. However, taking into account the fact that the cucumber juice underwent a tumultuous fermentation although its content was only with 15.09% higher, it seems that other chemical constituents of the raw materials are responsible for the above mentioned differences.

The initial content of sugars in cucumber juice was situated at the maximum limit determined by [19], while in the case of the white cabbage juice was close to that one determined by [20].

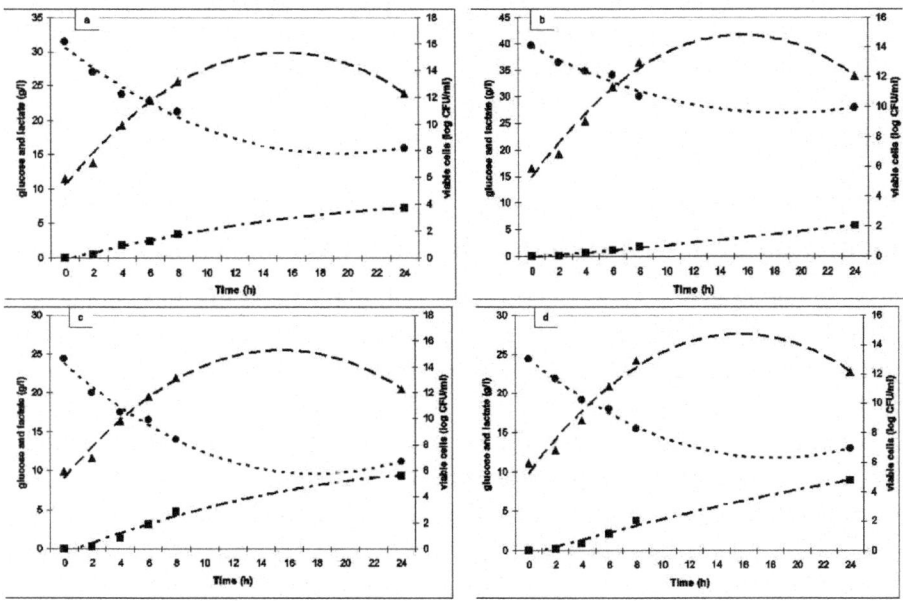

Figure 8: Correlation between the substrate consumption, lactate production and viable cells Cb (a), RCb (b), Cc (c) and CcO (d)● - glucose, ■ - lactate, ▲ - viable cells (points - experimental data, smooth lines - predicted values)

The metabolization of the reducing sugars after 24h of lactic acid fermentation of vegetable juices with *Lb. acidophilus* LA-5 ranged between 26.66% (Rb) to 54.09% (Cc). Relative close values were obtained by other authors in lactic acid fermentation of vegetable juices. Thus, the utilization of sugar during fermentation in a mixture of beetroot juice and carrot juice and different content of brewer's yeast autolysate with *Lb. plantarum* A112 and with *Lb. acidophilus* NCDO 1748 varied from 19.4 to 24.1% ([21]).

The tested pure culture, routinely used for dairy products, was found to be capable of growing on pure vegetable juices without nutrients added. In the batches obtained from cabbage, respectively cucumber, the maximum volumetric productivity was determined after 8 hours as follows: 19.25×10^{14} CFU/(l·h) for Cb, 11.9×10^{14} CFU/(l·h) for RCb, 18.6×10^{14} CFU/(l·h) for Cc and 10.25×10^{14} CFU/(l·h) for CcO respectively.

The relationship between the growth of *Lactobacillus acidophilus*, the substrate metabolization and the lactic acid accumulation is shown in Figure 8. The prediction functions of the values of the analyzed parameters in all the samples were defined as polynomial, the R squared being very close to unit.

Correlating the number of viable cells with the dynamics of the lactic acid, the values were lower until 6 hours in the red cabbage juice and cucumber juice with onion juice added respectively. The differences were lessened in the next period of the process. However, the final yield of the lactic acid production was better in the sample CcO, by 0.78, comparatively with 0.7 in the sample Cc.

Effect of Growth Factors on the Dynamics of the Lactic Acid Fermentation of the Carrot Juices by Lactobacillus Acidophilus La-5

Kinetic parameters such as the time to reach pH 5.0 and the maximum rate of acidification are important in terms of the shelf life of the fermented vegetable juices. These ones were differently modified by the presence of the amino acids or of the yeast extract at the initial moment of fermentation. From Table 2 we deduced that a highest influence on both $t_{pH \, 5.0}$ and v_{max} was exerted by cysteine, added to the juice in amount by 0.2% (w/v). Compared with the other supplements, the yeast extract had a relative good effect on the analyzed parameters. At the used concentrations, the behavior of valine and lysine seems to be unobservable from this point of view, excepting the poor effect of lysine on the maximum rate of acidification. Time to complete the fermentation ($t_{pH \, 4.2}$, h) ranged between 7.4 (YE) and 10.42 (Leu), trend that underline the statement that in the above mentioned experimental conditions *Lactobacillus acidophilus* growing faster.

Table 2: Effect of supplements on the kinetic parameters

Kinetic parameter	Supplement[s1])					
	Cys_1	Leu	Val	Lys	Cys_2	YE
Time-decreasing of $t_{pH\,5}$ [02])	1.28	0.85	0	0	1.69	1.1
Time-increasing of v_{ma} [x3])	0.82	0.84	0.98	1.05	1.1	1.05

MRS broth used for lactobacilli enumeration often incorporates L-cysteine to improve the recovery of these ones, especially due to the fact that *Lactobacillus acidophilus* LA-5 is micro-aerophilic. Cysteine, a sulfur containing amino acid, could provide amino nitrogen as a growth factor while reducing the redox potential. [22] reported that the incubation time to reach a pH of 4.5 was greatly affected by the addition of cysteine in yogurts made with different commercial cultures, although their viability was adversely affected in function of the amount of supplement and the type of the starter culture.

Lactic acid is the major metabolite of *Lactobacillus acidophilus*, influencing both the preservation of the fermented products and the sensorial characteristics of these ones. The effect of the amino acids and of the yeast extract on the dynamics of the lactic acid, assessed against the control, is underlined through the data from Table 3. The buffering capacity of the amino acids prevented a direct proportionality between the pH values and the lactic acid content.

Table 3: Time-increasing of lactic acid during 24h of lactic acid fermentation of carrot juices by*Lactobacillus acidophilus* LA-5

Time, h	Cys_1	Leu	Val	Lys	Cys_2	YE
2	8.737864	-12.6214	21.52778	-29.8611	107.6923	15.38462
4	17.66784	-23.6749	-5.55556	-2.77778	28.125	12.5
6	16.98113	-1.50943	11.71717	3.636364	1.818182	5.454545
8	20.63492	-1.5873	8.571429	1.428571	-11.1111	15.87302
24	0.925926	-0.92593	5.076142	4.568528	-14.433	11.34021

The values were expressed in percents by reporting the difference between sample and control to the control, at the same moment of time

Negative values shows that for the corresponding interval of time the supplements had not influence on the lactic acid production at the used levels.

Analyzing the whole process, only the samples with a minimum amount of cysteine added and those with yeast extract have been a great effect on the time-increasing of lactic acid. At the other opposite were found the samples with leucine added, this amino acid with non-polar hydrophobic chains clumsying the fermentation. From the viewpoint of increase the lactic acid content in the final stages of the process, the supplementation of the carrot juices with 0.2% (w/v) cysteine seems to be undesirable.

The beneficial effect of cysteine on the lactic acid accumulation in vegetable juices can occur due to its buffering capacity, which may diminish the toxic effects of organic acids on lactobacilli. Referring to the yeast extract, which contains more cell growth factors, being used generally as a source of assimilable nitrogen, vitamins and minerals, its influence at the level of 0.2%(w/v) on the time-increasing of lactic acid could be characterized as moderate. If some authors reported different maximum lactic acid concentration in media supplemented with yeast extract, several possible explanations include the strain of microorganism, the chemical composition of the substrate, the fermentation system, and generally the conditions employed during fermentation ([12]).

Effect of supplements on the performance of lactic acid production was evaluated based on lactic acid productivity and lactic acid yield, respectively on glucose ratio (Table 4).

The previous conclusion referring to the positive influence of the yeast extract and cysteine (in minimum amount) on the development of the lactic acid fermentation of vegetable juices is confirmed by the data from Table 4. Good values of lactic acid productivity were obtained after 24 h of fermentation in the samples with valine and lysine added, although in these ones the substrate consumption seems to be directed to the increasing of biomass, aspect emphasized by the average values of the lactic acid yield.

Table 4: Effect of supplements on lactic/acetic acid production after 48 h of fermentation[1])

Parameter	Cys_1	Leu	Val	Lys	Cys_2	YE
Lactic acid yield[2])	1.1	0.85	0.88	0.79	0.85	1.15
Lactic acid productivity[3])	1.01	0.99	1.06	1.05	0.7	1.13
Glucose conversion ratio[4])	1.1	0.9	1.2	1.05	0.92	1.25

The effect of supplements (amino acids and yeast extract) on the ascorbic acid dynamics is shown inFigure 9. L-Ascorbic acid (AA), also known as vitamin C, is a representative water-soluble vitamin possessing a variety of biological, pharmaceutical, and dermatological functions; it promotes collagen biosynthesis, provides photoprotection, causes melanin reduction, scavenges free radicals, and enhances immunity ([23]).

Due to the heat treatment applied with a view to destroy the epiphytic microbiota of the fresh vegetable juices, the losses occurred in the ascorbic acid content represented about 65%.

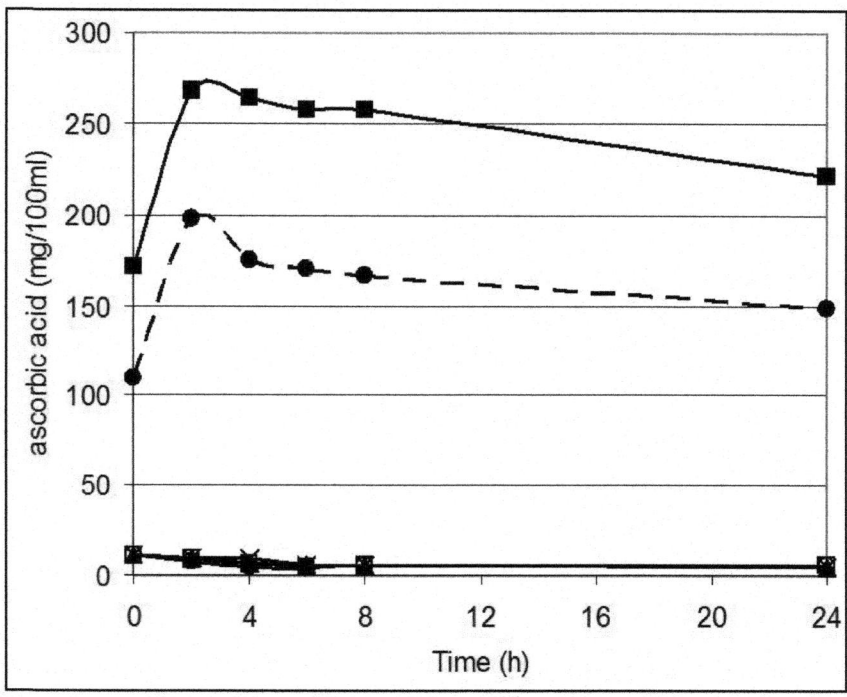

Figure 9: Time-course (0-24h) of the relative levels of ascorbic acid (●Cys_1, ■Cys_2, ○YE, □Leu, ▲Val, x Lis). The data shown are average values of two independent replicate experiments

The presence of ascorbic acid into vegetable juices submitted to fermentation by probiotic bacteria, especially by *Lactobacillus acidophilus* strains, is desired not only from the nutritional point of view, but also due to the fact that it could promote anaerobic conditions, acting as an oxygen scavenger. [24] have shown also that the fruit juices may be an alternative vehicle for the incorporation of probiotics because they are rich in nutrients and do not contain starter cultures that compete for nutrients with probiotics. Furthermore, fruit juices are often supplemented with oxygen scavenging ingredients such as ascorbic acid, thus promoting anaerobic conditions.

L-cysteine, a sulfur-containing amino acid known as a powerful reducing agent, caused the reduction of dehydroascorbic acid to ascorbic acid, which led a different behavior of the samples Cys_1 and Cys_2 by the others. The increase of this parameter was by 80% and 56.4% respectively, after 2h from the initial moment of fermentation. Subsequently, the analyzed parameter had the same diminishing tendency as in the other batches.

The losses occurred after 24h of lactic acid fermentation of carrot juices with *Lactobacillus acidophilus*LA-5 ranged from 48.39% (YE) to 61.9% (control). The possible reason could be the oxygen traces that cause the chemical oxidation of the vitamin C.

In order to evaluate the probiotic feature of the vegetable juices, the study of the effect of supplements on *Lactobacillus acidophilus* growth is from overwhelming importance, both during the lactic acid fermentation and during the storage of the final products.

Between the analyzed samples, those with yeast extract and 0.1% (w/v) cysteine added registered a higher increase of the number of viable cells till 14.4 - 14.5 log CFU/ml in the first 8h of the process. Concerning the yeast extract, the most possible explanation is due to an enhanced availability of minerals, which are growth promoters for *L. acidophilus* ([25]), while discussing the factors that affect the activity of endogenous probiotics, (26) mentioned that some of the growth promoters in cow milk were apparently cysteine-containing peptides.

Referring to the juices with leucine, lower values were determined comparative with the control during 24h, while in the samples with 0.2% (w/v) cysteine added the trend of the survival of lactobacilli was slow down in the period 6 - 8h, the level being by 13.5 and 13.6 log CFU/ml respectively. The last observation agrees with this one of [27], which have shown that the increasing of cysteine concentration improved the viability of *B. bifidum* in bio-yogurt, although it had no important effect on the viability of*Lactobacillus acidophilus*.

The batches supplemented with valine and lysine had occupied an intermediate position, the growth until 14.2 log CFU/ml after 8h of fermentation making from the utilization of these amino acids a promising variant in the future, with a view to optimize the conditions of the process unfolding. In the period 8 - 24h the number of viable cells decreased, as result of the lack of tolerance at lower pH of the analyzed strain.

The correlation between the most important parameters of the lactic acid fermentation of the carrot juices with *Lactobacillus acidophilus* LA-5 were evaluated using Pearson correlation analysis (significance level $p < 0.01$; confidence level of 99%).

Table 5: The Pearson coefficients for the experimental batches

Analytical variables	pH	lactic acid	glucose	viable cells	glycine	ascorbic acid
pH	1	-0.88^{9**}	0.82^{9**}	-0.94^{0**}	0.09^{9*}	-0.18^{4*}
lactic acid		1	-0.89^{1**}	0.84^{3**}	-0.20^{1*}	0.01^{6*}
glucose			1	-0.78^{9**}	0.09^{3*}	0.08^{4*}
viable cells				1	-0.06^{1*}	0.06^{6*}
glycine					1	-0.10^{3*}
ascorbic acid						1

The correlations are strong between pH and lactic acid, respectively pH and glucose, while a very strong relationship pH - viable cells could be considered (Table 5). A non-existent relationship between ascorbic acid / amino acids content (expressed as glycine) and the other analyzed parameters was determined.

A firm correlation between glucose and lactic acid was expected, but on the one hand it is known that the practical yield of sugars conversion to lactic acid of the strains of the group *Lb. acidophilus* is about 85%, while on the other hand the analysis does not include supplementary data referring to other factors that might be involved in the dynamics of the lactic acid fermentation of vegetable juices.

Factor Analysis (FA) is a multidimensional statistic method whose purpose is the analysis of the structure of mutual dependences of variables. The method is similar to the Principal Component Analysis (PCA) with the exception of the factor weights that are scaled ([28]).

Applying FA to the experimental data, the analytical variables were reduced to two principal components, which accounted for 59.72% (PC1) and respectively 18.95% (PC2) from the total variance. According to the component matrix, respectively to the values of the component loadings expressed by the first second principal components (rotation method: Varimax with Kaiser normalization), the most notable variables were pH and lactic acid (equal loading values by 0.954). Higher values were obtained also for viable cells (loading 0.939) and glucose (loading 0.933).

The combination of PC1 and PC2 (Figure 10) underlined the lack of correlation between amino acids content / ascorbic acid and all the other parameters taking into account both control and supplemented samples. While PC1 affected the dependent and independent variables involved in the progress of the lactic acid fermentation of vegetable juices, respectively in their probiotic feature, PC2 separated the variables which contribute to the nutritional characteristics of the final products.

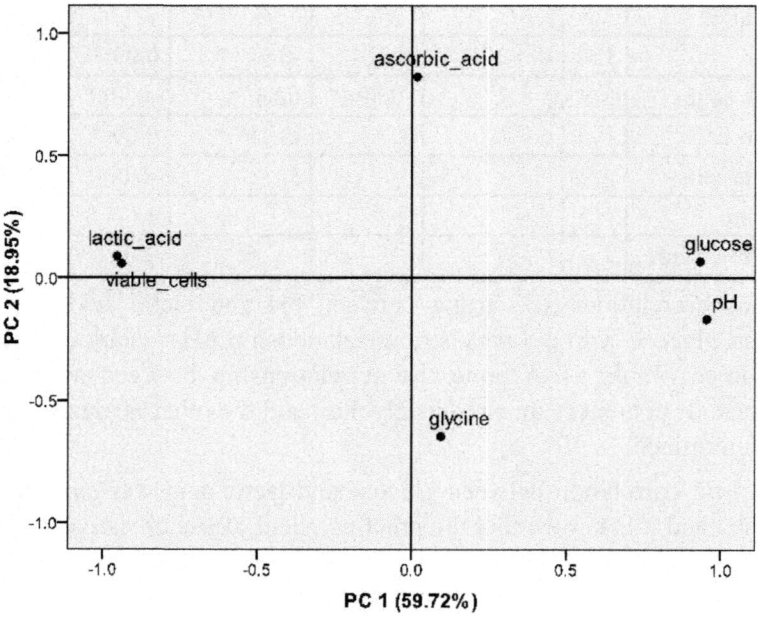

Figure 10: Component plot in rotated space

Applying PCA to the lactic acid fermentation of cabbage juices with various microorganisms, [29] established that the original 7 analytical variables were reduced also to 2 independent components that explained 88.2% from total variance of input data (PC1 66.9% and PC2 21.3%).

Cluster Analysis (CA) is a statistic method whose purpose is to join data into clusters with a view to increase their withingroup homogeneity. Usually, the FA is considered the first step of CA, with a view to reduce the data dimensionality. In order to better distinguish among experimental samples, the cluster method of the nearest neighbour was used. The distances between objects were measured as squared Euclidean distance. K-Means Cluster Analysis divided the experimental data into three groups, characterized by similar analytical properties, as follows:

- cluster 1: all the carrot juices (control samples and the batches with amino acids and yeast extract added) at the initial moment of fermentation, respectively at 2th h of fermentation. Supplementary, this cluster included the control and the sample with leucine at 4th h of fermentation (C_4 and Leu_4);

- cluster 2: all the carrot juices at 24th h of fermentation and the sample with lysine added at 8th h of fermentation ;
- cluster 3: the carrot juices with leucine and lysine added, respectively the control, at 6th and 8th h of fermentation (Leu_6, Leu_8, Lys_6, Lys_8, C_6, C_8); the carrot juices with cysteine, valine, respectively yeast extract from 4th to 8th h of fermentation (Cys_4 - Cys_8, (Val_4 - Val_8), (YE_4 - YE_8).

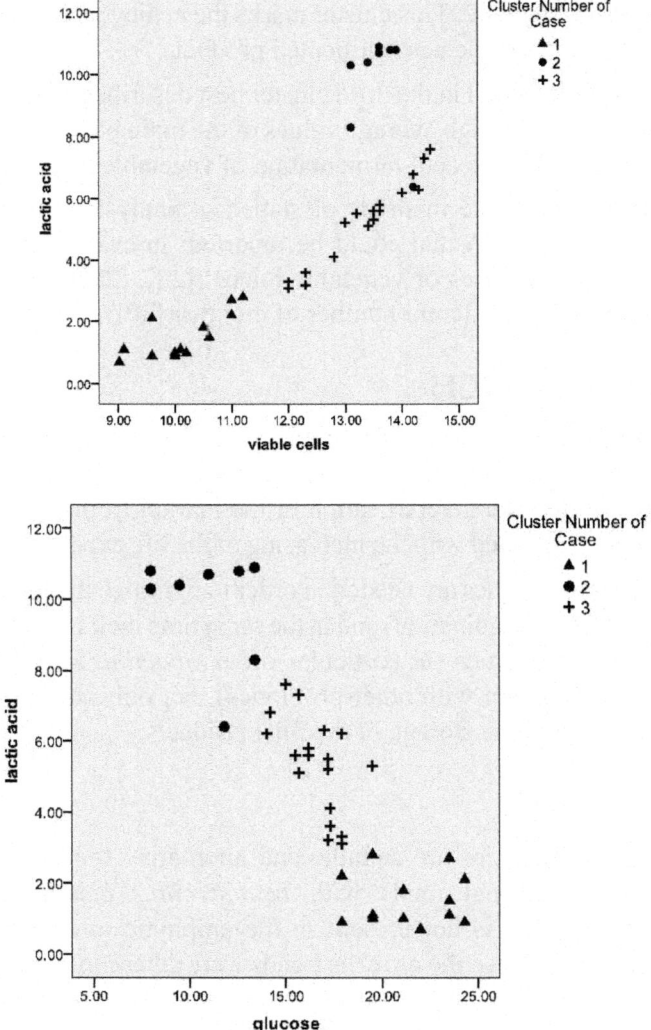

Figure 11: Clusters plotting in coordinate of two selected variables: lactic acid - viable cells and lactic acid - glucose

The clusters in axes of two selected variables (Figure 11) denote that the samples from the first cluster were marked with a higher content of substrate, null or very lower lactate amount and pH values more than 5. The corresponding time was both 0 and 2h (C, Val, Leu, and YE) or the entire interval 0 - 4h (Leu and control).

The samples at the final moment of fermentation and those with lysine after 8h of the process were included in the second cluster, characterized through lower or average values of glucose content, higher lactic acid amount and pH values close to 4.2. This cluster marks the achievement of the optimum characteristics of the lactic acid fermented products.

The samples included in the third cluster best describes a vigorous process, being characterized through average values of the main parameters involved in the dynamics of the lactic acid fermentation of vegetable juices.

The usefulness of the methods of statistical analysis is underlined by a lot of applications of CA that could be reported: in evaluation of analytical and sensory characteristics of vegetable juices ([28], [29]), in distinguishing between wines aged a different number of months ([30]).

FURTHER RESEARCH

The importance of consuming probiotic foods for the improvement of the quality of life increasingly more in the last years, being underlined by the scientific literature. The diversification of the market from this point of view could be strong correlated with the increasing of the life expectancy worldwide.

Our further researches are needed in order to optimize the level of nutrients (individually and in combination) and in the same time their influence on growth and viability of probiotics (in particular of*Lactobacillus acidophilus*, single strain or in combination with other probiotics), not only during fermentation but especially during the storage of the final products.

CONCLUSIONS

Different vegetable juices are suitable and alternative food matrices for the production of functional foods with *Lactobacillus acidophilus* LA-5, a probiotic strain which is not present in the epiphytic microbiota. Although some differences between the growths trends were determined, all the analyzed vegetables could be considered proper in order to obtain lactic acid fermented juices with a higher self-life. Application of Principal Component Analysis selected the most important parameters from analytical point of view: pH, lactic acid, biomass and viable cells, while the Cluster Analysis divided the experimental variables into three groups.

ACKNOWLEDGEMENT

The research was funded by Executive Unit for Financing Higher Education, Research, Development and Innovation (UEFISCDI) in the frame of the Project PN-II-ID-PCE-2008-2 (ID_1359). The microorganisms were kindly provided by Chr. Hansen, Romania.

REFERENCES

1. D. Granato, G. F. Branco, F. Nazzaro, A. G. Cruz, J. A. F. Faria, 2010Functional Foods and Nondairy Probiotic Food Development: Trends, Concepts and Products. Compr. Rev. Food Sci. Food Saf. 9292302

2. W. H. Holzapfel, P. Haberer, R. Geisen, J. Björkroth, U. Schillinger, (200, 2001Taxonomy and Important Features of Probiotic Microorganisms in Food and Nutrition. Am. J. Clin. Nutr. 73(suppl): 365S-73S.

3. T. Saito, 2004Selection of Useful Probiotic Lactic Acid Bacteria from the Lactobacillus acidophilus Group and their Applications to Functional Foods. Animal Sci. J. 75113

4. M. Pineiro, C. Stanton, 2007Probiotic Bacteria: Legislative Framework-Requirements to Evidence Basis. J. Nutr. 850S-853S.

5. K. Kailasapathy, J. Chin, 2000Survival and Therapeutic Potential of Probiotic Organisms with Reference to Lactobacillus acidophilus and Bifidobacterium spp. Immun. Cell Biol. 788088

6. J. Karovičová, Z. Kohajdová, 2003Lactic Acid Fermented Vegetable Juices. Hort Sci. (Prague), 304152158

7. E. Klewicka, 2010Antimutational Activity of Beetroot Juice. Food Technol. Biotechnol. 482229233

8. D. Altiok, 2004Kinetic Modelling of Lactic Acid Production from Whey, Ph D Thesis, Izmir Institute of Technology, Turkey

9. Jessica. J. Kious, Lactobacillus and Lactic Acid Production [online]. Energy Research Undergraduate Laboratory Fellowship Program (ERULF), National Renewable Energy Laboratory, Colorado, 2000Portable Document Format. Available from Internet:http://www.nrel.gov/docs/gen/fy01/NN0017.pdf.

10. L. Agarwal, K. Dutt, G. K. Meghwanshi, R. K. Saxena, 2008Anaerobic Fermentative Production of Lactic Acid using Cheese Whey and Corn Steep Liquor. Biotechnol. Lett. 30631635

11. Dave RI, Shah NP1997Viability of Yoghurt and Probiotic Bacteria in Yoghurts Made from Commercial Starter Cultures. Int. Dairy J. 73141

12. A. Amrane, 2005Analysis of The Kinetics of Growth and Lactic Acid Production for Lactobacillus Helveticus Growing on Supplemented Whey Permeate. J. chem. technol. biotechnol. 80345352

13. C. Kotzamanidis, T. Roukas, G. Skaracis, 2002Optimization of Lactic Acid Production from Beet Molasses by Lactobacillus delbrueckii NCIMB 8130. World j. microbiol. biotechnol. 18441448

14. Kim JW, Kim YS, Kyung KH2004Inhibitory Activity of Essential Oils of Garlic and Onion against Bacteria and Yeasts. J. food prot. 673499504

15. Kim JW, Huhi JE, Kyung SH, Kyung KH2004Antimicrobial Activity of Alk(En)Yl Sulfides Found in Essential Oils of Garlic and Onion. Food sci. biotechnol. 132235239

16. Y. Kohajdová, J. Karovičová, M. Greifová, 2007Analytical and Organoleptic Profiles of Lactic Acid-Fermented Cucumber Juice with Addition of Onion Juice. J. food nutr. res. 463105111

17. M. Rakin, J. Baras, M. Vukašinović, M. Maksimović, 2004The Examination of Parameters for Lactic Acid Fermentation and Nutritive Value of Fermented Juice of Beetroot, Carrot an Brewer's Yeast Autolysate, J. serbian chem. soc. 69 (8-9): 625-634.

18. R. Luedeking, E. L. Piret, 1959A Kinetic Study of the Lactic Acid Fermentation. Batch Process at Controlled pH, J. biochem. microbiol. technol. eng., 1393412

19. A. Amrane, Y. Prigent, 1999Analysis of Growth and Production Coupling for Batch Cultures of Lactobacillus Helveticus with Help of Anunstructurated Model. Process biochemistry, 34110

20. Z. Lu, H. P. Fleming, R. F. Mc Feeters, 2002Effects of Fruit Size on Fresh Cucumber Composition and the Chemical and Physical Consequences of Fermentation. J. food sci. 67829342939

21. Yoon KY, Woodams EE, Hang YD2005Production of Probiotic Cabbage Juice by Lactic Acid Bacteria. Bioresour. technol. 971214271430

22. Rakin MB, Baras JK, Vukašinović MS2005Lactic Acid Fermentation in Vegetable Juices Supplemented with Different Content of Brewer's Yeast Autolysate. Acta periodica technol. 367180

23. Dave RI1998Factors Affecting Viability of Yoghurt and Prbiotic Bacteria in Commercial Starter Cultures. PhD thesis, Victoria University of Technology, Werribee Campus, Victoria, Australia.

24. Jang KI, Lee HG2008Influence of Acetic Acid Solution on Heat Stability of L-Ascorbic Acid. Food sci. biotechnol. 17637641

25. Ding WK, Shah NP2008Survival of Free and Microencapsulated Probiotic Bacteria in Orange and Apple Juices. Int. food res. J. 15219232

26. A. Lourens-Hattingh, B. C. Viljoen, 2001Yogurt as Probiotic Carrier Food. Int. dairy j. 11117

27. A. Bezkorovainy, 2001Probiotics: Determinant of Survival and Growth in the Gut. Am. j. clin. nutr. 73(suppl): 399S-405S.

28. M. R. Bari, R. Ashrafi, M. Alizade, L. Rofegarineghad, 2009Effects of Different Contents of Yogurt Starter/Probiotic Bacteria, Storage Time and Different Concentration of Cysteine on the Microflora Characteristics of Bio-yogurt. Res. j. biol. sci. 4137142

29. J. Karovičová, Z. Kohajdová, 2002The Use of PCA, FA, CA for Evaluation of Vegetable Juices Processed by Lactic Acid Fermentation. Czech j. food sci. 20135143

30. Z. Kohajdová, J. Karovičová, 2004Optimisation of Method of Fermentation of Cabbage Juice. Czech j. food sci. 223950

31. Z. Li, Q. H. Pan, Z. M. Jin, J. J. He, N. N. Liang, C. Q. Duan, 2009Evolution of 49 Phenolic Compounds in Shortly-Aged Red Wines Made from Cabernet Gernischt (Vitis vinifera L. cv.). Food sci. biotechnol. 1810011012

Chapter 8

DYNAMIC STRESSES OF LACTIC ACID BACTERIA ASSOCIATED TO FERMENTATION PROCESSES

Diana I. Serrazanetti, Davide Gottardi, Chiara Montanari and Andrea Gianotti

Department of Food Science, Alma Mater Studiorum, University of Bologna, Bologna, Italy

Inter-Departmental Center of Industrial Agri-Food Research (CIRI Agroalimentare), Cesena, Italy

INTRODUCTION

Despite their negligible mass the microbial agents, starters and nonstarters, play a profound role in the characterization of the fermented foods in terms of chemical and sensorial properties. In fact, fermented foods may be defined as foods processed through the activity of microorganisms. Fermentation processes take a special place in the evolution of human cuisine, by altering the taste experience of food products, as well as extending the storage period. In particular, foods fermented with lactic acid bacteria (LAB) have constituted an important part of human diet and of fermentation processes (involving various foods, including milk, meat. vegetables and fruits) [1] since ancient times. They have played an essential role in the preservation of agricultural resources and in the improvement of nutritional and organoleptic properties of human foods and animal feed. Moreover, these organisms nowadays are increasingly used as health promoting probiotics, enzyme and metabolite factories and vaccine delivery vehicles [2].

It is interesting to outline how the changes of food characteristics during the fermentation process can be described as dynamic fluctuations of the food environment itself and, at the same time, stress source for the microorganisms involved [3, 4], such as LAB. In fact, whenever autochthonous bacteria are adapted and competitive in their respective environment, the environment can be described as stressful for LAB [5, 4]. The fermentation parameters, including temperature, water activity (Aw), oxygen, pH, as well as the concentration of starter cultures, affect the regulatory mechanism and the response mechanisms of LAB, as well as their effects on the final products properties [4].

When LAB are added to food formulations, several factors that may influence the ability of those microorganisms to survive, growth and become active in the new matrix have to be considered [6]. These factors include: 1) the physiological state of the LAB used as starters (whether the cells are from the logarithmic or the stationary growth phase); 2) the physical conditions of product ripening and storage (eg. temperature); 3) the chemical composition of the matrix (eg. acidity, available carbohydrates content, nitrogen source, mineral content, water activity and oxygen concentration); 4) possible interactions of the starter cultures with probiotics and other microorganisms naturally occurring or added to the system [6].

In figure 1 the main factors affecting the viability and the responses of LAB from production to storage are described [7].

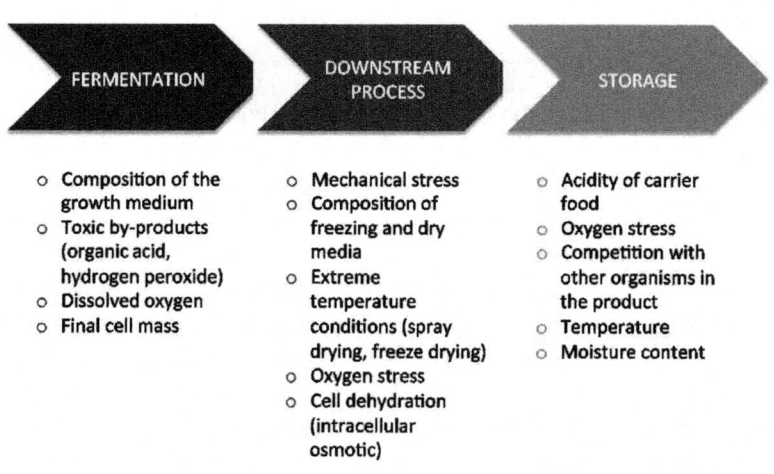

Figure 1: Factors affecting the viability and the responses of LAB to the various fermented foods production steps.

To better elucidate what happens to LAB during fermentation processes, we decided to use a model (defined "virtual food") that mimics various steps occurring during processing and that can affect LAB performances or viability.

LACTIC ACID BACTERIA AND STRESS: BASIC CONCEPTS

"Stress results from interactions between subjects and their environment that are perceived as straining or exceeding their adaptive capacities and threatening their well-being. The element of perception indicates that human stress responses reflect differences in personality, as well as differences in physical strength or general health" [8].

Stress has driven evolutionary changes (the development and natural selection of species over time). Thus, the species that adapted best to the causes of stress (stressors) have survived and evolved into the plant and animal kingdoms we now observe. The same evolutionary process regarded microorganisms. In fact, bacteria, irrespective of natural habitat, are exposed to constant fluctuations in their growth conditions. Consequently they have developed sophisticated responses, modulated by the re-modelling of protein complexes and by phosphorylation dependent signal transduction systems, to adapt and to survive to a variety of insults. To ensure survival to environmental adversities, bacteria may adapt to changes in their immediate vicinity by responding to the imposed stress. These responses are different and vast and depend on the microorganism nature and on the environmental stress and are accomplished by changes in the patterns of gene expression for those genes whose products are required to combat the deleterious [3]. In particular, cellular metabolic pathways are closely related to stress responses and the flux of particular metabolites to understand the hypothetically shifts and implications in the food systems has been studied in LAB [9-13, 4, 14, 15].

LAB are a functionally related group of organisms known primarily for their bioprocessing roles in food and beverages [16]. LAB play a crucial role in the development of the organoleptic and hygienic quality of fermented products. These microorganisms are used as starter cultures in many fermented products (i.e. beer, milk, dough, sausages and wine). Therefore, the reliability of starter cultures in terms of quality and functional properties (important for the development of aroma and texture), but also in terms of growth performance and robustness, has become essential for successful fermentations [17]. There have been some reports describing the physiological stress responses in LAB, particularly*Lactobacillus* species, which have a broad biodiversity [17-21, 13, 22, 4, 14, 15].

LAB evolved specific mechanisms to respond and to survive to environmental stresses and changes (stress-sensing system and defences). In fact, microorganisms could have specific regulators tailored to each of their regulated genes and adapt their expression according to environment. Stress defences are good examples of such integrated regulation systems. Bacterial stress responses rely on the coordinated expression of genes that alter different cellular processes (cell division, DNA metabolism, housekeeping, membrane composition, transport, etc.) and act in concert to improve the bacterial stress tolerance. The integration of these stress responses is accomplished by networks of regulators that allow the cells to react to various and complex environmental shifts. LAB respond to stress in a very specific way dependent on the species, on the strains and on the type of stress. The best-studied stresses are acid, heat, oxidative and cold stresses, although for the latter most of the studies focused on a specific family of proteins instead of analyzing the whole response [4].

Despite the extensive use of LAB, there is a paucity of information concerning the stress-induced mechanisms studied *in vivo* for improving the survival of these organisms during real food processing. A better knowledge of the adaptive responses of LAB is important because the fermentation processes often expose these microorganisms to adverse environmental conditions. LAB should resist to adverse conditions encountered in industrial processes, for example during starter handling and storage (freeze drying, freezing or spray-drying) and during the fermentation environment dynamic changes. These phenomena reinforce the need for robust LAB since they may have to survive and grow in different unfavorable conditions expressing specific functions (for example during stationary phase or storage) [17].

PRINCIPAL RESPONSES TO THE MOST COMMON STRESSES

Heat shock response: The effect of heat shock and the induction of a stress response in *Lactobacillus*spp. have been studied for *Lactobacillus delbrueckii* subsp. *bulgaricus* [23] and *Lactobacillus paracasei*[24, 25], *Lactobacillus acidophilus*, *Lactobacillus casei* and *Lactobacillus helveticus* [26], *Lactobacilluscollinoides* [27], *Lactobacillus sakei* [28], *Lactobacillus johnsonii* [29], *Lactobacillus rhamnosus* [30],*Lactobacillus plantarum* [31-33] and *Lactobacillus salivarius* [34]. The heat resistance of LAB is a complex process involving proteins with different roles in cell physiology, including chaperone activity, ribosome stability, stringent response mediation, temperature sensing and control of ribosomal functions [31]. The time taken to initiate the stress response is different for different treatments and different strains. The major problem encountered by cells at high temperature is the denaturation of proteins and their subsequent aggregation. In addition Earnshaw et al. [35],, Texeira et al. [36] and Hansen et al. [37] described also as response to heat stress the destabilization of macromolecules as ribosomes and RNA as well as alterations of membrane fluidity.

Heat stress response is characterized by the transient induction of general and specific proteins and by physiological changes. In every strain tested the involvement of Heat Shock Proteins (HSPs such as DnaK, GroEL and GroES during the heat stress was clear) [23-38]. The role of these stress proteins is complex; in fact, the bind substrate proteins in a transient non-covalent manner prevent premature folding and promote the attainment to the correct state *in vivo*. The resistance to heat stress is higher when the cells were previously exposed and adapted to this type of stress in the stationary phase, otherwise, when pre-adapted in exponential phase, the cells are more sensitive. In particular, the storage stability of the culture that was heat shocked after stationary phase was superior to that of culture heat shocked after log phase [34, 23, 30].

Cold shock response: It is very important to improve knowledge about LAB behavior in cold environment. In fact, during industrial processes, like in cheese ripening and refrigerated storage of fermented products, these microorganisms are subjected to different temperatures far below the optimal growth temperature. When LAB living cells are exposed to these cold environments, important physiological changes occur, such as decrease in membrane fluidity and stabilization of secondary structures of RNA and DNA, resulting in a reduced efficiency of translation, transcription and DNA replication. The response of microorganisms to these effects is termed cold-shock response during which a number of Cold Induced Proteins (CIPs) are synthesized. The roles of these proteins are at the levels of membrane fluidity, DNA supercoiling and transcription and translation. Few papers have described cold shock proteins and mechanisms in LAB, in particular they have focused on *Lactococcuslactis* and *L. plantarum* [39-42]. Kim et al. [39, 40] tested different LAB to evaluate cold shock effects on cryotolerance. Improved understanding of cold-shock-induced cryotolerance may contribute to the development of environmental conditions that allow improved viability/ activity of frozen or freeze-dried commercial LAB starter cultures. The results showed that, as with heat stress, there is also an improvement of the viability of the tested strains as concerning the cryotolerance after a cold shock. The process of freezing appeared to have different effects on different LAB as well as different effects on strains within the same genus. Moreover, the freezing response of the strains depends on the time of the cold shock process and the induction of cryotolerance appears to be dependent on the growth phase in which the cold shock took place [43-47].

Another interesting study regarding LAB response to sub-lethal cold stress was developed by Montanari et al.[14]. These Authors separated and quantified the cell cyclopropane fatty acids lactobacillic (C19cyc11) and dehydrosterculic (C19cyc9) to study the adaptive response to sub-lethal acid and cold stresses in *L. helveticus* and *Lactobacillus sanfranciscensis*. These microorganisms showed different fatty acids composition and environmental adaptation to short term cold and acidic stresses. In *L. helveticus* C19cyc11 dramatically increased after 2 h at 10°C and with the pH decrease, particularly in micro-aerobic conditions, in the presence of tween 80, and in anaerobic conditions. The increase of lactobacillic acid in *L. helveticus* is necessary to maintain the cell membrane in a suitable state of fluidity. Moreover, cyclopropane fatty acids confer resistance to ozonolysis, singlet oxygen and mild oxidative treatments [48, 49], suggesting a cross protection and response of LAB cell membrane to physicochemical stresses. A combined analysis of the genome-wide transcriptome and metabolism was performed with a dairy *Lactococcus lactis* subsp. *lactis* under dynamic conditions similar to the conditions encountered during the cheese-making process. Specific responses to acid and cold stresses were identified, but also the induction of unexpected pathways was determined. In particular, the induction of purine biosynthesis and prophage [50].

Oxidative stress response: LAB are facultative anaerobic microorganisms that have in common the reduction of part of pyruvate produced to lactate production in order to regenerate NAD+ from NADH formed during glycolysis. They do not require oxygen for growth and, in fact, a negative effect of oxygen on the development of these bacteria has often been observed. It was generally believed that these bacteria could under no condition use oxygen as the terminal electron acceptor [17]. However, many LAB have NADH oxidase and some can even express a functionally active respiratory chain in the presence of heme [51-57]. Respiration-competent LAB differ from the features of *Escherichia coli* and *Bacillus subtilis*, since they carry limited equipment for respiration. All respiring LAB carry genes encoding electron donor (NADH dehydrogenase) and a single electron acceptor (cytochrome bd oxidase) [58]. Addition of heme to the system activates respiration chain NADH oxidase activity, but none of the tested LAB synthesize heme [01].

When for some reasons the generation of free radicals is higher than the rate of their detoxification the cells are exposed to a constraint called "oxidative stress" [59]. For the food-associated LAB a still fragmented picture of the resistance mechanisms present emerges. Representatives of the different mechanisms have been described in different LAB [60-64]. Apart from the toxic effects of oxygen, aeration can induce important changes in the sugar metabolism of LAB. In fact, the presence of oxygen is a factor that greatly affects the outcome of a fermentation process. In general, LAB tolerate oxygen but grow better under nearly anaerobic conditions. However, in the presence of heme and oxygen LAB start respiration metabolism, by which the cell metabolism is reprogrammed so that pH, oxygen status, growth capacity and survival are markedly altered [56]. In the presence of oxygen and during the fermentation metabolism, H_2O_2 is formed. Numerous species of LAB contain peroxidase and/or catalase to prevent and eliminate these deleterious effects [17]. Concerning the prevention of reactive oxygen species (ROS) formation, the scope of the reactions is the eliminations of free oxygen. In a study on *L. helveticus* the fatty acids composition in the cell membrane changed in response to oxidative stress. In fact, the activity of oxygen consuming desaturase system increased to reduce the free radical damage to the cell [19]. Generally, the response to oxidative stress of LAB is similar, but also depends on the species, on the strains and, with regard to catalase action, on the bacterial density [4]. In *L. lactis* several genes have been identified and the respective encoded proteins have been shown to contribute to oxidative stress resistance. Moreover, the induction of these genes is growth phase-dependent (exponential or stationary) and their products confer multi-stress resistance [52]. General stress resistance mechanisms may also confer resistance to oxidative stress. In fact, in a model system several acid resistant mutants of *L. lactis* that appeared also more resistant to oxidative stress were isolated [64].

Acid stress response: Understanding the acid resistance mechanism used by LAB to survive to by-products of their own metabolism (i.e. homofermentative *L. lactis* converts 90% of metabolized sugar to lactic acid) and the response available in low-pH foods is of great importance. In LAB one of the most effective mechanisms for resistance in acid stress environment is the glutamate decarboxylase (GAD). In fact, few years ago, it was proposed that amino acid decarboxylase functions to control the pH of the bacterial environment by consuming hydrogen ions as part of carboxylation reaction [65]. LAB are also capable of inducing an Acid Tolerance Response (ATR) in response to mild acid treatments. The system induced includes pH homeostatis, protection and repair mechanisms. Genes and proteins, involved in pH homeostasis and cell protection or repair, play a role in acid adaptation, but this role can also extend to more general acid tolerance mechanisms. A more specific study was developed on the effects of lactic acid stress on *L. plantarum* by transcription profiling [66]. The difference, in terms of stress response, into the dissociated or undissociated forms of lactic acid has been highlighted. The toxicity of organic acids depends on their degree of dissociation and thus on the pH. For LAB end product inhibition by lactic acid could result in a disturbance of the regeneration of cofactor NAD+, especially under anaerobic conditions, in which the cell does not have the possibility of NAD+ regeneration by NADH oxidase. The response at membrane fatty acids level to acid stress was studied in *L. helveticus* and *L. sanfranciscensis* [14]. The relevant proportion of dodecanoic acid in the latter species under acid stress suggests that carbon chain shortening is the principal strategy of *L. sanfranciscensis* to modulate fluidity or chemico-physical properties of the membranes in the presence of acid stress. Moreover, a specific shift in leucine catabolic pathway at pH 3.6 was identified in *L. sanfranciscensis* [15]. In fact, the acid stress induced a metabolic shift toward overproduction of 3-methylbutanoic and 2-methylbutanoic acids, accompanied by sugar reduced consumption and primary carbohydrate metabolite production. The metabolites coming from branched chain amino acids (BCAAs) catabolism increased up to seven times under acid stress. While the overproduction of 3-methylbutanoic acid under acid stress can be attributed to the need to maintain redox balance, the rationale for the production of 2-methylbutanoic acid from leucine can be found in a newly proposed biosynthetic pathway leading to 2-methylbutanoic acid and 3 mol of ATP per mol of leucine. Leucine catabolism to 3-methylbutanoic and 2-methylbutanoic acids suggests that the switch from sugar to amino acid catabolism supports growth of *L. sanfranciscensis* in restricted environments such as sourdough, characterized by acid stress and recurrent carbon starvation.

Osmotic stress response: In the various applications in food and feed industry LAB can be exposed to osmotic stress when important amounts of salts or sugars are added to the product [17]. In fact, in most of the food habitats where lactobacilli live, they are confronted with salt [67] and sugar stress [68]. Study on the differences between salt and sugar osmotic stress revealed that the hyperosmotic conditions imposed by sugar stress are much less detrimental and only transient (transient osmotic stress), because the cells are able to balance the extra and the intracellular concentrations of lactose and sucrose [17]. Bacteria need to adapt to this change in their environment in order to survive [69], and they can do it by accumulating (by uptake or synthesis) compatible solutes, generally of organic origin, under hyperosmotic conditions [17]. The compatible solutes are defined as osmoprotectants. The main strategy to adapt to high osmolarity of non-halophilic bacteria is associated with the enhancement of the osmotolerance [68]. Moreover, the osmoprotectants can also stabilize enzymes and provide protection not only against osmotic stress but also against other type of stresses (high temperature, freezing and drying). The intracellular accumulation of compatible solutes prevents the loss of water caused by high external osmolarity and allows the maintenance of turgor [68]. The accumulation of carnitin, betain and proline was determined in LAB grown in MRS and complex diluted MRS medium (DMRS medium) [70]. Moreover, a specific response mechanism to osmotic stress was identified in a sourdough model system [13]. In particular, the growth of *L. sanfranciscensis* under osmotic stress resulted in a relevant accumulation of 3-methylbutanoic acid. Its synthesis is associated with the BCCAs., is NAD+ dependent and produces NADH during the reaction [71]. The accumulation of 3-methylbutanoic acid as predominant metabolite has been also observed in model systems simulating sourdough as a consequence of osmotic, acid or oxidative stress [12, 15].

High pressure stress response: High-pressure processing (HPP) or high pressure homogenization (HPH) are non-thermal processes capable of inactivating and eliminating pathogenic and food spoilage microorganisms in specific foods [11, 72], and it represents an exceptional stimulus for most mesophilic bacteria. Several proteins are induced after high pressure treatment and some of these have also been involved in the response to other various stresses [8]. The responses to HHP stress have been studied in particular on *L. sakei* and *L. sanfranciscensis* [73, 18]. These Authors suggested the presence of *de novo*protein synthesis as a consequence of HHP stress [73]. As concerning HPH several interesting studies on the responses on *Lactobacillus* spp., at the level of proteolytic and metabolic activities point of view have been conducted [11, 21, 22, 74]. HPH treatment positively affects the proteolytic activity of some of *Lactobacillus* strains, but the activation and the quantitative and qualitative changes of the metabolic activity appear to be the most promising results. The pre-treatment at different pressure was able to induce relevant changes in term of fermentation dynamics and metabolism with respect to the untreated cells [11]. The same approach was applied on *L. acidophilus* and *L. paracasei* to improve the technological performances of probiotic strains [21, 22, 74]. The sub-lethal treatment with HPH enhanced the capacity of some *in vitro* probiotic features (i.e. hydrophobicity and tolerance to simulated gastric acidity) in a strain dependant way. *L. paracasei* A13 enhanced cellular hydrophobicity and auto-aggregation capacity after HPH treatment at 50 MPa. On the contrary, the HPH treatment decreased these features in the other strains considered. Highest values of hydrophobicity were found for *L. acidophilus* DRU and its bile-resistant derivative *L. acidophilus* DRU+, while lower values were obtained for *L. paracasei* strain [74]. Moreover, the stress responses enable survival under more severe conditions, enhancing resistance to subsequent processing conditions [75]. HPH treatment at 50 MPa can favour the maintenance of cell viability during a refrigerated storage in buttermilk, a suitable medium to maintain the cell viability during refrigeration [76]. The increased viability can be attributed to the increased precocious availability of low molecular weight peptides and free fatty acids such as oleic acid [21, 22].

Competition and communication: Food fermentations are typically carried out by mixed cultures consisting of multiple strains or species [77]. Mixed-culture food fermentations are of primary economic importance. The performance of these cultures, consisting of LAB, yeasts, and/or filamentous fungi, is not the simple result of "adding up" the individual single-strain functionalities, but is largely determined by interactions at the level of substrates, exchange of metabolites and growth factors or inhibiting compounds [77].

General microbial interference is an effective non-specific control mechanism common to all populations and environments including foods. It represents the inhibition of the growth of certain microorganisms by other members of the habitat.

The mechanisms involved are common to all genera and include [78]:

- Nutrient competition,
- Generation of unfavorable environment,
- Competition for attachment/adhesion sites.

Most substrates for food fermentations have a highly heterogeneous physicochemical composition, which offers the possibility for the simultaneous occupation of multiple niches by "specialized" strains, for instance, through the utilization of different carbon sources. In these substrates, coexisting strains often interact through trophic or nutritional relations via multiple mechanisms [77].

Carbon sources are often present at high concentrations in food substrates, and therefore competition concerns the rapid uptake of nutrients and conversion into biomass. In dairy fermentations nitrogen is limiting, and initially organisms compete for the free amino acids and small peptides available. While in the later stages of fermentation, they compete for the peptides released by the actions of proteolytic enzymes [77].

In a cell-density-dependent quorum-sensing system, bacteria produce extracellular signaling molecules such as peptides or post-translationally modified peptides that act as inducers for gene expression when concentrations of these molecules exceed a certain threshold value [79]. These changes might eventually lead to competitive advantages for the population, more effective adaptation and responses to changing environmental conditions, or the co-ordination of interactions between bacteria and their abiotic and biotic environments [7]. In fact, microorganisms produce diffusible chemicals for the purpose of communication and it has been reported that the stress caused by the exposure of microbial cells to their own cell free conditioned media, containing metabolites and bioactive compounds including "quorum sensing" molecules, including 2(5H)-furanones, promotes cell differentiation, autolysis and overproduction of specific metabolites [12, 80, 9, 10]. In this way the microbial cultures used in food fermentations can also contribute (by "secondary" reactions and relations) to the formation of flavor and texture [81].

GENERAL STEPS REGARDING A VIRTUAL FERMENTED FOOD PROCESS

In the figure 2, the steps that mainly interest food fermentation are reported. A model virtual fermented food was identified to resume the common denominator of the fermented foods dynamics, particularly focused on the reciprocal influences between environmental fluctuation and LAB fermentation.

Whatever kind of food we want to produce, fermented or not, the first step of the process is the formulation: in this phase the main raw materials (meat, milk, fruit and vegetables or their derivatives) are mixed with other ingredients, that have different roles: salts or sugars to improve taste, spices to give specific sensorial quality and as antimicrobials, additives or other substances able to affect physical and structural properties, preservatives to improve microbial stability and shelf life. The addition of those ingredients can be perceived as stress. In fermented products, proper microorganisms, mainly yeasts and LAB, are also added as starter cultures, in order to start and lead the fermentation and to obtain a stable and standard final product. As a consequence, the microorganisms, naturally occurring or added as starter cultures, have to cope with a completely different system: in particular, naturally occurring microflora have to face the changes induced by the ingredients, while the starter cultures, deriving from growth media or added as lyophilized cultures, have to adapt to a real food system, where different sources of stresses are often present.

In particular, the first sub-lethal stress, which LAB face, regards the difference between the growth medium composition and the real food. Generally, LAB lyophilized cultures can be added to the ingredients after a reactivation and subsequently added to the product. This procedure identify the presence of a stress for the LAB cells. Starter cultures are added to the raw materials in large numbers and incubated under optimal conditions, but the adaptation to substrate or raw material is always necessary [82]. It is very important to consider the physiological state of the LAB before the inoculum. This state strongly depends on the time of harvesting of the culture (whether during the logarithmic or stationary phase of growth), on the conditions leading to transition to the stationary phase, on the treatment of the culture during and after harvesting and on the chemical composition of the environment. Therefore it is important during formulation and technological processes to consider also these factors, mainly for those products where microorganisms are added as starter cultures.

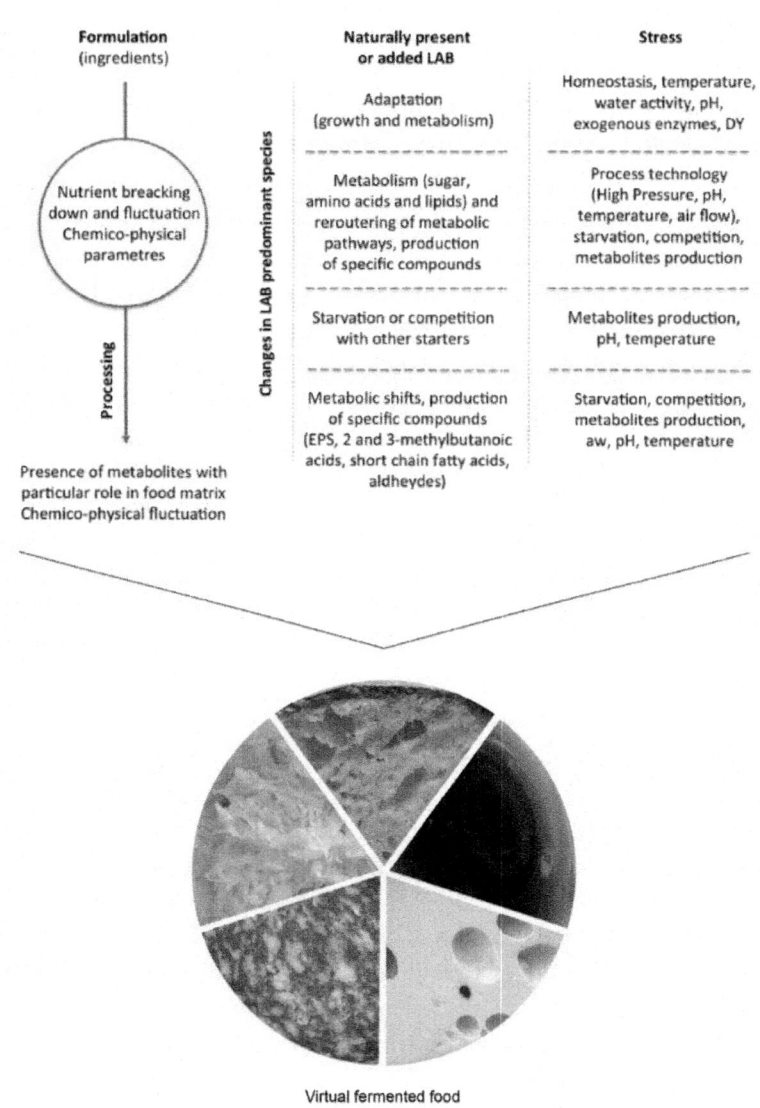

Figure 2: Fermented food model: reciprocal influences between environmental fluctuation and lactic acid bacteria fermentation.

The interaction between the starters and the ingredients and between the starters and the naturally present microbial population can trigger few important mechanisms that will influence the quality and the characteristics of the fermented product. Analogously, many food processes and formulations have been tested for safety by challenge test inoculating pathogen bacterial cells at different growth phases, and the results proved that cells grown to the stationary phase or adapted to various stresses have greater resistance than exponential cells [83].

Other ingredients usually added to obtain safe and stable products are food preservatives, including:

- Antioxidants,
- Anti-browning agents
- Antimicrobials.

These latter are arbitrarily classified into two groups: traditional or "regulatory approved" and naturally occurring [84]. The former includes acidifiers such as acetic acid, lactic acid and citric acid and antimicrobials such as benzoic acid and benzoates, propionate, nitrites and nitrates, sorbic acid and sorbates and sulfites. The latter includes compounds from microbial, plant and animal sources that are, for the most part, only proposed for use in foods as antimicrobials (e.g. lactoferrin, lysozyme, nisin). Throughout the ages, food antimicrobials have been used primarily to prolong shelf-life and preserve quality of foods through inhibition of spoilage microorganisms, while only few are used exclusively to control the growth of specific foodborne pathogens (e.g. nitrite, used for hundreds of years to inhibit growth and toxin production of *Clostridium botulinum* in cured meats). In food formulation antimicrobials are part of a multiple intervention system that involves the chemical along with environmental (extrinsic) and food related (intrinsic) stresses and processing steps. Some of these substances (for example lactic acid and citric acid) provoke a direct acidification of a food or food ingredient, and therefore challenge the microflora inducing and increase of acid resistance of the microflora itself. In fermented food the situation can be somewhat different, because the pH is gradually lowered by LAB creating a pH gradient, more likely than a sharp alteration in the pH due to direct acidification.

A good model describing the shock related to the inoculum of LAB in the raw complex material has been described during the production of fermented sausages [85]. The relatively high pH of raw meat rapidly decreases during the initial fermentation phase because organic acids, mainly lactate, are formed by LAB and the water activity is reduced during ripening, because of the addition of salt as well as drying. Furthermore, adjuvants, such as potassium or sodium nitrite and/or nitrate, are mostly added to optimize the fermentation process.

Generally strains used as starter cultures must tolerate these kinds of stresses and exhibit a high ecologic performance in the stressful food environment.

Genes related to stress response are induced when *L. sakei* is inoculated in the raw meat system [86]. In fact, ctsR, a gene that coded for a class III heat shock proteins repressor associated with the environmental stress response of Gram positive bacteria, increased its expression when *L. sakei* starts to adapt to the raw environment. This mechanism demonstrated that the sudden changes in the environment conditions are perceived as stress by*Lactobacillus* species. In particular, in the case of *L. sakei*, added to raw meat and spices, the principal stress response regarded high osmolarity and temperature shifts. Moreover, the presence of curing salt is regarded as one of the major hurdles in the initial phase of sausages fermentation. Because nitrite was found to be the effective for growth inhibition of pathogens, nitrite was also hypothesized as a stressor for *L. sakei* [85] and the exposure of this strain to stresses can induce changes in metabolic activities in a food environment [4]. The metabolic changes in *L. sakei* resulted in enhanced exploitation of available nutrients or increased activity of glycolytic enzymes, leading to the accelerated production of lactic acid by stress-treated *L. sakei* cells [85]. However, the exposition of *L. sakei* to low temperature and high osmolarity gives rise to the repression of phosphofructokinase and consequently to a decreased flux through the glycolytic pathway [87].

Moreover, it is important to consider that some ingredients can be also antimicrobials because of their own characteristics: in fact, if the recipe includes herbs and spices (aromatic plants, pepper), garlic and onions, an effect on microorganisms can be exerted by specific compounds characterizing these products, such as essential oils, terpenes and sulfur compounds [88].

Another essential aspect affecting the performances and metabolism of LAB are the intrinsic characteristics of raw materials that sometimes act in a synergic way with other ingredients. Considering for example fermented vegetables, the microflora of the starting fresh vegetables is typically dominated by Gram negative aerobic bacteria and yeasts, while LAB make up a minor portion of the initial population [89] and therefore they would not be able to start and lead a fermentation process. However, if anaerobic conditions are settled and salts are added, LAB can have a competitive advantage and induce spontaneous lactic acid fermentation. The growth of specific LAB is dependent on the chemical (substrate, salt concentration, pH) and physical (vegetable type, temperature) environments. As the environments change during fermentation, so can the dominant organisms, often leading to a specific and reproducible succession of bacteria.

In sauerkraut [89, 90] the presence of 1.8-2.2% of NaCl and a temperature of 18°C inhibits many strains of LAB, with the exception of *Leuconostoc mesenteroides* that initiates the fermentation; however this species is sensitive to acid conditions, so after a few days, when the concentration of lactic acid increases, *L. mesenteroides* is replaced by more acid resistant LAB such as *Lactobacillus brevis* and *L. plantarum*, able to further lower the pH up to 3-3.5, stabilizing the final product.

Considering olives fermentation is possible to outline the characteristics of the product affecting LAB: while the brine provides a good environment for LAB growth, with glucose, fructose and mannitol as the main source of fermentable sugars, the presence of high levels phenols (such as oleuropein) exert an antimicrobial activity, inhibiting some strains and selecting the types of organisms that predominate during the fermentation [91-93]. These LAB have to be resistant not only to phenols, but also to lye treatments and water washes, that can be performed during the processing and increase the initial pH, reducing also the nutrients content on the olive surface. The species able to face these kind of stresses usually belong to the genera *Pediococcus*, *Leuconostoc* and *Lactococcus*; after the first stage of fermentation, when the pH reaches 6, *L. plantarum* rapidly grows and dominate the fermentation, that goes on until the fermentable sugars are depleted. The viability and vigor of *L. plantarum* can be encouraged also by yeasts that are still present in this stage of fermentation and can produce vitamins [94].

Moreover, the presence of some gases can modify the growth performances of LAB. That is also influenced by the mixing step of the ingredients in some food processes (e.g. dough mixing). In fact, in bread making process, the continuous agitation of the dough can increase the microbes exposure to oxygen, and this can be a source of oxidative stress, mainly for LAB that are usually anaerobic or facultative anaerobic. Also in these cases the bacteria can react in different ways, activating metabolic and transcriptional responses in order to detoxify ROS, as previously described.

For the fermented vegetables, above reported, the rapid consumption of oxygen due to the presence of yeasts and aerobic bacteria in the first stage of fermentation has a positive effect on LAB. In fact, they are exposed only for a short time to oxidative stress and, due to their competitive advantage, they rapidly and intensively grow in the food system.

After formulation, the technological processes involving LAB include a fermentation process. It is reported that various beneficial phenotypic traits of LAB in food fermentations such as rapid acidification, selective proteolysis, tolerance of osmotic and stresses, resistance to ROS, and ability to thrive in nutrient poor conditions and at low temperatures are influenced by stress responses in various species of LAB [95, 96]. The knowledge of these mechanisms, and mainly of the stress responses activated by the fermentation process parameters can be useful in order to develop strains with optimal fermentation characteristics [83]. The first metabolic reaction regards the oxidation of carbohydrates (this reaction depends on the hetero-fermentative or homo-fermentative species involved) that give rise to acids, alcohols and CO_2. These metabolites are directly involved in flavor, aroma and texture of the product and in a second time can influence the production and the availability of other metabolites such as vitamins and antioxidant compounds [78]. Moreover, the LAB interactions with the ingredients increase also the digestibility and decrease the glycemic index, enhancing the healthy features of the fermented foods [97]. At the same time with carbohydrates oxidation, other metabolic mechanisms interest LAB cells such as proteolysis and lipolysis. The first reaction produces polypeptides with interesting characteristics as antimicrobial compounds, salt substitutes (the oligopeptides are able to increase the palatability of the system), and amino acids deriving aromatic compounds. On the other hand lipolysis produces medium chain fatty acids, with important antimicrobial properties. All these reactions (carbohydrates oxidation, lipolysis and proteolysis) generate precursors for other mechanisms in the cells and in the food matrix that give rise to the dynamic environment characteristics of fermented foods. It is important to outline that the compounds produced by the cells, metabolizing the substrate, can modify the system, producing also compounds that can stimulate the growth of symbiotic species or inhibit the growth of antagonistic microorganisms.

The conversion of carbohydrates to metabolites as acetic acid, lactic acid or CO_2 implies the acidification of the system. The contemporary pH decrease and the presence of sugar (osmotic stress) stimulate the exopolysaccharides (EPSs) production. In fact, in sourdough EPSs can be involved in acid tolerance of sourdough LAB [98]. EPSs are long-chain polysaccharides consisting of branched, repeating units of sugars or sugar derivatives. These sugar units are mainly glucose, galactose and rhamnose, in different ratios [99]. The presence of EPSs in the system can create a novel stress to the cells. The inclusion of cells within biofilm can increase their resistance to unfavorable environmental factors such as extreme temperature, low pH and osmolarity, the changes in the texture can induce in LAB also specific stress responses.

For example in yogurt production, the acidification by LAB implies proteins coagulation and thereby changes in the viscosity of the milk. In *L. bulgaricus*, during the acid adaptation present in the fermentation milk to obtain yogurt, some cellular changes were observed: the chaperones GroES, GroEL, HrcA, GrpE, DnaK, DnaJ, ClpE, ClpP and ClpL were induced and ClpC was repressed [100]. Some genes involved in the biosynthesis of fatty acids were induced (*fabH, accC, fabI*), while the genes involved in the mevalonate pathway of isoprenoid synthesis (*mvaC, mvaS*) were repressed [101, 102]. The changes in Aw value are depending not only on EPSs production by LAB after the exposition to acidic and osmotic stress, but also on the ingredients composition and on the step of fermentation.

Considering cheese, the Aw decreases during manufacture and ripening as a result of dehydration, salting, and production of water-soluble solutes from glycolysis, proteolysis, and lipolysis; the cheese Aw values range from 0.70 for extra hard cheeses to 0.99 for fresh, soft cheeses, such as cottage cheese, while semi-hard cheeses have Aw values of around 0.90. The cheese pH also decreases during manufacture and ripening [103]. The effects of different Aw and pH on *L. lactis* simulating cheese ripening have been analyzed [103]. The results evidenced that at low Aw, particularly at low pH, the growth and lactose utilization rates decreased and lactose fermentation to L-(1)-lactate switched to a pathway involving nontraditional saccharide products rather than the traditional lactococcal heterofermentative products.

In *L. plantarum* WCFS1 the addition of 300 mM and 800 mM of NaCl induced mild osmotic stress and osmotic stress respectively. In the presence of 800 mM of NaCl several genes showed an increased expression with respect to the control culture. In particular, those genes were associated with various stress responses in prokariotes, i.e. genes encoding Clp protease, an excinuclease, catalase (peroxide stress) and Dpr-like protein (peroxide stress). These differences in the gene expression were also identified in the presence of acid stress. These results suggest that lactic acid stress in *L. plantarum*WCFS1 also induces a more general stress response (as above described for different *Lactobacillus*species). An overlap between the stimulus for lactic acid and those for peroxide and UV radiation has also been reported for *L. lactis* [104, 66]. The response of *L. sanfranciscensis* to osmotic stress (saccarose 40%) gives rise to the overproduction of 3-methylbutanoic acid and gamma-decalactones when *L. sanfranciscensis* was co-inoculated with yeasts, simulating a sourdough environment. The production of lactones can be indicated as unfavourable environment for microbial growth and metabolism. In fact, these compounds have both particular aromatic and antimicrobial features [13].

The ability of the target strains to dominate the fermentation is related not only to the ingredients (as above described), but also to the fermentation conditions, mainly temperature and atmosphere. If the fermentation is not performed at the optimal growth temperature for the microorganisms, they could be unable to compete with naturally occurring microflora, and consequently the whole process could be compromised. On the contrary, some microbial species have developed specific thermal resistance mechanisms, and they can easily adapt to these unfavorable conditions without implications for the fermentation processes. Moreover, the adaptation to thermal stresses often leads to tolerance to other stresses, in a mechanism usually define "cross protection", as reported for *L. lactis* [105]. The ability of commercial *L. lactis* ssp. *lactis* and *L. lactis* ssp. *cremoris* to withstand freezing at –60°C for 24 h was significantly improved by a prior 25 min heat shock at ~40°C or by a 2 h cold shock at 10°C, opening interesting perspectives for the production on resistant starter cultures, both frozen or lyophilized [105].

Other Authors with regard to different stresses reported the "cross protection" mechanism: for example the mechanisms of multiple adaptations to hops of two different strains of *L. brevis* have been characterized [106]. Hop resistance of lactobacilli requires multiple resistance mechanisms. This is consistent with the stress conditions acting on bacteria in beer, which mainly consist of acid stress and the antimicrobial effect of the hop compounds, in addition to ethanol stress and starvation. The effect of interaction of acid stress and presence/absence of oxygen in the system on *L. helveticus* and *L. sanfranciscensis*, in particular on their cell membrane composition, has been reported [14]. Upon acid stress the level of cyclopropane fatty acids increased at the expense of the level of long-chain unsaturated fatty acids. *L. helveticus* and *L. sanfranciscensis*, exposed to acid sub lethal stress demonstrated the same increase in cyclopropane fatty acids. In particular, *L. helveticus* presented higher concentration of C19cyc11 at pH 4 and pH 3, while *L. sanfranciscensis* presented more C19cyc9 at pH 3 in microaerophilic condition without tween 80, at pH 3.6 in anaerobiosis with tween 80, and at pH 4 in anaerobiosis without tween 80. These results demonstrated the same behavior in front of multiple stresses by LAB membrane [106, 14]

Consider the atmosphere, i.e. the presence or not of oxygen, as another important variable during fermentation, it is known that oxygen can inhibit the growth of LAB, especially in the first stages. However, the food system is usually a consortium of different microorganisms: for example in bakery products and in fermented sausages the fermentation is carried out both by yeasts and LAB; the formers can therefore consume the amount of oxygen present in the mix, allowing the growth of LAB. The same thing happens for fermented vegetables, where naturally occurring Gram negative bacteria and yeast rapidly remove the oxygen, promoting the rapid predominance of Lactobacilli. Some secondary metabolites such as bacteriocins can play a role in LAB performances and metabolism, affecting also the total population and ecology of fermented foods [107, 108]. Bacteriocins are antimicrobial peptides or proteins produced by bacteria that can be active on different microorganisms, depending on their structure. LAB belonging to the genera *Lactococcus, Pediococcus, Lactobacillus,Leuconostoc, Carnobacterium, Propionibacterium* are known to produce bacteriocins with both narrow and broad inhibitory spectra [109]. The use of functional LAB starter cultures (eg. bacteriocinogenic starter cultures), well adapted to the environment and the process conditions applied, may contribute to the development of better controllable and more efficient production processes [110]. An example can be nisin, a peptide produced by *L. lactis* ssp. *lactis*, that has a narrow spectrum affecting primarily only Gram-positive bacteria and their spores, including lactic acid bacteria, *Bacillus*, *Clostridium*, *Listeria*, and *Streptococcus*. However some LAB such as *Streptococcus thermophilus* and *L. plantarum* are able to produce the enzyme nisinase, which neutralizes the antimicrobial activity of the peptide [111]. Therefore these LAB could be suitable for a co-fermentation with *L. lactis*.

Another interesting case of bacteriocin production, as a consequence of oxidative stress and carbon dioxide exposure, has been reported [110]: oxidative stress and carbon dioxide are involved in the production of a specific bacteriocin, amylovorin L, by *Lactobacillus amylovorus*, able to inhibit other LAB species. During traditional sourdough fermentation, a decrease in redox potential of the rather firm mixture occurs. The oxygen initially present is consumed by *Candida* spp. or converted into hydrogen peroxide or water, thereby creating microaerophilic or anaerobic environment in which the growth of the desired LAB is favored. While in a large-scale sourdough type II fermentation currently the use of dough mixture with high dough yield is exploited. This sourdough has to be stirred to liberate part of the carbon dioxide produced to prevent running over. During mixing, oxygen is incorporated into the dough. Also, the development of yeast and hence the production of carbon dioxide is favored in continuously stirred sough mixtures with high water content. Elevation of the airflow rates leading to oxidative stress conditions resulted in an enhanced specific amylovorin L production. Growth in the presence of carbon dioxide also increased the specific bacteriocin production. Mild aeration or a controlled supply of oxygen as well as growth in an environment containing high amounts of carbon dioxide might thus contribute to the competitiveness of *L. amylovorus* DCE471 in a sourdough ecosystem [110]. The production of plantaricin A by *L. plantarum* was also demonstrated in relation to a quorum sensing mechanism [79].

Another example of the influence of the process on LAB metabolism has been widely described [112]. These Authors monitored the evolution of the gene expression of *L. plantarum* IMDO 130201 during a sourdough process. In particular, the genes and the metabolites related to acidic stress were analyzed. It is interesting to highlight that during the pH decrease (production of lactic acid by *L. plantarum*) the genes coding for plantaricin production had higher levels of expression at low pH values, indicating that the bacteriocin production was activated under acid stress conditions by *L. plantarum* IMDO 130201 strain. The presence of the pheromone plantaricin A (PlnA) in a system inoculated with *L. plantarum*DC400 was also reported [79]. Biosynthesis of PlnA was variously stimulated depending on the microbial partner. In fact, *L. sanfranciscensis* DPPMA174 induced the highest synthesis of PlnA, which, in turn, determined lethal conditions for it. The proteome of *L. sanfranciscensis* DPPMA174 responded to the presence of PlnA. The up-regulation of 31 proteins related to stress response, amino acid metabolism, energy metabolism, membrane transport, nucleotide metabolism, regulation of transcription and cell redox homeostasis was found. At the same time, other proteins such as cell division protein (FtsZ), glutathione reductase (LRH_11212) and response regulator (rrp11) were down-regulated. These results demonstrated a hypothetically and interesting waterfall of events all related with stresses response and with the typical fermentation products dynamics (Figure 3). At the same time, the low pH values implied a poor expression of the genes involved in carbohydrate degradation in*L. plantarum* IMDO 130201. The bacterium was directed toward survival at low pH by amino acid conversions rather than by relying on growth [112]. The same behavior was identified in *L. sanfranciscensis* LSCE1 response to pH 3.6 [15]. Under the adopted experimental conditions, which did not produce any decrease in viability of *L. sanfranciscensis* LSCE1, the acid stress, within 2 h, was accompanied by a reduction of the carbohydrate metabolism, as shown by the decrease of ethanol, acetate, and lactate. This mechanism suggests the existence of a switch from sugar to amino acid catabolism that supports survival and growth also in specific and restricted environments, such as sourdoughs, characterized by acid stress and recurrent carbon starvation. Under the acid conditions (pH 3.6) and in the presence of specific nutrients 3-methylbutanoic acid was the predominant metabolite among those detected by solid phase micro-extraction gas chromatographic analysis and mass spectrometry (GC-MS-SPME), released after 2 h of acid stress exposure [15]. The acid stress implied less carbohydrate utilization and ethanol, lactate, and acetate production, but high amino acids catabolism that confers a different and characteristic metabolites pattern. Stress resistance assume great importance as one of the adaptation factors to gastrointestinal tract of probiotic strains as reported in a detailed review [113].

STRESS RESISTANCE OF PROBIOTIC LAB

There are two main categories of factors that contribute to the optimal functioning of probiotic lactobacilli: factors that allow optimal adaptation to the new niches that they temporarily encounter in the host (adaptation factors) and factors that directly contribute to the health-promoting effects (probiotic factors) [113].

Adaptation factors include stress resistance, active metabolism adapted to the host environment, and adherence to the intestinal mucosa and mucus.

In fact, probiotic lactobacilli encounter various environmental conditions upon ingestion by the host and during transit in the gastro intestinal tract (GIT). They need to survive to: 1) the harsh conditions of the stomach secretion generating a fasting pH of 1.5, increasing to pH 3 to 5 during food intake; 2) the bile excreted by liver in small intestine represents another challenge for bacteria entering the GIT. Bile salts also seem to induce an intracellular acidification so that many resistance mechanisms are common for bile and acid stress. Indeed, the protonated form of

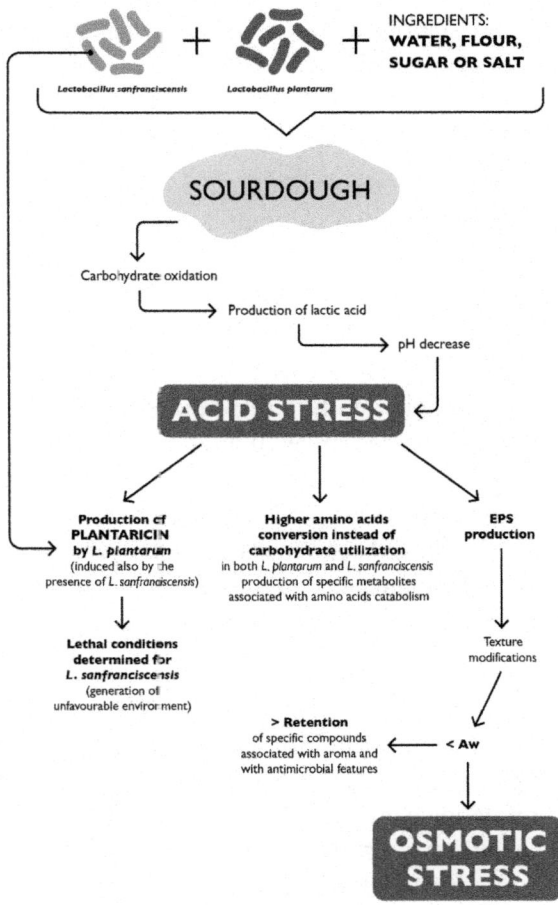

Figure 3: Sourdough fermentation dynamics. Case of possible parallel phenomena interesting acid and osmotic stress.

bile salts is thought to exhibit toxicity through intracellular acidification in a manner similar to those of organic acids like the lactic acid produced by the lactobacilli themselves. For a detailed overview of acid, bile, and other stress resistance mechanisms of lactobacilli, the reader is referred to more extensive review [113]. 3) In analogy to the stresses encountered by intestinal pathogens, they also encounter oxidative and osmotic stress in GI tract. 4) Interactions with other microbes and 5) Interactions with cells of the host immune system and the various antimicrobial products that they produce can also impose a serious threat for the probiotic microbes. Analogously to what described in food LAB, the phenomenon of cross-adaptation is often observed, i.e., that adaptation to one stress condition also protects against another stress factor, implying some common mechanisms. In this respect, also for probiotic LAB non-actively-growing stationary-phase cells are generally more resistant to various stressors than early-log-phase cells.

Maintaining Integrity of the Cell Envelope

The different macromolecules constituting the cell membranes and cell walls of lactobacilli have been shown to contribute to maintaining cell integrity during stress to various degrees. For example, low pH caused a shift in the fatty acid composition of the cell membrane of an oral strain of *L. casei*. Similarly, bile salts have been shown to induce changes in the lipid cell membrane of *Lactobacillus reuteri*CRL1098.

The role of EPS in acid and bile resistance is less clear. However, EPS production has not been studied in detail after exposure to bile. In fact, to our knowledge, phenotypic analyses of dedicated*Lactobacillus* mutants affected in EPS biosynthesis genes have not yet been performed. Homopolysaccharides (HoPSs) from *L. reuteri* have been reported to have a more established role in stress resistance by the maintenance of the cell membrane in the physiological liquid crystalline phase under adverse conditions.

Repair and Protection of DNA and Proteins

A number of proteins that play a role in the protection or repair of macromolecules such as DNA and proteins also seem to be essential for acid and bile resistance. Intracellular acidification can result in a loss of purines and pyrimidines from DNA. Bile acids have also been shown to induce DNA damage and the activation of enzymes involved in DNA repair. Perhaps even more vital in the general stress response are chaperones that intervene in numerous stresses for important tasks such as protein folding, renaturation, protection of denatured proteins, and removal of damaged proteins.

Two-Component and Other Regulatory Systems

Mechanisms to specifically sense the presence of certain stress factors and regulate gene expression in response to these stimuli are also crucial for bacterial survival under adverse conditions. Although these mechanisms are not well characterized for lactobacilli, they often involve two-component regulatory systems (2CRSs). 2CRSs allow bacteria to sense and respond to changes in their environment after receiving an environmental signal through transmembrane sensing domains of the histidine protein kinase (HPK).

METHODOLOGICAL APPROACHES TO STUDY THE EFFECTS OF STRESS ON LAB

The study of stress responses by LAB is getting closer and closer to the different "omic" fields: genomic, proteomic and metabolomic. Other traditional approaches regarding the membrane cells composition and modifications, both from a structural (cellular fatty acids composition by gas-chromatographic method) and morphological (membrane and wall modification by electronic microscopy) point of view are still used.

Genes implicated in LAB stress responses are numerous and the levels of characterization of their actual role and regulation differ widely between species. The studies concerning stress responses in LAB sometimes benefit from the knowledge already acquired in other bacteria. For example, parts of the studies on heat response have been focused on specific genes because of their major role demonstrated in other microorganisms [17]. The cheapest and easiest way to study a stress response in LAB is to follow some specific genes related to stresses such as heat shock, salts and acids [114, 115]. This type of study is useful especially if the entire genome sequence of some LAB is still unknown. However, nowadays the study of whole trascriptome (the total set of RNAs) is one of the most exhaustive ways to study modifications of gene expression as a result of a stress condition. The transcriptome of a cell contains information about the biological state of the cell and the genes that play a role under specific circumstances. The principal technique used to study the trascriptome is microarray [116].

DNA microarray technology has been used in numerous experiments to analyze gene expression: one example is the evaluation of the general stress response of *B. subtilis* [117] or the investigation of the transcription profiles of *L. plantarum* grown in steady-state cultures that varied in lactate/lactic acid concentration, pH, osmolarity [66, 104]. This approach is useful also to study the behaviour of bacteria in a real food system. Hüfner et al. [5] studied the global transcriptional response of *L. reuteri* to sourdough environment, showing a significant changes of mRNA levels for 101 genes involved in diverse cellular processes, from carbohydrate and energy metabolism, to cell envelope biosynthesis, exopolysaccharide production, stress responses, signal transduction and cobalamin biosynthesis.

The gene expression dynamics of *L. casei* during fermentation in soymilk when grown up to lag phase, late logarithmic phase, or stationary phase were also studied. Comparisons of different transcripts close to each other revealed 162 and 63 significantly induced genes, in the late logarithmic phase and stationary phase, whose expression was at least threefold up-regulated and down-regulated, respectively. Approximately 38.4% of the up-regulated genes were associated with amino acid transport and metabolism, followed by genes/gene clusters involved in carbohydrate transport and metabolism, lipid transport and metabolism, and inorganic ion transport and metabolism [118].

The study of trascriptome is a good approach that gives a good overview of the changes that can occur inside a stressed bacterium. A limitation of this technique is that it is expensive and requires that the genome sequences of the organisms under study should be available for designing the oligonucleotides for the microarray [119].

A different but, at the same time, related point of view regards the study of proteins and proteome. The most common method to obtain this information is to extract total proteins and separate them by a sodium dodecyl sulphate polyacrylamide gel electrophoresis (SDS-PAGE) followed by a western blotting (in the first case) or a two dimesional electrophoresis (2D-E) analysis (in the second case). Also in this case if the study is focused on a single protein, it is necessary to know before the characteristic of the target protein to optimize the analytical conditions. 2D-Electrophoresis can provide more than 10000 detectable protein spots in a single gel run. Thus, proteins with post-translational modifications (PTMs), such as processing, phosphorylation and glycosylation, can be easily detected as separate spots. A spot separated by 2D-E theoretically consists of an almost homogeneous protein, and thus can be identified following digestion with a sequence-specific protease by peptide mass fingerprinting (PMF) approaches, typically using matrix-assisted laser desorption ionization (MALDI)- time-of-flight (TOF) mass spectrometers. The same level of automation is also available for proteomic approaches involving tandem mass spectrometry (MS-MS) analysis, extremely useful when studying organisms with incomplete or partial genomic information [120].

This kind of approach was used to investigate the cell surface proteins of a typical strain of *L. casei* in response to acidic growth conditions [121]. They demonstrated that growth of *L. casei* under acidic conditions caused molecular changes at the cell surface in order to accomplish an adaptive strategy, resulting in slower growth at low pH. Moreover, the proteomic approach was useful to study the heat shock response respectively on *L. helveticus* PR4 and *L. plantarum* [26, 31]. The cold adaptation of *Lactococcus piscium* strain CNCM I-4031 was studied with the same approach [122]. This analysis could be also performed to compare the effects that new technologies produce on bacteria comparing with the normal stress conditions. In fact, the HHP stress response of *L. sanfranciscensis* was compared with cold, heat, salt, acid and starvation stresses responses [18].

Due to increasingly available bacterial genomes in databases, proteomic tools have recently been used to screen proteins expressed by microorganisms in food, in order to better understand their metabolism *in situ*. While up to now the main objective has been the systematic identification of proteins, the next step will be to bridge the gap between identification and quantification of these proteins [123]. Proteomics has also been used to analyse the proteins released during the ripening of Emmentaler cheese. In an innovative study, proteomics was used to prepare a reference map of the different groups of proteins found in cheese [124]. These authors were able to categorize these proteins into five classes: those involved in proteolysis, glycolysis, stress response, nucleotide repair and oxidation-reduction. In addition, information was obtained regarding the peptidases released into the cheese during ripening process. This study enabled the Authors to differentiate between the various casein degradation mechanisms present, and to suggest that the streptococci within the cheese matrix are involved in peptide degradation and together with the indigenous lactobacilli contribute to the ripening process. Using proteomics these Authors were able to get a greater understanding of the microbial succession involved in the ripening of Emmentaler cheese, which information could not have been obtained using other protein separation techniques. This example illustrates the power of proteomics as a tool for analyzing the composition of a complex mixture of proteins and peptides [119].

The global identification of stress-induced proteins in a given organism has technical limitations. Membrane proteins, for example, are rarely detected by this method. Secondly, it may be that changes in membrane proteins composition result from long-term adaptation processes, while short-term responses may primarily be accounted for the activation (and/or stabilization) of proteins already present. The latter hypothesis is valid especially in the case of transport systems, although for some of the systems studied a transcriptional induction has also been observed [17]. The use of this technique is not as widespread as that of DNA microarrays due to the challenges associated with the purification and separation of complex mixtures of proteins found in cell extracts. At the same time the study of the only transcriptome should take into consideration that a lot of post-transcriptional processes may act on RNA (ex. RNA interference, polyadenilation ecc) [125].

As reported above, the stress responses of LAB are studied also through the analysis of membrane composition, structure and integrity. Not unexpectedly, in fact, the cell membrane plays an important role in stress resistance. First of all, the membrane itself can change in adaptation to environmental conditions and these changes contribute to the protection of the bacteria [17]. The adaptive response to sub-lethal acid and cold stresses in *L. helveticus* and *L. sanfranciscensis* has been analyzed (as described above) [14]. The extraction and identification by GC-MS of lipid fatty acids and free fatty acids could give an overview of the membrane fluidity state. In the same article they developed a gas chromatographic method to separate and quantify the cell cyclopropane fatty acids lactobacillic (C19cyc11) and dehydrosterculic (C19cyc9) demonstrating different responses of the strains tested in terms of cyclopropane fatty acids production, probably due to the different original optimal environment. The comparison between the wild type and the acid-resistant mutant *L. casei* LBZ-2 evidenced in the latter higher membrane fluidity, higher proportions of unsaturated fatty acids, and higher medium chain length. In addition, cell integrity analysis showed that the mutant maintains a more intact cellular structure and lower membrane permeability after environmental acidification [126].

The last but not least approach used to study the stress response of LAB is the metabolic one. The study of the metabolites released, as a consequence of the stress exposure, can contribute to the understanding of the mechanisms that regulate the microbial interactions and the metabolic alterations induced by stress conditions. Moreover, these approaches can be exploited to identify which technological conditions induce microorganisms to produced desirable metabolites [4, 15].

With this perspective the use of GC-MS-SPME as a potent and easy tool to study the generation of volatile metabolite compounds such as flavoring molecules or aroma precursors was widely adopted [9,11-13, 15] and contributed to rationalize the process and optimize the products. In particular, the effects of HPH on different species of *Lactobacillus* involved in dairy product fermentation and ripening, monitoring the changes in volatile compounds as indicators of metabolic profiles has been studied [11]. Analysing the oxidative and heat stresses in *L. helveticus* two new 2[5H]-furanones released by this strain both as a possible signalling molecules and as possible important flavouring compounds has been identified by GC-MS-SPME [9]. On the contrary the study of non-volatile metabolites can be performed by normal chromatographic technique (HPLC), especially for amino acids and sugars [15], or by Fast Protein Liquid Chromatography (FPLC) separation for peptides, followed by a mass spectrometry identification [127]. An NMR approach to evaluate the effects on the growth of *L. plantarum* raising the medium molarity by high concentrations of KCl or NaCl and iso-osmotic concentrations of non-ionic compounds was performed [128].

Since all the techniques described above, if used alone, do not allow a total comprehension of stress responses, a lot of studies are trying to combine two or more approaches together. Combined transcriptomic and proteomic analyses were used to evaluate the glucose-limited chemo-stat in*Enterococcus faecalis* V583 [129] or to study the effect of bile salts in the growth of *L. casei* [130]. A combined physiological and proteomic approach, instead, was followed to unravel lactic-acid-induced alterations in *L. casei* [131].

Therefore it is possible to understand, from the references above, that techniques used to study the stress responses of LAB are taking more and more "omic" approach. This comports an accumulation of a huge number of data that it is not easy to manage and to compare. For this reason the use of new programs of data analysis is required. One of these approaches could be the use of heat maps, a technique born as a tool to understand microarray results [66]. Nowadays it could be useful also to manage the data from other fields: in fact, a heat maps was used to show the correlation between metabolites produced, the relative gene expression of specific genes and stress conditions [15]. The same useful tool, combined with other statistical analyses, has been also applied [132].

CONCLUSION

It is known that LAB can adapt to stress with different mechanisms widely studied in model and real systems. An overview of those responses has been described and reported in this chapter.

Stress not only induces changes enabling better survival, but also different performances in a system. In fermented food, the knowledge of the mechanisms that regulate LAB metabolic changes and their effects gain importance especially when those responses can be exploited in order to improve the food properties [4]. In particular, fermented foods are dynamic systems subjected to continuous evolution of their physico-chemical characteristics. The complex fluctuation of the food environment itself, during processing, is stress source for every microorganism involved and the changes that affect the fermented food habitats, can be perceived by LAB as stress.

In this chapter examples of the dynamic fluctuation effect on LAB metabolism have been described in order to outline that every reaction can cause a waterfall of metabolic events influencing the sensorial quality, the shelf-life and the bioactive compounds production of fermented foods.

The subjects of those events are LAB, indicating the importance of metabolism of these microorganisms in food. The cell physiology is crucial to ensure that cells are well suited to survival during downstream processes and that they exhibit high performances.

The production and exploitation of naturally adapted strains can be interesting for companies because of the absence of ethical and legal concerns. The adapted strains are not considered genetically modified microorganisms (GMOs) and therefore they can be applied in food processing without legal restrictions and, more important, without affecting the consumer perception, currently (in Europe) not ready to introduce in his diet foods produced with GMOs.

Individual stresses used in food processing and preservation may render probiotic LAB more resistant to further and different stresses, including those encountered in the human body, e.g. those encountered during gastro-intestinal passage (pH of the stomach, exposure to bile salts in small intestine etc.). A positive correlation has been recently observed between EPS production and resistance to bile salt and low pH stress in *Bifidobacterium* species isolated from breast milk and infant faeces [128].

This knowledge can open interesting perspectives to improve at the same time the performances of LAB, the quality of fermented food and the health-promoting properties of the LAB used.

Moreover, it will be interesting to identify the gastrointestinal tract also as a complex and dynamic system in which LAB need to adapt to adverse conditions, responding with metabolic shifts provided with interesting technological an healthy features.

The "omics" technologies could be particularly useful for identifying the mechanism leading to LAB stress responses. These approaches could also help to identify the mechanisms for cell fitness and stress adaptation that will be needed to develop more generic and science based technologies [7].

ACKNOWLEDGEMENT

We thank Prof.ssa Maria Elisabetta Guerzoni for her enormous scientific support and Luca Vagnini for his graphic abilities (http://www.lucavagnini.com).

REFERENCES

1. B Poolman, J Ruhdal, A Gruss, 2011LAB Physiology and energy metabolism. In: Ledeboer A, Hugenholtz J, Kok J, Konings W, Wouters J. Thirty years of research on lactic acid bacteria.Rotterdam: 24 Media Labs. 77101

2. A Ledeboer, J Hugenholtz, J Kok, W Konings, J Wouters, 2011In: Ledeboer A, Hugenholtz J, Kok J, Konings W, Wouters J. Thirty years of research on lactic acid bacteria.Rotterdam: 24 Media Labs. pp. v-vi.

3. J Marles-wright, R Lewis, 2007Stress response of bacteria. Curr. Opin. Struct. Biol. 17755760

4. D. I Serrazanetti, M. E Guerzoni, A Corsetti, R. F Vogel, 2009Metabolic impact and potential exploitation of the stress reactions in lactobacilli. Food Microbiol. 26700711

5. E Hüfner, R. A Britton, S Roos, H Jonsson, C Hertel, 2008Global transcriptional response of Lactobacillus reuteri to the sourdough environment. Syst. Appl. Microbiol. 31323338

6. K. J Heller, 2001Probiotic bacteria in fermented foods: product characteristics and starter organisms. Am. J. Clin. Nutr. 73:374S-379S.

7. C Lacroix, S Yildirim, 2007Fermentation technologies for the production of probiotics with high viability and functionality. Curr. Opin. Biotechnol. 18176183

8. D. I Serrazanetti, 2009Effects of acidic and osmotic stresses on flavor compounds and gene expression in Lactobacillus sanfranciscensis. Ph.D. thesis. University of Teramo, Teramo, Italy.

9. M Ndagijimana, M Vallicelli, P. S Cocconcelli, F Cappa, F Patrignani, R Lanciotti, M. E Guerzoni, 2006H Two, furanones as possible signaling molecules in Lactobacillus helveticus. Appl. Environ. Microbiol. 7260536061

10. L Vannini, M Ndagijimana, P Saracino, P Vernocchi, A Corsetti, M Vallicelli, F Cappa, P. S Cocconcelli, M. E Guerzoni, 2007New signaling molecules in some Gram-positive and Gram-negative bacteria. Int. J. Food Microbiol. 1202533

11. R Lanciotti, F Patrignani, L Iucci, P Saracino, M. E Guerzoni, 2007Potential of high-pressure homogenization in the control and enhancement of proteolytic and fermentative activities of some Lactobacillus species. Food Chem. 102542550

12. M. E Guerzoni, P Vernocchi, M Ndagijimana, A Gianotti, R Lanciotti, 2007Generation of aroma compounds in sourdough: effects of stress

exposure and lactobacilli-yeasts interactions. Food Microbiol. 24139148

13. P Vernocchi, M Ndagijimana, D. I Serrazanetti, A Gianotti, M Vallicelli, M. E Guerzoni, 2008Influence of starch addition and dough microstructure on fermentation aroma production by yeasts and lactobacilli. Food Chem. 10812171225

14. C Montanari, S. L Sado-kamdem, D. I Serrazanetti, F. X Etoa, M. E Guerzoni, 2010Synthesis of cyclopropane fatty acids in Lactobacillus helveticusand Lactobacillus sanfranciscensis and their cellular fatty acids changes following short term acid and cold stresses. Food Microbiol. 27493502

15. D. I Serrazanetti, M Ndagijimana, S. L Sado, A Corsetti, R. F Vogel, M Ehrmann, M. E Guerzoni, 2011Acid stress-mediated metabolic shift inLactobacillus sanfranciscensis LSCE1. Appl.Environ. Microbiol. 7726562666

16. T. R Klaenhammer, R Barrangou, B. L Buck, M. A Azcarate-peril, E Altermann, 2005Genomic features of lactic acid bacteria effecting bioprocessing and health. FEMS Microbiol. Rev. 29393409

17. van de Guchte MSerror P, Chervaux C, Smokvina T, Ehrlich SD, Maguin E (2002Stress responses in lactic acid bacteria. Antonie Leeuwenhoek. 82187216

18. S Hörmann, C Scheyhing, J Behr, M Pavlovic, M Ehrmann, R. F Vogel, 2006Comparative proteome approach to characterize the high-pressure stress response of Lactobacillus sanfranciscensis DSM 20451T. Proteomics. 618781885

19. M. E Guerzoni, R Lanciotti, P. S Cocconcelli, 2001Alteration in cellular fatty acid composition as a response to salt, acid, oxidative and thermal stresses in Lactobacillus helveticus. Microbiol. 14722552264

20. M Pavlovic, S Hörmann, R. F Vogel, M. A Ehrmann, 2008Characterisation of a piezotolerant mutant of Lactobacillus sanfranciscensis. Z. Naturforsch. 63791797

21. P Burns, F Patrignani, D. I Serrazanetti, G. C Vinderola, J. A Reinheimer, R Lanciotti, M. E Guerzoni, 2008Probiotic crescenza cheese containingLactobacillus casei and Lactobacillus acidophilus manufactured with High Pressure-Homogenized Milk. J Dairy Sci. 91500512

22. F Patrignani, P Burns, D. I Serrazanetti, G Vinderola, J Reinheimer, R Lanciotti, M. E Guerzoni, 2009Suitability of high pressure-homogenized milk for the production of probiotic fermented milk containing Lactobacillus paracasei and Lactobacillus acidophilus. J. Dairy Res. 519

23. G Gouesbert, G Jan, P Boyaval, 2002Two-dimensional electrophoresis study of Lactobacillus delbrueckii subsp. bulgaricus thermotolerance. Appl. Environ. Microbiol. 6810551063

24. C Desmond, G. F Fitzgerald, C Stanton, R. P Ross, 2004Improved stress tolerance of GroESL-overproducing Lactococcus lactis and probioticLactobacillus paracasei NFBC 338. Appl. Environ. Microbiol. 7059295936

25. B. M Corcoran, R. P Ross, G. F Fitzgerald, P Dockery, C Stanton, 2006Enhanced survival of GroESL-overproducing Lactobacillus paracasei NFBC 338 under stressful conditions induced by drying. Appl. Environ. Microbiol. 7251045107

26. Di Cagno RDe Angelis M, Limitone A, Fox PF, Gobbetti M (2006Response of Lactobacillus helveticus PR4 to heat stress during propagation in cheese whey with a gradient of decreasing temperatures. Appl. Environ. Microbiol. 7245034514

27. J. M Laplace, N Sauvageot, A Harke, Y Auffray, 1999Characterization of Lactobacillus collinoides response to heat, acid and ethanol treatments. Appl. Microbiol. Biotechnol. 51659663

28. G Schmidt, C Hertel, W. P Hammes, 1999Molecular characterisation of the dnaK operon of Lactobacillus sakei LTH681. Syst. Appl. Microbiol. 22321328

29. R Zink, C Walker, G Schmidt, M Elli, D Pridmore, R Reniero, 2000Impact of multiple stress factors on the survival of dairy lactobacilli. Sci. Aliment. 20119126

30. J Prasad, P Mcjarrow, P Gopal, 2003Heat and osmotic stress responses of probiotic Lactobacillus rhamnosus HN001 (DR20) in relation to viability after drying. Appl. Environ. Microbiol. 69917925

31. M De Angelis, M Gobbetti, 2004Environmental stress responses in Lactobacillus: a review. Proteom. 4106122

32. A Bucio, R Hartemink, J. W Schrama, J Verreth, F. M Rombouts, 2005Survival of Lactobacillus plantarum 44a after spraying and drying in feed and during exposure to gastrointestinal tract fluids in vitro. J. Gen. Appl. Microbiol. 51221227

33. C Castaldo, R. A Siciliano, L Muscariello, R Marasco, M Sacco, 2006CcpA affects expression of the groESL and dnaK operons in Lactobacillus plantarum. Microb. Cell Fact. 5:35.

34. G. E Gardiner, O Sullivan, E Kelly, J, Auty MAE, Fitzgerald GF, Collins JK, Ross RP, Stanton C (2000Comparative survival rates of human-

derived probiotic Lactobacillus paracasei and L. salivarius strains during heat treatment and spray drying. Appl. Environ. Microbiol. 6626052612

35.   R. G Earnshaw, J Appleyard, R. M Hurst, 1995Understanding physical inactivation processes: combined preservation opportunities using heat, ultrasound and pressure. Int. J. Food Microbiol. 28197219

36.   P Teixeira, H Castro, C Mohacsi-farkas, R. J Kirby, 1997Identification of sites of injury in Lactobacillus bulgaricus during heat stress. Appl. Microbiol. 83219226

37.   P. J Hansen, M Drost, R. M Rivera, F. F Paula-lopes, A. I-K. a. t. a. n. a. n. i. t Y. M Krininger, C. E Chase, CC (2001Adverse impact of heat stress on embryo production: causes and strategies for mitigation. Theriogenol. 5591103

38.   M Kilstrup, S Jacobsen, K Hammer, F. K Vogensen, 1997Induction of heat shock proteins DnaK, GroEL, and GroES by salt stress in Lactococcus lactis. Appl. Environ. Microbiol. 6318261837

39.   Kim, , Dunn NW (1997) Identification of a cold shock gene in lactic acid bacteria and the effect of cold shock on cryotolerance. Curr. Microbiol. 35:59-63.

40.   Kim, , Khunajakr N, Dunn NW (1998) Effect of cold shock on protein synthesis and on cryotolerance of cells frozen for long periods in Lactococcus lactis. Cryobiol. 37:86-91.

41.   J. A Wouters, B Jeynov, F. M Rombouts, W. M De Vos, O. P Kuipers, T Abee, 1999Analysis of the role of 7 kDa cold-shock proteins of Lactococcus lactis MG1363 in cryoprotection. Microbiol. 14531853194

42.   S Derzelle, B Hallet, K. P Francis, T Ferain, J Delcour, P Hols, 2000Changes in cspL, cspP, and cspC mRNA abundance as a function of cold shock and growth phase in Lactobacillus plantarum. J. Bacteriol. 18251055113

43.   F Fonseca, C Béal, G Corrieu, 2001Operating Conditions That Affect the Resistance of Lactic Acid Bacteria to Freezing and Frozen Storage. Cryobiol. 43189198

44.   G Zhang, M Fan, Y Li, P Wang, Q Lv, 2012Effect of growth phase, protective agents, rehydration media and stress pretreatments on viability ofOenococcus oeni subjected to freeze-drying Afr. J Microbiol. Res. 614781484

45.   L Bâati, C Fabre-gea, D Auriol, P. J Blanc, 2000Study of the cryotolerance of Lactobacillus acidophilus: effect of culture and freezing conditions on the viability and cellular protein levels. Int. J. Food. Microb. 59241247

46. J. M Panoff, B Thammavongs, M Guéguen, 2000Cryotolerance and cold stress in lactic acid bacteria. Sci. Aliment. 20105110

47. J. M Panoff, B Thammavongs, J. M Laplace, A Hartke, P Boutibonnes, Y Auffray, 1995Cryotolerance and Cold Adaptation in Lactococcus lactissubsp. lactis IL1403. Cryobiol. 32516520

48. D. W Grogan, J. E Cronan, 1986Characterization of Escherichia coli mutants completely defective in synthesis of cyclopropane fatty acids. J. Bacteriol. 166:872.

49. D Grogan, J Cronan, 1997Cyclopropane ring formation in membrane lipids of bacteria. Microbiol. Mol. Biol. Rev. 61429441

50. S Raynaud, R Perrin, M Cocaign-bousquet, P Loubiere, 2005Metabolic and Transcriptomic Adaptation of Lactococcus lactis subsp. lactis biovardiacetylactis in Response to Autoacidification and Temperature Downshift in Skim Milk. Appl. Environ. Microbiol. 7180168023

51. G Wolf, E. K Arendt, U Pfähler, W. P Hammes, 1990Heme-dependent and hemeindependent nitrite reduction by lactic acid bacteria results in different N-containing products. Int. J. Food Microbiol. 10323329

52. P Duwat, B Cesselin, S Sourice, A Gruss, 2000Lactococcus lactis, a bacteriamodel for stress responses and survival. Int. J. Food Microbiol. 558386

53. T Rochat, J. J Gratadoux, A Gruss, G Corthier, E Maguin, P Langella, van de Guchte M (2006Production of a heterologous nonheme catalase byLactobacillus casei: an efficient tool for removal of H2O2 and protection of Lactobacillus bulgaricus from oxidative stress in milk. Appl. Environ. Microbiol. 7251435149

54. Y Yamamoto, C Poyart, P Trieu-cuot, G Lamberet, A Gruss, P Gaudu, 2005Respiration metabolism of Group B Streptococcus is activated by environmental haem and quinone and contributes to virulence. Mol. Microbiol. 56525534

55. K Vido, H Diemer, A Van Dorsselaer, E Leize, V Juillard, A Gruss, P Gaudu, 2005Roles of thioredoxin reductase during the aerobic life ofLactococcus lactis. J. Bacteriol. 187601610

56. L Rezaiki, B Cesselin, Y Yamamoto, K Vido, E Van West, P Gaudu, A Gruss, 2004Respiration metabolism reduces oxidative and acid stress to improve long-term survival of Lactococcus lactis. Mol. Microbiol. 5313311342

57. D Lechardeur, B Cesselin, A Fernandez, G Lamberet, C Garrigues, M Pedersen, P Gaudu, A Gruss, 2011Using heme as an energy boost for

lactic acid bacteria. Curr. Opin. Biotechnol. 22143149

58. Brooijmans RJWde Vos WM, Hugenholtz J (2009Lactobacillus plantarum WCFS1 Electron Transport Chains. Appl. Environ. Microbiol. 7535803585

59. V. I Lushchak, 2001Oxidative stress and mechanisms of protection against it in bacteria. Biochem. Moscow. 66476489

60. C Hertel, G Schmidt, M Fischer, K Oellers, W. P Hammes, 1998Oxygendependent regulation of the expression of the catalase gene katA of Lactobacillus sakei LTH677. Appl. Environ. Microbiol. 6413591365

61. A Miyoshi, T Rochat, J. J Gratadoux, Le Loir Y, Costa Oliveira S, Langella P, Azavedo 2003Oxidative stress in Lactococcus lactis. Genet. Mol. Res. 2:348-359.

62. J. M Bruno-bárcena, J. M Andrus, S. L Libby, T. R Klaenhammer, H. M Hassan, 2004Expression of a heterologous manganese superoxide dismutase gene in intestinal lactobacilli provides protection against hydrogen peroxide toxicity. Appl. Environ. Microbiol. 7047024710

63. A Jänsch, M Korakli, R. F Vogel, M. G Ganzle, 2007Glutathione reductase from Lactobacillus sanfranciscensis DSM20451T: contribution to oxygen tolerance and thiol exchange reactions in wheat sourdoughs. Appl. Environ. Microbiol. 7344694476

64. F Rallu, A Gruss, S. D Ehrlich, E Maguin, 2000Acid- and multistress-resistant mutants of Lactococcus lactis: identification of intracellular stress signals. Mol. Microbiol. 35517528

65. P. D Cotter, C Hill, 2003Surviving the acid test: responses of Gram-positive bacteria to low pH. Microbiol. Mol. Biol. Rev. 67429453

66. B Pieterse, R. J Leer, Schuren FHJ, van der Werf MJ (2005Unravelling the multiple effects of lactic acid stress on Lactobacillus plantarum by transcription profiling. Microbiol. 15138813894

67. M Piuri, C Sanchez-rivas, S. M Ruzal, 2005Cell wall modifications during osmotic stress in Lactobacillus casei. J. Appl. Microbiol. 988495

68. M Piuri, C Sanchez-rivas, S. M Ruzal, 2003Adaptation to high salt in Lactobacillus: role of peptides and proteolytic enzymes. J. Appl. Microbiol. 95372379

69. S Jordan, M. I Hutchings, T Mascher, 2008Cell envelope stress response in Gram-positive bacteria. FEMS Microbiol. Rev. 32107146

70. Kets EPWTeunissen PJM, Bont JAM (1996Effect of Compatible Solutes on Survival of Lactic Acid Bacteria Subjected to Drying. Appl. Environ.

Microbiol. 62259261

71. D. E Ward, C. C Van Der Wejden, M. J Van Der Merwe, H. V Westerhoff, A Claiborne, J. L Snoep, 2000Branched-chain α-keto acid catabolism via the gene products of the bkd operon in Enterococcus faecalis: A new, secreted metabolite serving as a temporary redox sink. J Bacteriol. 18232393246

72. K. M Considine, A. L Kelly, G. F Fitzgerald, C Hill, R. D Sleator, 2008High-pressure processing- effects on microbial food safety and food quality. FEMS Microbiol. Lett. 28119

73. A Jofré, M Champomier-verges, P Anglade, F Baraige, B Martín, M Garriga, M Zagorec, T Aymerich, 2007Protein synthesis in lactic acid and pathogenic bacteria during recovery from a high pressure treatment. Res. Microbiol. 158512520

74. G Tabanelli, 2011Use of sub-lethal high pressure homogenization (HPH) treatments to enhance functional properties of Lactic acid bacteria probiotic strains. Ph.D. thesis. University of Bologna, Bologna, Italy.

75. H. J Chung, W Bang, M. A Drake, 2006Stess response of Escherichia coli. Compr Rev Food Sci F. 55264

76. P Burns, G Vinderola, F Molinari, J Reinheimer, 2008Suitability of whey and buttermilk for the growth and frozen storage of probiotic lactobacilli. Int. J. Dairy Technol. 61156164

77. S Sieuwerts, de Bok FAM, Hugenholtz J, van Hylckama Vlieg JET (2008Unraveling Microbial Interactions in Food Fermentations: from Classical to Genomics Approaches. Appl. Environ. Microbiol. 7449975007

78. E Caplice, G. F Fitzgerald, 1999Food fermentations: role of microorganisms in food production and preservation. Int. J. Microbiol. 50131149

79. Di Cagno RDe Angelis M, Calasso M, Vincentini O, Vernocchi P, Ndagijimana M, De Vincenzi M, Dessì MR, Guerzoni ME, Gobbetti M (2010Quorum sensing in sourdough Lactobacillus plantarum DC400: induction of plantaricin A (PlnA) under co-cultivation with other lactic acid bacteria and effect of PlnA on bacterial and Caco-2 cells. Proteomics. 1021752190

80. M. C Lorenz, N. S Cutler, J Heitman, 2000Characterization of Alcohol-induced Filamentous Growth in Saccharomyces cerevisiae. Mol. Biol. Cell 11183199

81. P. J Hansen, M Drost, R. M Rivera, F. F Paula-lopes, A. I-K. a. t. a.

n. a. n. i. t Y. M Krininger, C. E Chase, CC (2001Adverse impact of heat stress on embryo production: causes and strategies for mitigation. Theriogenology. 5591103

82. G Giraffa, 2004Studying the dynamics of microbial populations during food fermentation. FEMS Microbiol. Rev. 28251260

83. E. A Johnson, 2002Microbial Adaptation and Survival in Foods. In: Microbial Stress Adaptation and Food Safety. Ed. A.E. Yousef and V.K Juneja. CRC Press

84. P. M Davidson, 2001Chemical preservatives and natural antimicrobial compounds, 593627In: Food Microbiology: Fundamentals and Frontiers, 2nd ed. M.P. Doyle.

85. E Hüfner, C Hertel, 2008Improvement of raw sausage fermentation by stress-conditioning of the starter organism Lactobacillus sakei. Curr. Microbiol. 57490496

86. E Hüfner, T Markieton, S Chaillou, Crutz-Le Coq AM, Zagorec M, Hertel C (2007Identification of Lactobacillus sakei genes induced during meat fermentation and their role in survival and growth. Appl. Environ. Microbiol. 7325222531

87. A Marceau, M Zagorec, S Chaillou, T Mera, M. C Champomier-verges, 2004Evidence for involvement of at least six proteins in adaptation ofLactobacillus sakei to cold temperatures and addition of NaCl. Appl. Environ. Microbiol. 7072607268

88. S Kamdem, F Patrignani, M. E Guerzoni, 2007Shelf-life and safety characteristics of Italian Toscana traditional fresh sausage (Salsiccia) combining two commercial ready-to-use additives and spices. Food Control. 18421429

89. M Schneider, 1988Zur mikrobiologie von sauerkraut bei der vergärung in verkaufsfertige kleinbehältern. Dissertation Hohenheim University, Stuttgart.

90. C. S Pederson, M. N Albury, 1969The sauerkraut fermentation. New York State Agricultural Experiment Station Technical Bulletin 824, Geneva, New York.

91. J. L Ruiz-barba, A Garrido-fernandez, R Jiménez-diaz, 1991Bactericidal action of oleuropein extracted from green olives against Lactobacillus plantarum. Let. Appl. Microbiol. 126568

92. G Ciafardini, V Marsilio, B Lanza, N Possi, 1994Hydrolysis of oleuropein by Lactobacillus plantarum strains associated with olive fermentation. Appl. Environ. Microbiol. 6041424147

93. P Lavermicocca, F Valerio, S. L Lonigro, M De Angelis, L Morelli, M. L Callegari, C. G Rizzello, A Visconti, 2005Study of Adhesion and Survival ofLactobacilli and Bifidobacteria on Table Olives with the Aim of Formulating a New Probiotic Food.Appl. Environ. Microbiol. 7142334240

94. J. L Ruiz-barba, R Jiménez-diaz, 1995Availability of essential B-group vitamins to Lactobacillus plantarum in green olive fermentation brines. Appl. Environ. Microbiol. 6112941297

95. O Sullivan, E Condon, S (1997Intracellular pH is a major factor in the induction of tolerance to acid and other stresses in Lactococcus lactis. Appl. Environ. Microbiol. 6342104215

96. J. W Sanders, G Venema, J Kok, 1999Environmental stress responses in Lactococcus lactis. FEMS Microbiol. Rev. 23483501

97. M. E Guerzoni, A Gianotti, D. I Serrazanetti, 2011Fermentation as a tool to improve healthy properties of bread". In V. R. Preedy, R. R. Watson, & V. B. Patel, (Eds.), Flour and breads and their fortification in health and disease prevention (385393London, Burlington, San Diego: Academic Press, Elsevier.

98. M. G Gänzle, C Schwab, 2009Ecology of exopolysaccharide formation by lactic acid bacteria: sucrose utilisation, stress tolerance, and biofilm formation. In: Ulrich, M. (Ed.), Bacterial Polysaccharides- Current Innovation and Trends. Horizon Press.

99. A. D Welman, I. S Maddox, 2003Exopolysaccharides from lactic acid bacteria: perspectives and challenges. Trends Biotechnol. 21269274

100. E. M Lim, S. D Ehrlich, E Maguin, 2000Identification of stress-inducible proteins in Lactobacillus delbrueckii subsp. bulgaricus. Electrophoresis. 2125572561

101. F Streit, G Corrieu, C Beal, 2007Acidification improves cryotolerance of Lactobacillus delbrueckii subsp. bulgaricus CFL1. J. Biotechnol. 128659667

102. F Streit, J Delettre, G Corrieu, C Beal, 2008Acid adaptation of Lactobacillus delbrueckii subsp. bulgaricus induces physiological responses at membrane and cytosolic levels that improves cryotolerance. J. Appl. Microbiol. 10510711080

103. S. Q Liu, R. V Asmundson, P. K Gopal, R Holland, V. L Crow, 1998Influence of Reduced Water Activity on Lactose Metabolism by Lactococcus lactis subsp. cremoris at Different pH Values. Appl. Environ. Microbiol. 6421112116

104. A Hartke, S Bouche, J. C Giard, A Benachour, P Boutibonnes, Y Auffray, 1996The lactic acid stress response of Lactococcus lactis subsp. lactis. Curr. Microbiol. 33194199

105. J. R Broadbent, C Lin, 1999Effect of heat shock or cold shock treatment on the resistance of Lactococcus lactis to freezing and lyophilization. Cryobiology. 3988102

106. J Behr, M. G Gänzle, R. F Vogel, 2006Characterization of a highly hop-resistant Lactobacillus brevis strain lacking hop transport. Appl. Environ. Microbiol. 72648392

107. A Atrih, N Rekhif, M Michel, G Lefebvre, 1993Detection and characterization of a bacteriocin produced by Lactobacillus plantarum C19. Can. J. Microbiol. 3911731179

108. R Jiménez-diaz, R. M Rios-sanchez, M Desmazeaud, J. L Ruiz-barba, J. C Piard, 1993S Plantaricin, and T, two new bacteriocins produced byLactobacillus plantarum LPCO10 isolated from a green olive fermentation. Appl. Environ. Microbiol. 5914161424

109. T. R Klaenhammer, 1988Bacteriocins of lactic acid bacteria. Biochim. 70337349

110. P Neysens, L De Vuyst, 2005Carbon dioxide stimulates the production of amylovorin L by Lactobacillus amylovorus DCE 471, while enhanced aeration causes biphasic kinetics of growth and bacteriocin production. Int. J. Food Microbiol. 105191202

111. P. M Davidson, M. A Harrison, 2002Microbial Adaptation to Stresses by Food Preservatives, in Microbial Stress Adaptation and Food Safety. Ed. A.E. Yousef and V.K Juneja. CRC Press

112. G Vrancken, L De Vuyst, T Rimaux, J Allemeersch, S Weckx, 2011Adaptation of Lactobacillus plantarum IMDO 130201, a Wheat Sourdough Isolate, to Growth in Wheat Sourdough Simulation Medium at Different pH Values through Differential Gene Expression. Appl. Environ. Microbiol. 7734063412

113. S Lebeer, J Vanderleyden, De Keersmaecker SCJ (2008Genes and Molecules of Lactobacilli Supporting Probiotic Action. Microbiol. Mol. Biol. R. 72728764

114. M. P Jobin, D Garmyn, C Divies, J Guzzo, 1999Expression of the Oenococcus oeni trxA gene is induced by hydrogen peroxide and heat shock. Microbiol. 14512451251

115. V Capozzi, M. P Arena, E Crisetti, G Spano, D Fiocco, 2011The hsp 16 Gene of the Probiotic Lactobacillus acidophilus Is Differently Regulated

by Salt, High Temperature and Acidic Stresses, as Revealed by Reverse Transcription Quantitative PCR (qRT-PCR) Analysis. Int. J. Mol. Sci. 1253905405

116. B Pieterse, 2006Transcriptome analysis of the lactic acid and NaCl-stress response of Lactobacillus plantarum Ph.D. thesis. Wageningen University, Wageningen, The Netherlands

117. A Petersohn, M Brigulla, S Haas, J. D Hoheisel, U Volker, M Hecker, 2001Global Analysis of the General Stress Response of Bacillus subtilis. J. Bacteriol. 18356175631

118. J. C Wang, W. Y Zhang, Z Zhong, A. B Wei, Q. H Bao, Y Zhang, T. S Sun, A Postnikoff, H Meng, H. P Zhang, 2012Transcriptome analysis of probiotic Lactobacillus casei during fermentation in soymilk. J. Ind. Microbiol. Biotech. 39191206

119. S Anandan, 2006Genomic and Proteomic Approaches for Studying Bacterial Stress Responses. In: Advances in Microbial Food Safety. V. Juneja, J.P. Cherry and M.H Tunick eds. ACS Books.

120. G Renzone, D Aambrosio, C Arena, S Rullo, R Ledda, L Ferrara, L Scaloni, A (2005Differential proteomic analysis in the study of prokaryotes stress resistance. Ann. Ist. Super. Sanità. 41459468

121. M. H Nezhad, M Knight, M. L Britz, 2012Evidence of changes in cell surface proteins during growth of Lactobacillus casei under acidic conditions. Food. Sci. Biotech. 21253260

122. M Garnier, S Matamoros, D Chevret, M. F Pilet, F Leroi, O Tresse, 2010Adaptation to Cold and Proteomic Responses of the Psychrotrophic Biopreservative Lactococcus piscium Strain CNCM I-4031. Appl. Environ. Microbiol. 7680118018

123. J Jardin, D Mollé, M Piot, S Lortal, V Gagnaire, 2012Quantitative proteomic analysis of bacterial enzymes released in cheese during ripening. Int. J. Food Micro. 1551928

124. V Gagnaire, M Piot, B Camier, Vissers JPC, Gwenael J, Leonil J (2004Survey of bacterial proteins released in cheese: a proteomic approach. Int. J. Food Microbiol. 94185201

125. M Güell, E Yus, M Lluch-senar, L Serrano, 2011Bacterial transcriptomics: what is beyond the RNA horiz-ome? Nat. Rev. Microbiol. 9658669

126. C Wu, J Zhang, M Wang, G Du, J Chen, 2012Lactobacillus casei combats acid stress by maintaining cell membrane functionality. J. Ind. Microbiol. Biotechnol. Accepted

127. R Coda, C. G Rizzello, D Pinto, M Gobbetti, 2012Selected Lactic Acid

Bacteria Synthesize Antioxidant Peptides during Sourdough Fermentation of Cereal Flours, Appl. Environ. Microbiol. 7810871096

128. E Glaasker, Tjan FSB, Ter Steeg PF, Konings WN, Poolman B (1998Physiological Response of Lactobacillus plantarum to Salt and Nonelectrolyte Stress. J. Bacteriol. 18047184723

129. I Mehmeti, E. M Faergestad, M Bekker, L Snipen, I. F Nes, H Holo, 2012Growth Rate-Dependent Control in Enterococcus faecalis: Effects on the Transcriptome and Proteome, and Strong Regulation of Lactate Dehydrogenase. Appl. Environ. Microbiol. 78170176

130. C Alcántara, M Zúniga, 2012Proteomic and transcriptomic analysis of the response to bile stress of Lactobacillus casei BL23. Microbiol. Accepted

131. C Wu, J Zhang, W Chen, M Wang, G Du, J Chen, 2011A combined physiological and proteomic approach to reveal lactic-acid-induced alterations inLactobacillus casei and its mutant with enhanced lactic acid tolerance. Appl. Microbiol. Biotechnol. 93707722

132. S Jozefczuk, S Klie, G Catchpole, J Szymanski, A Cuadros-inostroza, D Steinhauser, J Selbig, L Willmitzer, 2010Metabolomic and transcriptomic stress response of Escherichia coli. Mol. Sys. Biol. 6:364.

Chapter 9

NIGERIAN INDIGENOUS FERMENTED FOODS: PROCESSES AND PROSPECTS

Evans C. Egwim, Amanabo Musa, Yahaya Abubakar and Bello Mainuna

[1]Biochemistry Department, Federal University of Technology, Minna, Niger State, Nigeria

INTRODUCTION

The deliberate fermentation of foods by man predates written history and is possibly the oldest method of preserving perishable foods. Evidence suggests that fermented foods were consumed 7,000 years ago in Babylon (Battcock and Aza-Ali, 1998). Scientist speculates that our ancestors possibly discovered fermentation by accident and continued to use the process out of preference or necessity. Preserving by fermentation not only made foods available for future use, but more digestible and flavourful. The nutritional value produced by fermenting is another benefit of fermenting.

Fermented foods are generally produced using plant or animal ingredients in combination with fungi or bacteria which are either sourced from the environment, or carefully kept in cultures maintained by humans. Just as living organisms cover the surface of the earth, fermentation microbes cover the surface of the organisms. Wild yeasts are found living on grapes (Chamberlain et al. 1997), and bacteria line the human digestive tract.

Fermented foods, whether from plant or animal origin, are an intricate part of the diet of people in all parts of the world. Fermented food plays a very important role in the socio-economics of developing countries. Each nation has its own types of fermented food, representing the staple diet and the raw ingredients available in that particular place. It makes major contributions to the protein requirements of the rural population. The preparation of many indigenous or "traditional" fermented foods and beverages remains a household art today.

PURPOSE AND BENEFITS OF FOOD FERMENTATION

The primary benefit of fermentation is the conversion of sugars and other carbohydrates to usable end products. According to Steinkraus (1995), the traditional fermentation of foods serves several functions, which includes: enhancement of diet through development of flavour, aroma, and texture in food substrates, preservation and shelf-life extension through lactic acid, alcohol, acetic acid and alkaline fermentation, enhancement of food quality with protein, essential amino acids, essential fatty acids and vitamins, improving digestibility and nutrient availability, detoxification of anti-nutrient through food fermentation processes, and a decrease in cooking time and fuel requirement.

Nutritional Benefits

Fermentation can produce important nutrients or eliminate anti- nutrients. Food can be preserved by fermentation, since fermentation uses up food energy and creates conditions unsuitable for spoilage microorganisms. For instance, in pickling, the acid produced by the dominant organism inhibits the growth of all other microorganisms.

Fermenting makes foods more edible by changing chemical compounds, or predigesting, the foods for us. There are extreme examples of poisonous plants like cassava that are converted to edible products by fermenting. Some coffee beans are hulled by a wet fermenting process, as opposed to a dry process (Battcock and Aza-Ali, 1998).

Reduction in anti-nutritional and toxic components in plant foods by fermentation was observed in a research which showed " Cereals, legumes, and tubers that are used for the production of fermented foods may contain significant amounts of antinutritional or toxic components such as phytates, tannins, cyanogenic glycosides, oxalates, saponins, lectins, and inhibitors of enzymes such as alpha-amylase, trypsin, and chymotrypsin.

- These substances reduce the nutritional value of foods by interfering with the mineral bioavailability and digestibility of proteins and carbohydrates. In natural or pure mixed-culture fermentations of plant foods by yeasts, molds, and bacteria, antinutritional components (e.g. phytate in whole wheat breads) can be reduced by up to 50%; toxic components, such as lectins in tempe and other fermented foods made from beans, can be reduced up to 95%.(Larsson and Sandberg, 1991)

Fermentation increases nutritional values of foods, and allows us to live healthier lives. Here are a few examples:

- The sprouting of grains, seeds, and nuts, multiplies the amino acid, vitamin, and mineral content and antioxidant qualities of the starting product (Wigmore, 1986).

- Fermented beans are easier for the bodies to digest, like the proteins found in soy beans that are nearly indigestible until fermented (Katz, 2003).

- Fermented dairy products, like, cheese, yogurt, and kifir, can be consumed by those not able to digest the raw milk, and aid the digestion and well-being for those with lactose intolerance and autism.

- Porridge made from grains allowed to ferment increases the nutritional values so much that it reduces the risk of disease in children (Battcock and Aza-Ali, 1998).

- Probotic supplements (beneficial bacterial cultures for microbial balance in the body) are capable of fighting cancer and other diseases.

- Vinegar is used to leach out certain flavours and compounds from plant materials to make healthy and tasty additions to the meals.

Health Benefits

Fermented food, enjoyed across the globe, conveys health benefits through lactic acid fermentation. The fermentation process can transform the flavour of food from the plain and mundane to a mouth-puckering sourness enlivened by colonies of beneficial bacteria and enhanced micronutrients.

- Studies have revealed that *Lactobacillus rhamnosus* and *L. reuteri* which are common organisms in Nigerian fermented foods like ogi and kunun- zaki could colonize the vagina, kill viruses, and reduce the risk of infections, including bacterial vaginosis (Reid *et al.*, 2001a; Cadieux *et al.,*2002). The potential therapeutic effects of Lactic Acid Bacteria (LAB) and *ogi*, including their immunostimulatory effect, are due primarily to changes in the gastrointestinal (GI) microflora to suppress the growth of pathogens. Increase in population of LAB in the intestinal or vagina reduces the cause of bacterial vaginosis, which is a major risk factor for the contraction of HIV (Reid, 2002a). It also reduces the occurrence of gonorrhoea, chlamydia, and other sexually transmitted diseases (Reid *et al.,* 2001b) and diarrhoea (Adebolu *et al.,* 2007).

- All lactic acid producing bacteria (E.g*Lactobacillus acidophilus, L.bulgaricus, L. plantarum, L. caret, L. pentoaceticus, L. brevis and L. themophilus*) produces high acidity during fermentation. The lactic acid they produce is effective in inhibiting the growth of other bacteria that may decompose or spoil the food. Despite their complexity, the whole basis of lactic acid fermentation centres on the ability of lactic acid bacteria to produce acid, which then inhibits the growth of other non-desirable organisms. Other compounds are important as they improve particular testes and aromas to the final products. The *L. mesenteroides* initiates growth in vegetables more rapidly over a range of temperatures and salt concentrations than any other lactic acid bacteria. It produces carbon dioxide and acids which rapidly lower the pH and inhibit the development of undesirable microorganism.

- Over 200 species of bacteria live in gut of humans. These microbes help break down food in the intestines, aid in the digestion process, help fight off disease, and boost the immune system. A good balance of intestinal flora is very important to the overall health. If we eat nothing but overly processed and hard to digest foods, then the fermentation process occurs within the GIT resulting into gas, bloating, diarrhoea, and constipation might possibly lead to other diseases like cancer. However, providing the body with predigested foods such as fermented foods will help the existing microbes within to do the job they need to do.

- Fermentation is not only a way to preserve certain foods, in some cases it actually adds to the nutrient value of it. Fermented vegetables contain more vitamin C and fermented milk products have ample amounts of B vitamins. The bioavailability of these vitamins also increases with fermentation. Probiotics, or "good bacteria" are also formed through the process of fermentation. Fermented soy products contain more vitamin B_{12}(Chung et al, 2010)

- The desirable bacteria cause less deterioration of the food by inhibiting the growth of the spoiling types of bacteria. Some fermenting processes lower the pH of foods preventing harmful microorganisms to live with too acidic an environment. Controlled fermentation processes encourage the growth of good bacteria which starves, or fights off, the bad microbes.

- The fermentation process can be stopped by other means of preserving, such as, canning (heating), drying, or freezing. Heat (pasteurization, 63°C), and low temperatures (freezing, 0°C or below) stops the fermenting process by slowing, or killing, the preferred microorganisms, and other bacteria. A few undesirable bacteria are not killed by either means, and continue to grow. When the beneficial bacteria are gone, the unfavorable bacteria take over, growing exponentially! This causes rotting, disease, illness, and inedible foods. When the good guys are present and happy, the food remains edible.

- Phytates (phytic acid) are the storage form of phosphorus [a mineral] bound to inositol [a B vitamin] in foods high in fiber (all plant foods), and particularly the fiber of raw whole grains, legumes, seeds, and nuts. Although these foods have high phosphorus content, the phosphates in phytates are not released by human digestion. Phytates, particularly in such raw foods as bran, are a concern because they can bind a portion of the iron, zinc, and calcium in foods, making the minerals unavailable for absorption. When bread is leavened (fermented) by yeast, enzymes degrade phytic acid, and phytates pose no problem. Enzymes, called phytases, destroy phytates during fermentation processes such as: the yeast-raising of dough, Even a small amount of phytates in food can reduce iron absorption by half (by 50%), but the effect is less marked if a meal is supplemented with ascorbic acid (Vitamin C) which also helps the absorption of zinc and calcium.

- Fermented food, enjoyed across the globe, conveys health benefits through lactic acid fermentation. The fermentation process can transform the flavor of food from the plain and mundane to a mouth-puckering sourness enlivened by colonies of beneficial bacteria and enhanced micronutrients. While fermented food like yogurt, sauerkraut and kefir are well-known many other lesser-known foods also benefit from the lactic acid fermentation process. Indeed, virtually every food with a complex or simple sugar content can be successfully fermented.

- Born of both necessity and practicality, lactic acid fermentation proved to be not only an efficient method of preserving food for our ancestors, but also a critical one. Indeed, fermented food like sauerkraut, cheese, wine, kvass, soured grain porridge and breads often sustained tribes and villages during harsh winters when fresh food simply wasn't available let alone plentiful.

- In many societies including our own where yogurt has been heralded as a health food since the 19th century, fermented food has gained a reputation for its beneficial effects on immunity, intestinal health and general well-being. Modern researchers are just beginning to understand what the sages of old were tuned in to: fermented food conveys clear and calculable health benefits to the human diet. Lactic acid fermentation in and of itself enhances the micronutrient profile of several foods.

Detoxification

Detoxification of anti-nutrients through food fermentation processes. The renewal of anti-nutrient from the Nigerian fermented food is an important step in ensuring that the fermented food is safe to eat. Many fermentation foods contain naturally accruing toxins and anti-nutritional compounds. These can be removed or detoxified by the action of micro-organism during fermentation for instance, the fermentation process that produces the Sudanese product, kawal, removes the toxins from the leaves of*Cassia obtusitfolia* and fermentation is an important step in insuring that the fermentation foods are safe to eat.

Removing cyanide by fermentation: Cassava contains naturally occurring chemicals, cyanogenic glycoside. When eaten raw or improperly processed, this substances releases cyanide into the body, which can be fatal, correct processing removers this chemicals. The cassava is first peeled (as about 60-70% of the poison is in the peel) and then soaked in stagnant water or fermented in sacks for about three days. It is sometimes grated or rasped as this helps to speed up the fermentation process. At the beginning of the fermentation, *Geotricum candidia* acts on the cassava. This helps to make the product acidic, which finally kills off the microorganisms as they cannot exist in such a medium. A second strain of microorganisms (*corynebacteriumlactis*) which can tolerate the acidic environment then take over and by the third day 90-95% of the dangerous chemicals would have been hydrolyzed. The cassava also develops its characteristic flavour. The product is then sieved and the fine starch particles are fried in an iron pan over aflame or with some oil. During this process most, if not all the remaining toxins are given off. The liquor from a previous fermentation is used as a starter, thereby reducing the period of fermentation to about 6-8hours.

NIGERIAN FERMENTED FOODS

Fermented Tubers

These include mainly cassava and yam used in the production of foods such as garri, fufu, lafun and elubo etc.

Nigeria is one of the leading producers of cassava in the world with an annual production of 35-40 million metric tons. Over 40 varieties of cassava are grown in Nigeria and cassava is the most important dietary staple in the country accounting for over 209 of all food crops consumed in Nigeria (IITA,2004). Cassava tubers are rich in starch [20-30%] and with possible exception of sugar cane; cassava is considered the highest producer of carbohydrates among crop plants. Despite its vast potentials, the presence of the two gynogenic glycosides, lineman calculating for 93% of total content (Okafor *et al.*, 1984) and lotaustralin or methyl linamarin, hydrolysis by the enzymes linamarase to release toxic HCN, is the most important problem limiting cassava utilization. Generally cassava contains 10-500 mg HCN/KG of root depending on the variety, although much higher levels, exceeding 1000 mg HCN/kg, may be present in unusual cases. Cassava varieties are frequently described as sweet or bitter. Sweet cassava varieties are low in cyanogens with most of the cyanogens present in the peels. Bitter cassava varieties are high in cyanogens that tend to be evenly distributed throughout the roots. Environmental (soil, moisture, temperature) and other factors also influence the cyanide content of cassava. Low rainfall or drought increase cyanide level in cassava tools due to water stress on plant. Apart from acute toxicity that may result in death, consumption of sub-lethal dose of cyanide from cassava production over a long period of time results in chronic cyanide toxicity. That increases the prevalence of goiter and cretinism in iodine deficient area. Symptoms of cyanide poisoning from consumption of cassava with high level of cyanogens include vomiting, stomach pains, dizziness, headache, weakness and charkha. Chronic cyanide toxicity is also associated with several pathological conditions including konzo, an irresistible paralysis of the legs reported in eastern, central and southern Africa. And tropical ataxic neuropathy, reported in west Africa, characterized by lesion of the skin, mucous membranes, optics and auditory nerves, spinal cord and peripheral nerves and other symptoms. Without the benefits of modem science, a process for detoxifying cassava roots by canvassing potentially toxic roots into garriand fufu was developed, presumably empirically in West Africa.

Gari

Gari is a creamy-white, granular flour with fermented flavour and a slightly sour taste made from fermented, gelatinized fresh cassava tubers. Gari is widely known in Nigeria and other West African countries. It is commonly consumed either by being soaked in cold water with sugar, coconut, roasted groundnuts, dry fish, or boiled cowpea as complements or as a paste made with hot water and eaten with vegetable sauce. There are basically three types of gari

- Rough-sour gari which is preferred for soaking with sugar and sometimes roasted peanut or coconut.

- Medium gari is usually cooked by adding to boiling water and stirred. This is usually eaten with stew or soup.

- Smooth gari which could be mixed with pepper and other spicy ingredients. A small amount of warm water and palm oil is added and mixed with the hand to soften. This type of gari is served with fried fish.

- Fufu

Fufu is a fermented white paste made from cassava it is ranked next to gari as an indigenous food of most Nigerians in the South. Fufu is made by sleeping whole or cut peeled cassava roots in water to ferment for maximum of three days, during the steeping, fermentation decrease the pH, softens the roots and help to reduces the potentially toxic cyanogenic compound(Agbor-Egbe and Lape Mbome, 2006)

• Lafun

Lafun is a fibrous powdery form of cassava similar to fufu in Nigeria. The method of producing lafun is different from that of fufu in the traditional preparation; fresh cassava roots are cut into chucks and steeped for 3-4 days or until the roots become soft. The fermented roots are peeled, broken up into small pieces and sun dried on mats, flat rocks, cineol flours, or the roots of houses. The dried pieces are milled into flour. Alternatively, chips are made directly from fresh roots, cut into chucks and sun dried. Drying takes 2-4 days, depending on weather. Unlike fufu, the fiber is the related root for lafun are dried along with the mash and later sieved out. The flour is made into dough with boiling water before consumption. When properly stored, it has a shelf life of six months or more.

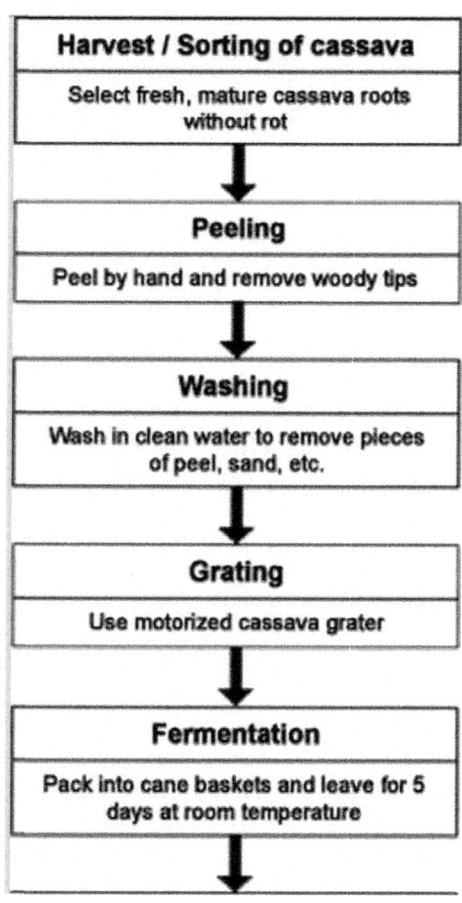

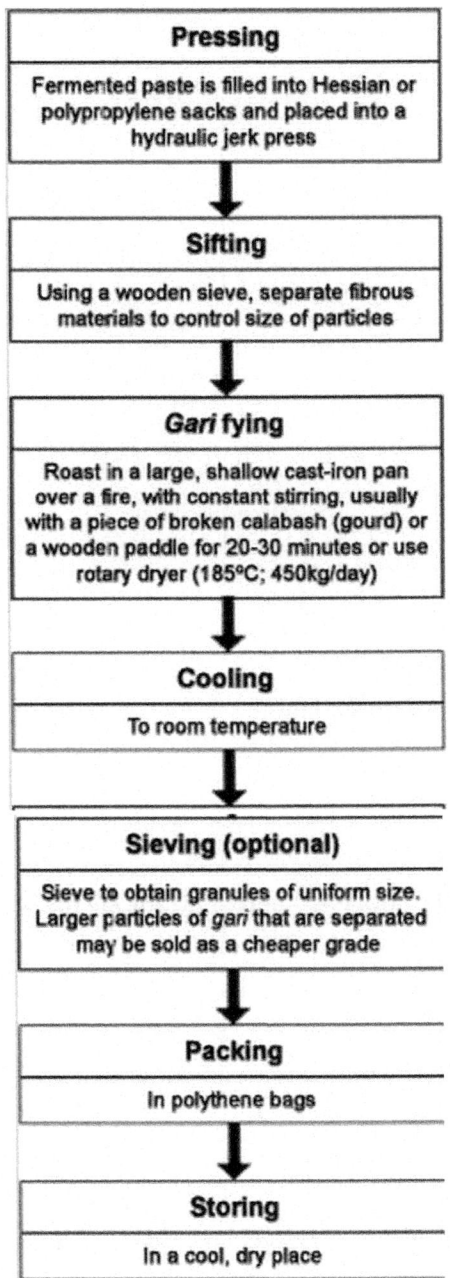

Figure 1: Process flow chart for Gari

Matured cassava roots without rot
↓
Peel and remove skin
↓
Wash in clean water to remove sand
↓
Soak roots in water in a bowl for 48-72 hours.
↓
Fufu mash is allowed to be concentrated before decaling
↓
Fermented paste is filled into polypropylene sacks and placed in a jerk press.
↓
Motorized grater is used to pulverize cake into smaller particle and increases surface area for easy drying.
↓
Dry, using rotary dryer (184⁰C, 450kg/day)
↓
Cool at room temperature
↓
Mill to obtain powder and pack in polyether bags
↓
Store in a cool, dry place

Figure 2: Process flow chart for fufu

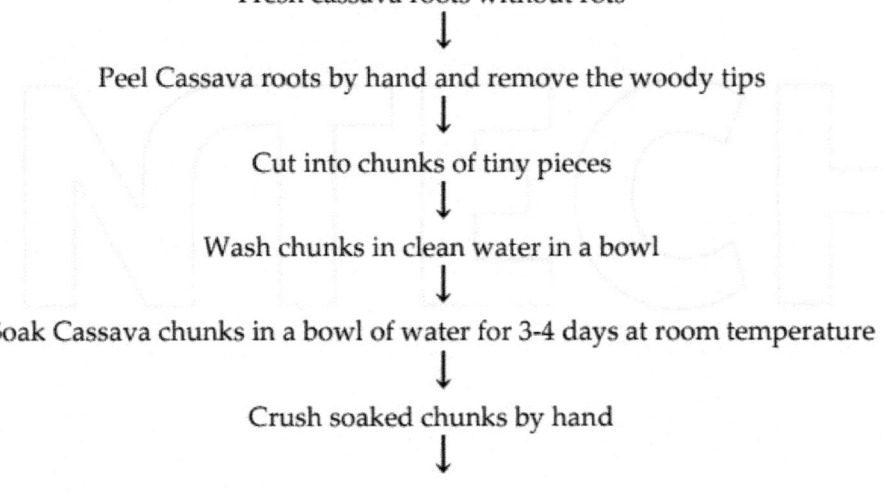

Fresh cassava roots without rots
↓
Peel Cassava roots by hand and remove the woody tips
↓
Cut into chunks of tiny pieces
↓
Wash chunks in clean water in a bowl
↓
Soak Cassava chunks in a bowl of water for 3-4 days at room temperature
↓
Crush soaked chunks by hand
↓

Figure 3: Process flowchart for the production of lafun.

Fermented Cereals

Cereals which include maize (*Zea mays*), Sorghum (*Sorghum bicolor*), millet (*Peninselum americarum*) and acha etc. are used in the production of gruels which is used as complementary food for babies and serves as breakfast for adults.

Maize, millet, rice and sorghum cereals provide mainly carbohydrates and low quality protein. The generation and fermentation of cereals enhance the availability of elemental iron, the deficiency of which is responsible for the high incidence of anaemia in tropical countries. It is estimated that about 50% of perishable food commodities including fruits, vegetables, roots and tubers and about 30% of food commodities including maize, sorghum millet, rice and cowpeas are lost after harvest in Nigeria. Nigerian in fact experience a slower growth rate and weight gain during the weaning period than during breastfeeding, due primarily to the poor nutritional qualities of traditional Nigerian complementary food such as "Ogi" which are mainly produced from cereal fermentation. Apart from their poor nutritional qualities, traditional Nigerian cereal based gruels used as complementary foods have high paste viscosity and require considerable dilution before feeding; a factor that further reduces energy and nutrient density.

Although nutritious and safe complementary foods produced by food multinationals are available in Nigeria, they are far, too expensive for most families. The economic situation in these country necessitate the adoption of simple, inexpressive processing techniques that result in quality improvement and that can be carried out at household and community levels for the production of nutritious, safe and affordable complementary foods which is the leading cause of protein-energy malnutrition in infants and preschool children in Nigeria.

Ogi

This is an example of traditional fermented food, it is a staple cereal of Yorubas of Nigeria and is the first native food given to babies at wearing. It is produced generally by soaking corn grains in warm water for one to two days followed by wet milling and sieving through a screen mesh. The sieved material is allowed to sediment and fermented, and is marketed as wet cake wrapped in leaves. Various food dishes are made from the fermented cakes or ogi. During the steeping corn, *Corynebacterium* spp. become prominent and appears to be responsible for the diastolic action necessary for the growth of yeast and lactic acid bacteria. Along with the corn in bacteria, *S. cerevisiae, Enterobacter cloacae* and *L. plantarum* have been found to be prominent in traditional ogi fermentation.

In Nigeria, the first weaning food is called pap, akamu, ogi, or koko and it is made from maize (zea mays), millet (*Pennisetum americanum*), or guinea corn (*Sorghum spacers*). In Anambra state most mothers introduce the thin gruel at three to six months of age. The baby is fed on demand with a spoon or a few mothers used the traditional forced hand-feeding method. After the successful introduction the thin gruel, other staple foods in the family menu are given to the child. These food include yam (*Dioscoria spp*) rice (*oriza sativa*) gari (fermented cassava grit) and cocoyam (*xanthosoma sagitifolum*), which may be eaten with soup. These foods are usually mashed thinned or pre-chewed. As soon as a child can chew, he or she is given pieces of from the family pot.

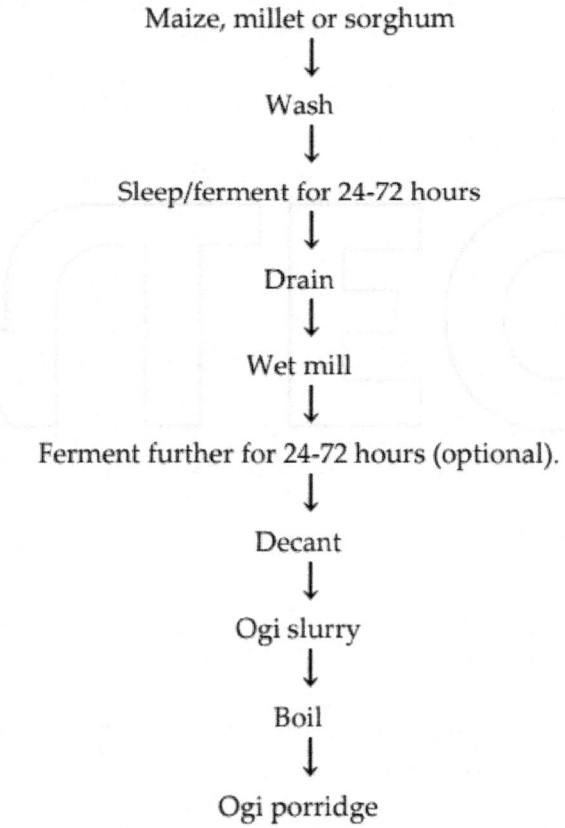

Maize, millet or sorghum
↓
Wash
↓
Sleep/ferment for 24-72 hours
↓
Drain
↓
Wet mill
↓
Ferment further for 24-72 hours (optional).
↓
Decant
↓
Ogi slurry
↓
Boil
↓
Ogi porridge

Figure 4: Process flow chart for ogi

Masa

Masa (waina) is a fermented puff batter of rice, millet, maize, or sorghum cooked in a pan with individual cup like depressions. It resembles the Indian idli in shape and dosa in test. Masa is consumed in various forms by all groups in the northern States of Nigeria and other North African countries (Mali, Burkina Faso, Niger and Chad.) It is the principal ingredients of a variety of cereal-based foods and is a good source of income for the women who prepare the traditional product for sale. Though, masa is as popular as a Nigeria ogi, it has received very little attention. The problem of masa, apart from the short shelf keeping quality, is that of low protein content and inconsistence in the use of varied cereals and spices has resulted in variation in the quality of the products.

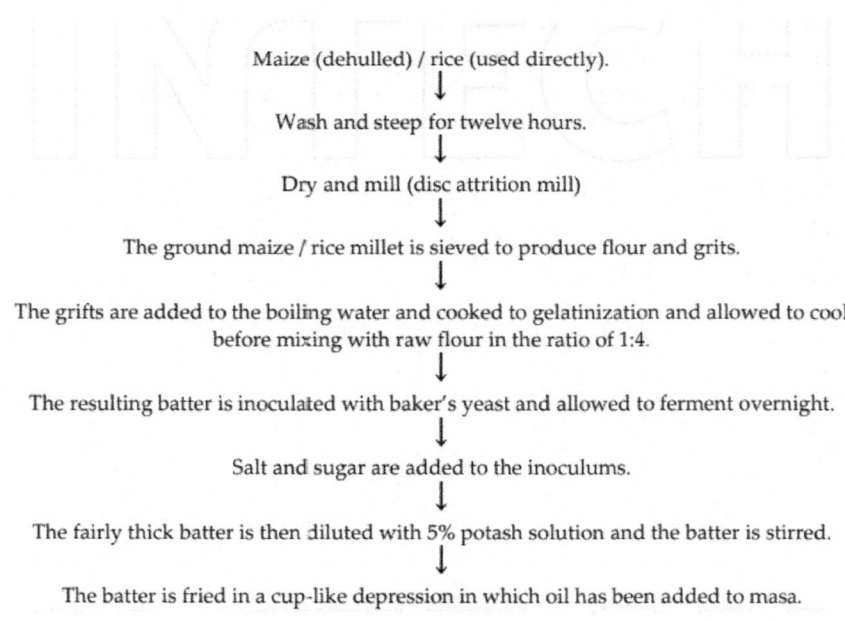

Maize (dehulled) / rice (used directly).
↓
Wash and steep for twelve hours.
↓
Dry and mill (disc attrition mill)
↓
The ground maize / rice millet is sieved to produce flour and grits.
↓
The grifts are added to the boiling water and cooked to gelatinization and allowed to cool before mixing with raw flour in the ratio of 1:4.
↓
The resulting batter is inoculated with baker's yeast and allowed to ferment overnight.
↓
Salt and sugar are added to the inoculums.
↓
The fairly thick batter is then diluted with 5% potash solution and the batter is stirred.
↓
The batter is fried in a cup-like depression in which oil has been added to masa.

Figure 5: Process flow chart for the production of masa.

The addition of cowpea, groundnut or soybeans flour into masa during preparations improves the nutritional quality of masa. Groundnut-maize enriched masa could be a source of protein to the consumer particularly in developing countries like Nigeria where cost of feeding on animal sourced protein is unaffordable. The high calorie content of groundnut-maize masa could be due to the high fat content of the added paste. The decrease in the weight of masa with addition of groundnut paste could be due to increase in the oil content in the paste which has been proofed to be relatively lighter. Masa formulation containing millet or rice blended with cowpea or groundnut was prepared and sodium concentrations were high. Significant improvements in lysine (9-75%), threonine (16-25%) and Isoleucine(10-28%) were observed from masa samples. The biological value (81-93%), apparent digestibility (82-88%) and net protein utilization (74-79%) of all masa samples showed improved nutritional qualities.Supplementedmasa was nutritionally better than masa made from millet or rice alone.

Pito

Pito is the traditional beverage drink of the Benins in the Mid- West part of Nigeria. It is however popularly consumed throughout Nigeria owing to its refreshing nature and low price. Pito is also widely consumed in Ghana. The preparation of pito involves soaking the cereal grains (maize, sorghum, or combinations of both) in water for two days, followed by malting and allowing them to sit for five days in basket lined with moistened banana leaves. The malted grains are ground mixed with water and boiled. The resulting mash is allowed to cool and later filtered through a fine mash, allowed to cool and later filtered through a fine mesh basket. The filterate thus obtained is allowed to stand overnight or until it assumes a slightly sour flavour, following which it is boiled to concentrate.

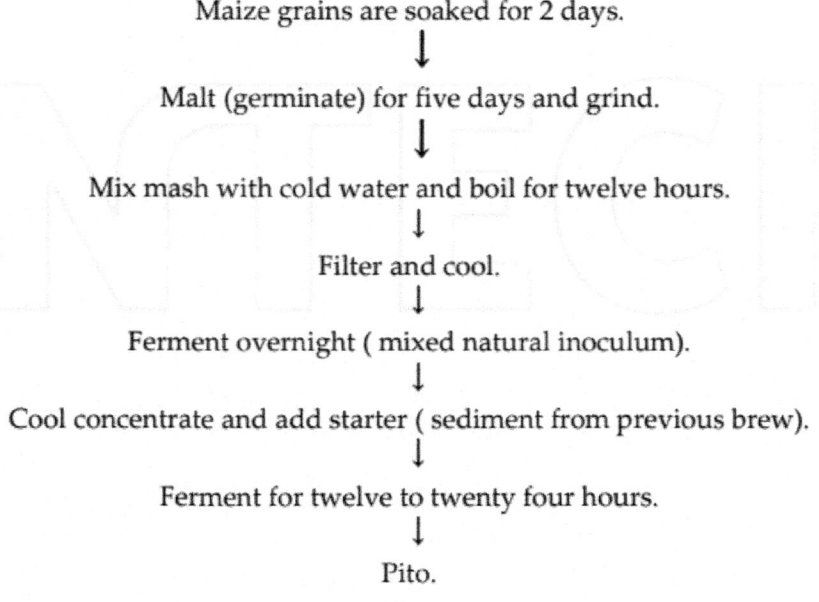

Maize grains are soaked for 2 days.
↓
Malt (germinate) for five days and grind.
↓
Mix mash with cold water and boil for twelve hours.
↓
Filter and cool.
↓
Ferment overnight (mixed natural inoculum).
↓
Cool concentrate and add starter (sediment from previous brew).
↓
Ferment for twelve to twenty four hours.
↓
Pito.

Figure 6: Process flowchart for production of pito.

A starter from the previous brew is added to the cool concentrate which is again allowed to ferment overnight. Pito, the product obtained thus is dark brown liquid which varies in taste from sweat to bitter. It contains lactic acid, sugars and amino acids and has an alcoholic content of about 3%. Organisms responsible for souring include *Geotricum candidum, and Lactobacillus species,* while *Candida species*are responsible for the alcoholic fermentation.

Burukutu

This is a popular alcoholic beverage of vinegar-like flavour, consumed in Northern Guinea Savannah region of Nigeria, in the republic of Benin and Ghana. The preparation of burukutu involves steeping sorghum grains in water overnight, following which excess water drained. The grains are then spread out onto a mat or tray, covered with banana leave and allowed to germinate. During the germination processes, the grains were watered on alternate days and turned over at intervals. Germination continues for four to five days until the plumule attain a certain length. The malted grains are spread out in the sun to dry for one to two days, following which the dried malt is ground to powder. Garri (a farinaecious fermented cassava product) is added to the mixture of the ground malt and six parts water. The resulting mixture is allowed to ferment for two days, following which it is boiled for two days. The resulting product is cloudy alcoholic beverage.

The pH of the fermenting mixture decreases from about 6.4 to 4.2 within 24 hours of fermentation and decreases further to 3.7 after 48 hours. At the termination of the 2-days maturing period, *Acetobacter species and Candida species* are dominant microorganisms. Boiling prior to maturation eliminates lactic acids and other yeast. Fully matured burukutu beer has an acetic acid content which varies between 0.4 to 0.6%.

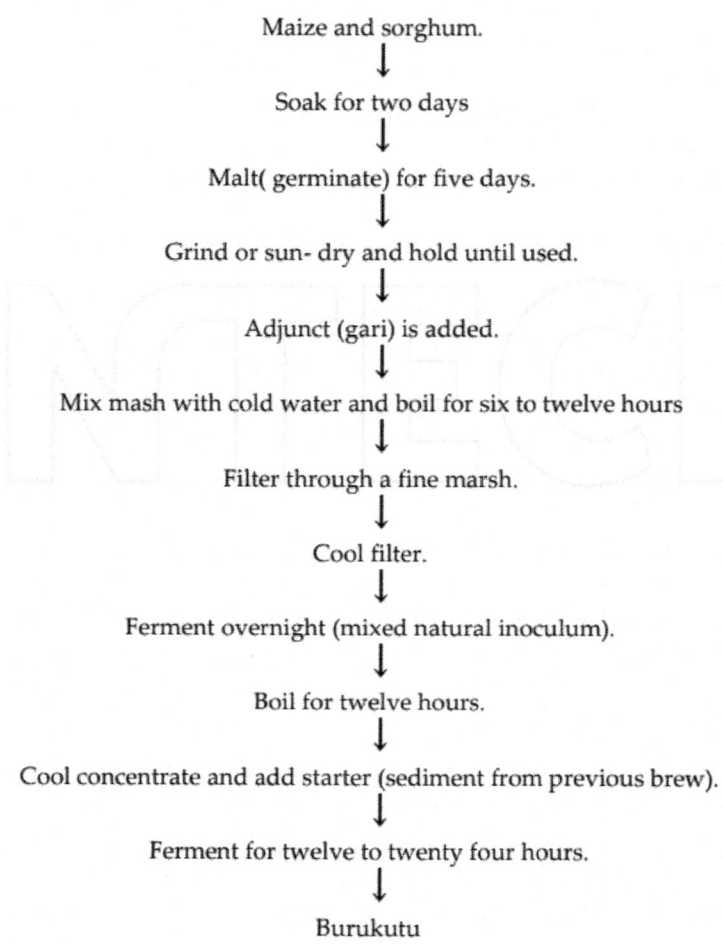

Maize and sorghum.
↓
Soak for two days
↓
Malt(germinate) for five days.
↓
Grind or sun- dry and hold until used.
↓
Adjunct (gari) is added.
↓
Mix mash with cold water and boil for six to twelve hours
↓
Filter through a fine marsh.
↓
Cool filter.
↓
Ferment overnight (mixed natural inoculum).
↓
Boil for twelve hours.
↓
Cool concentrate and add starter (sediment from previous brew).
↓
Ferment for twelve to twenty four hours.
↓
Burukutu

Figure 7: Process flowchart for production of burukutu.

Kunun-zaki.

Kunun-zaki is a non-alcoholic fermented beverage widely consumed in Northern part of Nigeria. This beverage is however becoming more widely consumed in southern Nigeria,owing to its refreshing qualities. Kunun-zaki is consumed anytime of the day by both adult and children as breakfast drink, food supplement. It is a refreshing drink usually used to entertain visitors, appetizers and is commonly used / served at social gathering. Although, there are various types of kunun processed and consumed in Nigeria which include; kunun-zaki, kunun-gyada, kunun-jiko, and amshau and kunun-gayamba. However, kunun-zaki is mostely consumed. The traditional process for the manufacture of kunun-zaki involves the steeping of millet grains, wet milling with spices (ginger,cloves and pepper),wet sieving and partial gelatinization of the slurry, followed by the addition of sugar and bottling.

The fermentation which occurs briefly during steeping of the grains in water for 8 to 48 hours period involves mainly lactic acid bacteria and yeast. Storage studies revealed that the product has a shelf-life of about 24hours at ambient temperature, which was extended to 8 days by pasteurization at 60 °C for 1 hour and stored under refrigeration conditions.

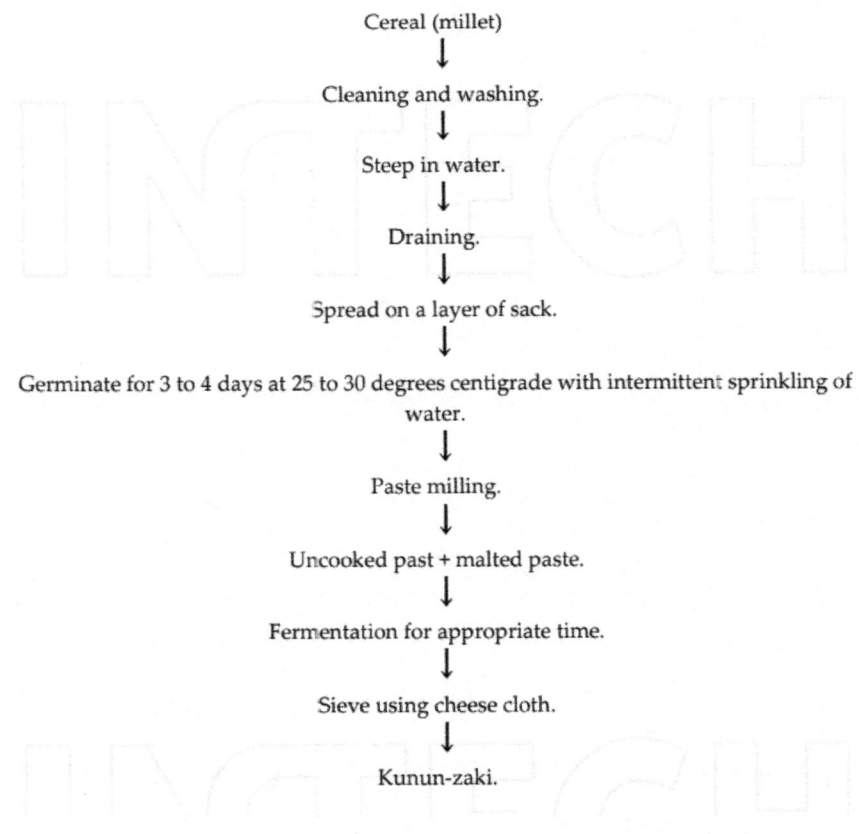

Cereal (millet)
↓
Cleaning and washing.
↓
Steep in water.
↓
Draining.
↓
Spread on a layer of sack.
↓
Germinate for 3 to 4 days at 25 to 30 degrees centigrade with intermittent sprinkling of water.
↓
Paste milling.
↓
Uncooked past + malted paste.
↓
Fermentation for appropriate time.
↓
Sieve using cheese cloth.
↓
Kunun-zaki.

Figure 8: Flowchart for the traditional processes of kunun-zaki.

FRUITS AND VEGETABLES

Ogiri

This is a condiments gotten from the fermentation of castor oil seeds. The raw castor oil seed are boiled for two hours until the seed changes colour to brown. The seeds are dehaulled, rinsed in clean water. The boiled seeds are boiled again for one more hour. It is then cooled and wrapped with enough banana leaves, which is then packed in a clean container with cover to ferment at room temperature.

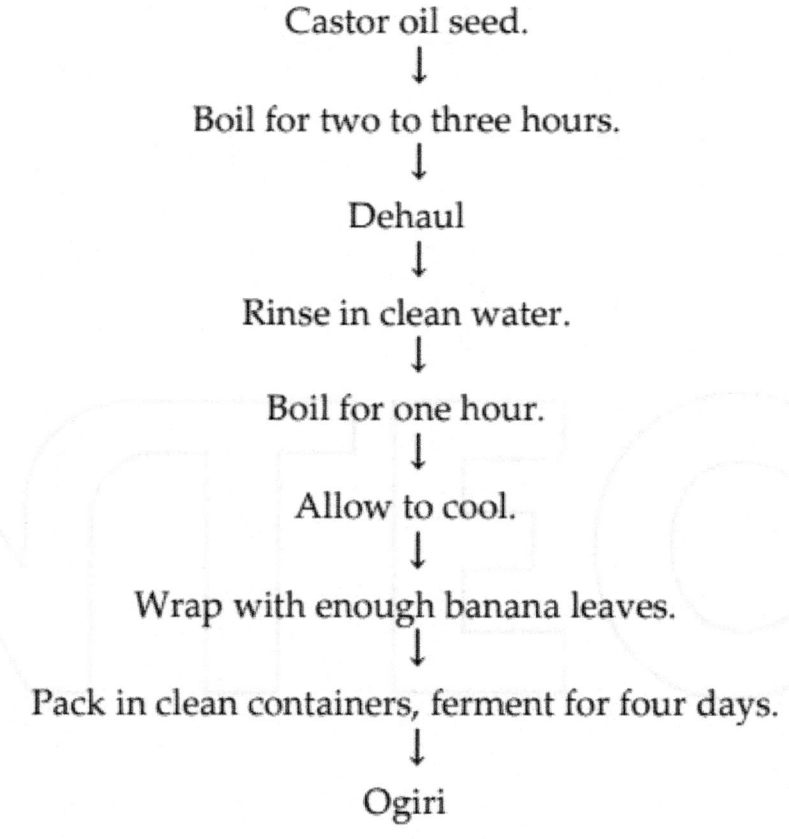

Castor oil seed.

↓

Boil for two to three hours.

↓

Dehaul

↓

Rinse in clean water.

↓

Boil for one hour.

↓

Allow to cool.

↓

Wrap with enough banana leaves.

↓

Pack in clean containers, ferment for four days.

↓

Ogiri

Figure 9: flow chart for the production of ogiri

LEGUMES (LOCUST-BEANS,AFRICAN OIL BEANS, SOYA BEANS) IRU, DADAWA UGBA, AFIYO, DANGWUA)

Dadawa/Iru

This is one of the most important food condiments in Nigeria and many countries of West and Central Africa. It is used in much the same way as bouillon cubes are used in the Western world as nutritious flavouring additives along with cereal grains sauce and may serve as meat substitute. Dadawa (Iru) is prepared from the seeds cf African locust beans, thus are rich in fat (39 to 40%) and protein (31 to 40%)(Achi,2005) and contributes significantly to the energy intake, protein and vitamins, especially riboflavin, in many countries of West and Central Africa. Dadawa or iru is made from locust-bean (*Parkia biglobosa*) seed, a leguminous tree found in the Savannah region of Africa, Southeast Asia and South America. Dadawa is produced by a natural un-inoculated solid –substrate fermentation of the boiled and dehaulled cotyledon, the major fermenting organisms are the *Bacillus* and *Staphylococcus*.The beans mass after fermentation is sun-dried and moulded into round balls or flattened cakes. Due to the high protein content, it has a great potential as a key protein source and basic ingredient for food supplement.

Dadawa fermentation is very similar to that of okpehe prepared from the seeds of *Prosopis africana*, ogiri prepared from melon seeds (*Citrullus vulgaria*) and castor oil bean (*Ricinus cummunis*). Although, the organisms involved in the fermentation of these foods condiments and others have been identified, this has marginal effects as the industrial or commercial production is concerned. Traditionally fermentation of African locust beans involves boiling the beans for twelve hours in excess water, until they are very soft to allow for hand dehaulling after which the separated cotyledon is boiled for another two hours to soften it. The cotyledon is then wrapped with enough banana leaves (*Musa saplentum*) and packed with cover to ferment at room temperature.

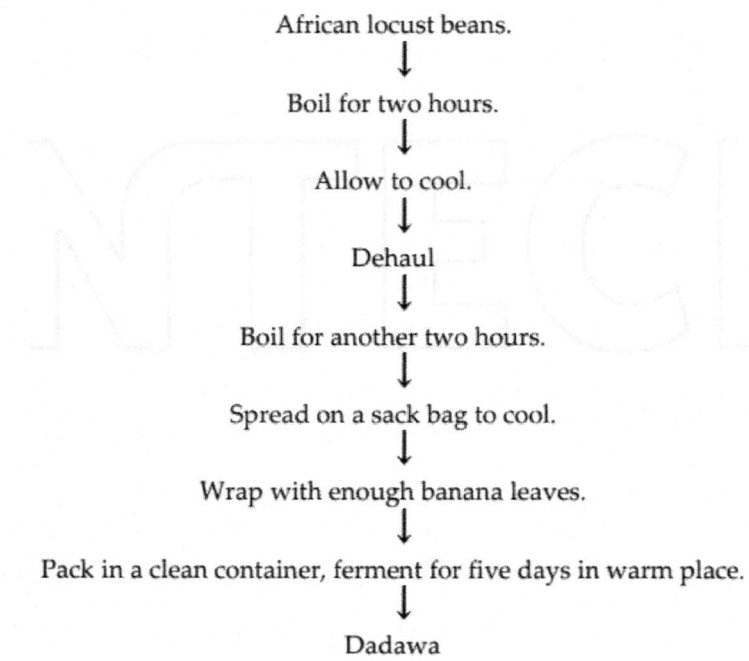

African locust beans.

↓

Boil for two hours.

↓

Allow to cool.

↓

Dehaul

↓

Boil for another two hours.

↓

Spread on a sack bag to cool.

↓

Wrap with enough banana leaves.

↓

Pack in a clean container, ferment for five days in warm place.

↓

Dadawa

Figure 10: A Flow chart for the production processes of dadawa

Other biochemical changes that occur during dadawa fermentation include the hydrolysis of indigestible oligosaccharide present in African locust beans notably stachyose and raffinose, to simple sugars by alpha and beta galactosidase, the synthesis of B-vitamins(thiamin and riboflavin),vitamin C and the reduction of anti-nutritional factors(oxalates and phytates). An improved process for industrial production of dadawa involves dehaulling African locust bean with ball(disc) mill, cooking in a pressure retort for one hour inoculating with *Bacillus subtilis* culture, drying the fermented beans and milling into a powder. Cadbury Nigeria Plc. In 1991 introduced cubed dadawa but it failed to make the desired market impact and it is withdrawn. It would appear that consumers preferred the granular product to the cubed product.

Ugba

Fermented African Oil bean seed, (*Panthaclethra macrophylla benth*) *Ugba,* is an indigenous fermented food and a popular staple among the eastern part of Nigeria. It is rich in protein and other minerals and is obtained by solid-state fermentation of the African Oil bean seed.

It is gotten traditionally from the fermentation of oil bean seed. It contains up to 44% protein, which comprise of at least 17 of the 20 amino and protein digestibility and utilization increases with fermentation (Okechukwu *et al*, 2012). The oil bean seeds are boiled for three hours, dehaulled and cooked, the cooked seeds are then sliced (0.5 to 1mm thickness) and boiled again for two hours, drained, rinsed thrice in water and steeped in cold water for four hours so as to eliminate the bitter taste. The sliced beans are wrapped with enough banana leaves (*Musca sapientum*), packed in a clean container and covered to ferment at room temperature.

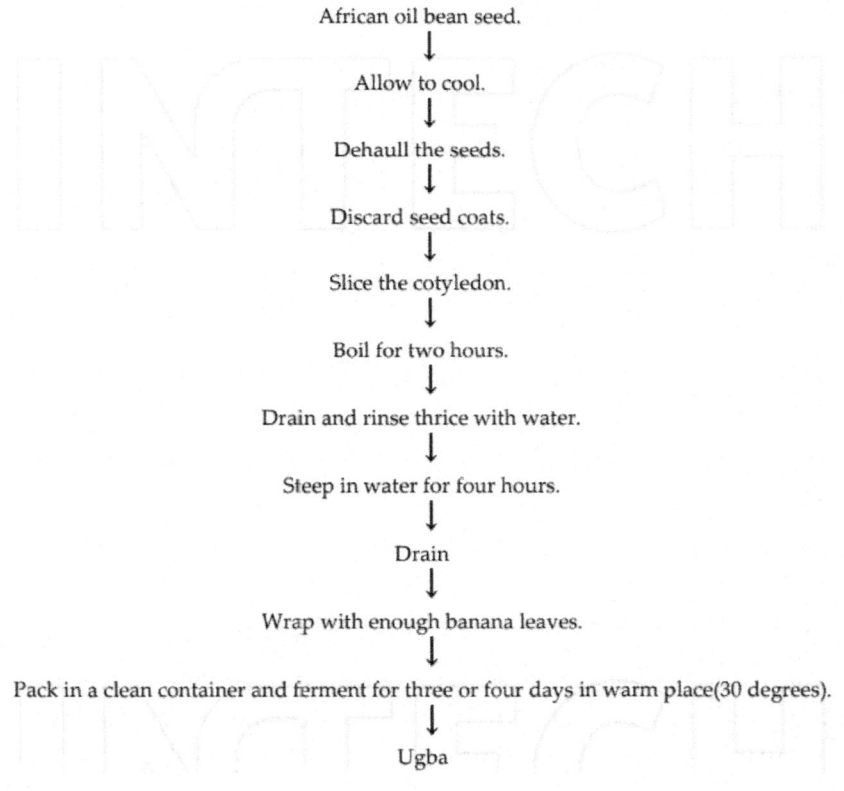

African oil bean seed.
↓
Allow to cool.
↓
Dehaull the seeds.
↓
Discard seed coats.
↓
Slice the cotyledon.
↓
Boil for two hours.
↓
Drain and rinse thrice with water.
↓
Steep in water for four hours.
↓
Drain
↓
Wrap with enough banana leaves.
↓
Pack in a clean container and ferment for three or four days in warm place(30 degrees).
↓
Ugba

Figure 11: A flow chart for the production process of ugba

Afiyo (Okpehe)

Afiyo as is called by the Hausas or Okpehe as known by the Idomas in Benue state is a fermented food flavouring condiment most popular in the middle belt of Nigeria. It is produced from *Prosopis Africana,* which is a leguminous oil seed, fermented in most part of Benue, Niger, Kaduna states and northern parts of Kwara state. This fermented product of *Prosopis Africa* is a strong smelling mash of sticky browned seed and fermentation is moist solid state by chance inoculation, supposedly by various species of micro-organisms.

Various fermented foods have been recorded and these are highly placed condiments while some serve as main meals. Of the thousands of legumes, less than twenty are used extensively in use. Those in common use include peanuts, soy beans, locust beans, oil beans, cowpea and lentils etc.

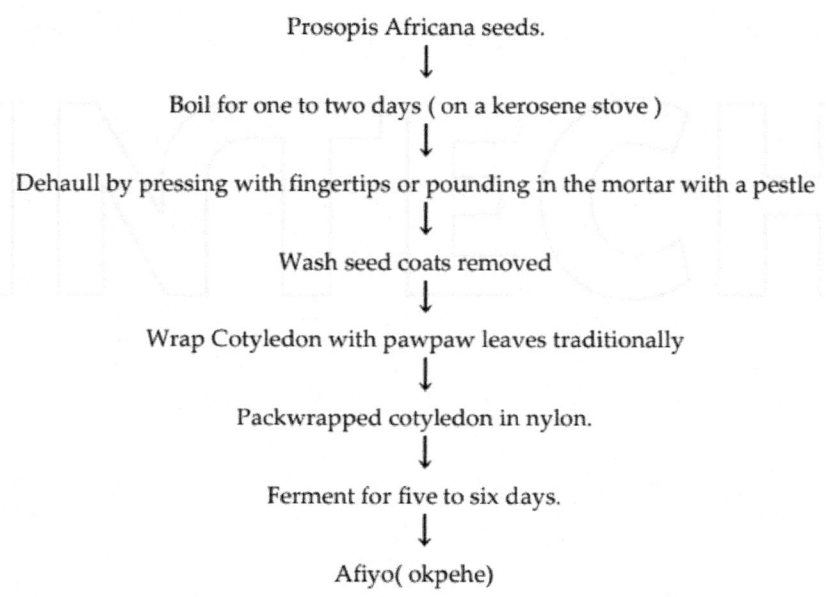

Prosopis Africana seeds.
↓
Boil for one to two days (on a kerosene stove)
↓
Dehaull by pressing with fingertips or pounding in the mortar with a pestle
↓
Wash seed coats removed
↓
Wrap Cotyledon with pawpaw leaves traditionally
↓
Packwrapped cotyledon in nylon.
↓
Ferment for five to six days.
↓
Afiyo(okpehe)

Figure 12: Flow chart for the traditional preparation of okpehe.

Generally, the concentration of amino acids increases during the production of condiments (dadawa, ogiri, ugba) as the fermentation day increases, and it reaches a peak at day four, day three, and day two respectively. This progressive increase in concentration of amino acids in condiments is due to the decrease in total protein as fermentation progresses, which may be attributed to the effect of protease enzymes which result in the hydrolysis of protein molecules to small molecules such as amino acids, such protease activities in the fermenting oil seed increases digestibility than the seed.

Reducing sugar concentration increases with days of fermentation and reaches a peak at the day five, day four, day three for African locust bean, African oil bean seed and castor oil seed respectively. The increase in reducing sugar is due to the hydrolysis of carbohydrates in the presence of certain enzymes such as amylases and galactase. This phenomenon is expected since microbial amylase hydrolyses higher carbohydrates (polysaccharides and disaccharides) to reducing sugars which are then readily digestible by humans. Similarly, galactose softens the texture of the seeds and liberates sugar for digestion. The reduction in the amino acids and the reducing sugar concentration may be due to the presence of some micro-organism that feed on amino acids and reducing sugars.

ANIMAL PRODUCTS (MILK, MEAT) NONO, CHEESE, KILISHI

Nono

Nono is a fermented food drink derivatives gotten from cow milk. As a general practice among Fulani Herdsmen, the milking is done between the third and sixth months of lactation. Until the third month, the calves are left to consume milk. Cows are only milked at night and since no milking is possible during the day calves roam with the dam. Milk, if left untreated, spoils within a short time due to microbial activity; thus, processing of milk improves its storage and diversifies its use.

Traditionally, nono is prepared by inoculating freshly drawn cow milk with a little of the leftover as starter and then is allowed to ferment for twenty four hours at room temperature. During fermentation, some of the lactose is converted to the lactic acid. At the end of the fermentation period, the milk butter is removed by churning for further use and the remaining sour milk, nono is a delicious and refreshing beverage. Most of the organisms involved in the fermentation process are usually of three main groups; bacteria, yeast, and mould. Of these, *Lactobacilli*(*L. acidophilus and L.bulgaris*), *Lactococci species* (*L. cremoni, and L.lactis*), *Streptococcus thermophilus, Leuconostoc species and Saccharomyces species*seems to be the most prominent, each giving the product a characteristic flavour.

Nono has yoghurt-like taste (sharp acid taste), and is therefore usually taken with sugar, and fura which is made up of millet flour compressed in balls and cooked for about twenty to forty minutes. The cooked fura is crumbled in a bowl of nono (now called fura de nunu). Nono is an excellent source of protein, rich in essential amino acids and a good source of calcium, phosphorus and vitamin A, B, C, E and B complex. However, like other milk products, it is poor in ascorbic acid and iron. Wives of pastorialist usually process fresh milk into various traditional milk product. These include nono,(sour milk), kindirimo(sour yoghurt), maishanu (local butter), cuku (Fulani cheese) and wara (Yoruba cheese). These products are usually hawked around the local markets in certain towns. These products are usually only available within the walking distance of Fulani settlement. For the same reason, these products are also more readily available in the northern states of the country.

Production of kilishi and other processed meats of interest.

Suya (esire or balangu), banda (kundi) and kilishi are the most important traditional processed meats in Nigeria and other West African countries including Chad, Niger and Mali.

Banda is a salted, smoke-dried meat product made from chunks of cheap, low quality meat from various types of livestock including donkeys, horses, camel, buffallo and wild life. The meat chunk is pre-cooked before smoking/ kiln drying or sun-drying. The traditional smoking/ kiln for banda, usually an open top. Fifty-gallon of oil drum fitted with layers of wire mesh that hold the product, and fired from the bottom. Banda is a poor product,stone-hard and dark I and n colour. Unlike banda, suya, and kilishi are made by roasting, spiced, salted, slices,/ strips of meat(usually beef).

Kilishiis is different from suya in that a two stage sun-drying process proceeds to roasting. Consequently, kilishi has a low moisture content (6-14%) than suya (25-35%) and a longer shelf- life. A variety of spices and other dried ingredients are used in kilishi processing including ginger(*Zingiberofficnale*), chillies (*Capsicufrtescens*), melegueta pepper (*Aframomum melegueya*), onion(*Allium cepa*) and defatted peanut(*Arachis hypogea*), cake powder, kilishi consist of about 46% meat and 54% non-meat ingredients with defatted powder accounting for about 35% of the ingredients formulation. Other traditional processed meat products in Nigeria include ndariko and jirge.

The summary of micro organism associated with Nigerian fermented foods is shown in table 1

Table 1: Role of fermented focd in detoxification

SUBSTRATE	MICROORGANISM	PRODUCT
Cassava	*Leuconostoc sp Geotrichum candidum Pseudomonas* sp. *Scclecotrichum graminisBacteriodes* sp. *Tallospora asperaActinomyces* sp. *Passalora bacilligera Corynebacterium* sp. *Varicosporium* species *Lactobacillus* sp. *Culicidospora gravida Diplococcium spicatum*	Gari
Yam	*Streptococcus* sp. *Articulospora inflate Lactobacillus* sp. *Aspergillus niger Listeria* sp. *Aspergillus rapens Aspergillus flavus Lemonniera aquatic*	Elubo-isu
African locust beans *Parkia filicoida*	*Lactobacillus* sp. *Rhizopus stolonifer Streptococcus* sp. *Aspergillus fumigatus Pediococcus* sp. *Triscelophorus monosporus Bacillus* sp., *Coryneform bacteria*	Iru
Castor seed *Ricinus communis*	*Bacillus* spp., *Pseudomonas* spp. *Micrococcuss* spp., *Streptococcus*	Ogiri-igbo
Fluted pumpkin seeds *Telferia ocidentalis*	*Bacillus* spp., *E. coli*, *Staphylococcus* Spp., *Pseudomonas*	Ogiri ugu

SUBSTRATE	MICROORGANISM	PRODUCT
African oil beans *Pentaclethra macrophylla*	*Bacillus subtilis, Staphylococcus* spp., *Micrococcus* spp., *Corynebacterium* spp.	Ugba / Ukpaka
Soya bean	*B. subtilis, B. licheniformis, B.megaterium, Staphylococcus epidermidis, Micrococcus* spp.	Okpiye/Okphehe
African yam beans *Stenophylis stenocarpa*	*Bacillius subtilis, B. licheniformis, B. pumilis, Staphylococcus* sp.	Owoh
Melon seed *Citrulus vulgaris*	*Bacillus subtilis, B. megateruim, B. firmus E.coli,. Proteus, Pediococcus, Alcaligenes* spp., *Pseudomonas aeruginosa*	Ogiri-egusi
Cereals: maize, sorghum, millet	*Saccharomyces cerevisiae, Lactobacillus* spp., *Fusarium* spp.	Ogi Agidi
Milk	*Lactobacillus* spp. *Lactococcus* spp. *Streptococcus* spp. *Pediococcus* spp. *Leuconostoc* spp. *Propionibacter* spp.	Wara (Nigerian cheese)
Grain flour	*Saccaromyces cerevisiae*	Bread
Cereals (Millet, sorghum, maize, rice)	*Lactobacillius plantarum, L. fermentum and Lactococcus lactis*	Kunun-zaki

Mycotoxin Detoxification

Food and feeds are often contaminated with a number of toxins either naturally or through infestation by micro-organisms such as moulds, bacteria and virus. Certain moulds often produce secondary toxic metabolites called mycotoxins. These include fumonisins, ochratoxins A, zearalenone and aflatoxins. Several methods are available for degrading toxins from contaminated food. For example, using alkaline ammonia treatment to remove mycotoxins from food. However, these methods are harsh to food as they involve the use of chemicals which are potentially harmful to health or may impair or reduce the nutritional value of foods. Cooking foods does not remove mycotoxins either as most of them are heat stable. Detoxification of mycotoxins in foods through LAB fermentation has been demonstrated over the years (Biernasiak et.al., 2006). Using LAB fermentation for detoxification is more advantageous in that it is a milder method, which preserve the nutritive value and flavour of de-contaminated food. In addition to this, LAB fermentation irreversibly degrades mycotoxins without leaving any toxic residues. The detoxifying effect is believed to be through toxin binding effect.

In mycotoxin detoxification, LAB fermentation has also been successfully used to detoxify cassava toxins (cyanogens) following fermentation of cassava food product. In addition to cyanogens detoxification, cassava fermentation contributes to the preservation and improvement of flavour and aroma of cassava ferment. Although cooking has been used as a method of cyanogens detoxification, it has a number of problems as it leaves residual cyanogens in processed cassava, which exist as glucosides, cyanohydrins or free cyanide which are equally toxic as their parent compounds in uncooked food.

In a review, Bankole and Adebanjo (2003), found that the level of aflatoxin B1, B2 and

G1 were significantly higher in corn from the high incidence area for human hepatocellular carcinoma and the average daily intake of aflatoxin B1 from the high risk area was 184.1 g/kg aflatoxin. Udoh *et al*(2000) reported 33% of maize sample from ecological zones of Nigeria contaminated with aflatoxins.

Fermented maize (Ogi) is a staple cereal in Nigeria and it is a popular weaning food in most rural communities in Nigeria. Oluwafemi and Ikeowa (2005) have reported that in fermenting maize to ogi, aflatoxin B1 was reduced by about 50% after 72hours of fermentation. Maize as well as other Nigerian cereals are also important raw materials for both local and commercial beer brewing. Oluwafemi and Taiwo (2003) have shown that the role of *S. cerevisiae* in reducing the pH from 5.2 to 3.7 during fermentation is important in detoxifying aflatoxin B during beer fermentation.

Cyanide Detoxification

Processing of cassava roots improves palatability, reduces or eliminates potential toxicity, transforms raw cassava into other preservable forms which are more beneficial to man. Fermentation is by far the most common method of processing the cassava crop in Africa (Okafor *et al.*, 1984). The rate of detoxification of cyanide by traditional fermentation is shown in figure 13. Fermentation is a very effective way for detoxification of cyanide in cassava with r^2 of 98%.

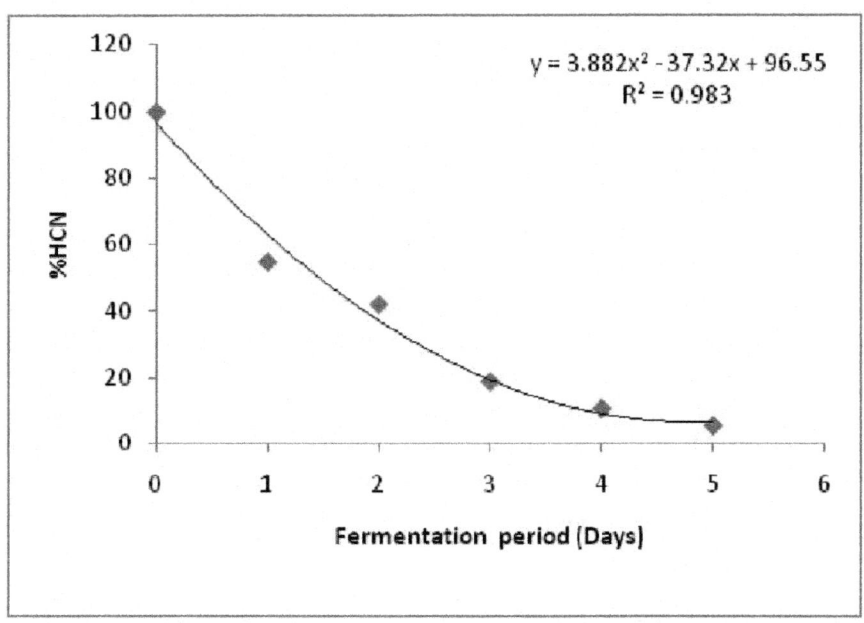

Figure 13: Effect of fermatation on %HCN

Loop Fermentation

Loop fermentation is achieved by using starter culture from already fermented product to inoculate a fresh barge of fermentation process. Ohenhen and Ikenebomeh(2007) have shown that that loop fermentation can prolong the shelf-life of ogi from about 40days, obtained by traditional fermentation method to well over 60days. We have observed in our laboratory that by using loop fermentation technique in gari processing, fermentation was completed in three days as against five days by the conventional traditional fermentation. Cyanide content also reduced to about 3% with loop fermentation. With a second loop (double loop) ie using products of a first loop fermentation to inoculate a fresh process, fermentation in gari production was completed in 2day with only about 2.6% cyanide remaining. The explanation is that the organisms are "trained" to better utilize the compounds in the fermenting substrate When the fermenting substrate in the double loop was acidified by squeezing some limes (citrus) juice into it before inoculating, cyanide content was 0% after 3days. The summary is shown in Fig. 14.

Detoxification of Phytates, Tannins and Oxalates

The anti-nutrients including Phytates, tannins and oxalates interferes with mineral absorption and palatability of the cereals so detoxification is vital to enhance their nutrient value and organoleptic properties. Several detoxification methods are available, including decortication, malting, fermentation and alkali treatment (Osuntogun *et al.*. 1989; Banda-Nyirenda and Vohra, 1990). Yeast fermentationhas proved very effective in the detoxification of antinutrients. The table below summarizes the effectiveness of yeast in detoxification of different anti-nutrient.

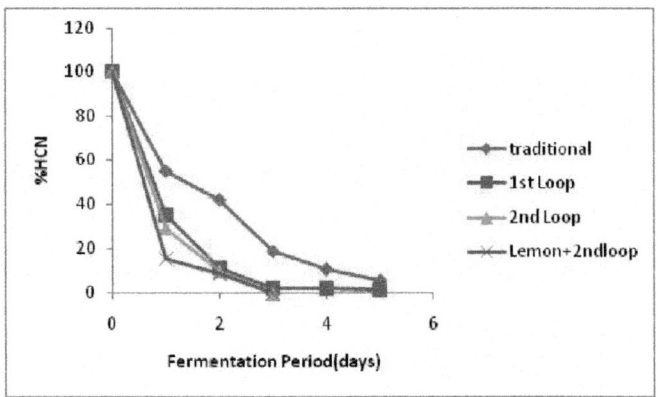

Figure 14: Effect of fermentation loop on Percentage residual HCN

Onyesom *et al*(2008) have also shown that fermentation of cassava to fufu with lemon grass reduces cyanide to less than 1% after 5days.

Phytic acid is well documented to block absorption of not only of phosphorus, but also other minerals such as calcium, magnesium, iron and zinc. It also negatively affects the absorption of lipids and protein. One reason this is true is because phytic acid also inhibits enzymes that are needed to digest foods such as pepsin (which helps break down protein), amylases (convert starch into sugar for digestion) and trypsin (also used in protein digestion). While whole grains have a much higher mineral content than processed grains, the full benefit of that nutrition is lost if phytic acid blocks the nutrients from being absorbed. It is well known that most cereals and legumes contain high levels of these ant-nutrients. It is also common knowledge that most of the Nigerian staples as in other developing countries constitute mainly cereals and legumes. It is therefore important that these foods staples are fermented as well as improve on the traditional fermentation techniques.

Table 2: Overview of yeasts activities in degradation of anti-nutrients

Activity	Yeast species	Effects
Biodegra-dation of phytate	Saccharomyces cerevisiae, Saccharomyces kluyveri, Schwanniomyces castel-lii, Debaryomyces castellii, Arxulaadeninivorans, Pichia anomala, Pichia rhodanensis, Pichia spartinae, Cryptococcus laurentii, Rhodotorula graci-lis, Torulaspora delbrueckii, Kluyveromyces lactis Candida krusei (Issatchenkia orientalis) and Candida spp.	Nutritional importance, i.e., bioavailability of divalent min-eralssuch as iron, zink, calcium and magnesium
Folate biofortifica-tion	S. cerevisiae Saccharomyces bayanus, Saccharomyces paradoxus, Saccharomyces pastorianus, Metschnikowia lochheadii, Debaryomyces melissophi-lus, Debaryomyces vanrijiae var. vanrijiae, Debaryomyces hansenii, Pichia philogaea, Kodamaea anthophila, Wick-erhamiella lipophilia, Candida cleridarum and Candida dro-sophilaeCandida milleri and T. delbrueckii Saccharomyces ex-iguous andCandida lambicaP. anomala and Candida glabrata Kluyveromyces marxianus and C. krusei (I. orientalis)	Prevention of neural tube defects in the foetus, megalo-blastic anaemia and reduction of the risk for cardiovascular disease, cancer and Alzheim-er's disease
Degrada-tion of mycotoxins	. cerevisiae sppPhaffia rhodo-zyma and Xanthophyllomyces dendrorhous	Antitoxic in some degree

CONCLUSION

Developing countries like Nigeria require food processing technologies that are appropriate, suitable for tropical regions and affordable to rural and urban economies. Fermentation techniques are one of such technologies that have been developed indigenously for a wide range of food products. These include root crops, cereals, legumes, fruit and vegetables, dairy, fish and meat. As a unit operation in food processing, fermentation offers various advantages, including, improved food safety, improved nutritional values, enhance flavour and acceptability, reduction in anti-nutrients, detoxification of toxigenic compounds, enhanced shelf-life and improved functional properties.

The present review has shown that Nigerian fermented food and food products can be developed into medium or large scale level for standard commercial products. However, there is the need to further optimize the processes

REFERENCES

1. A. O Achi, 2005Traditional fermented protein condiments in Nigeria. African Journal of Biotechnology 416121621

2. T. T Adebolu, A. O Olodun, and B. C Ihunweze, 2007Evaluation of ogi liquor from different grains for antibacterial activities against some common pathogens. Afr. J. Biotech. 6911401143

3. O. O. I Agarry, O Nkama, Akoma (2010Production of Kunun-zaki (A Nigerian fermented cereal beverage) using starter culture. International Research Journal of Microbiology 1018025

4. T Agbor-egbe, Lape Mbome I. (2006The effects of processing techniques in reducing cyanogen levels during the production of some Cameroonian cassava foodsJournal of Food Composition and Analysis; 19354363

5. D. B. C Banda-nyirenda, and P Vohra, 1990Nutritional improvement of tannin containing sorghums (Sorghum bicolor) by sodium bicarbonate. Cereal Chem. 67533537

6. S. A Bankole, A Adebanjo, 2003Aflatoxin contamination of dried yam chips marketed in Nigeria.Tropical Science 43 (3-4).

7. J Biernasiak, M Piotrowska, and Z Libudzisz, 2006Detoxification of mycotoxins by probiotic preparation for broiler chickens. Mycotoxin Research 22, (4): 230-235.

8. P Cadieux, J Burton, and C. Y Kang, 2002Lactobacillus strains and vaginal ecology.JAMA; 28719401941

9. G Chamberlain, J Husnik, and R. E Subden, 1997Freeze-desiccation survival in wild yeasts in the bloom of icewine grapesCanadian Instituteof Food Science and Technology Journal 30435439

10. Chung Shil KwakMee Sook Lee, Se In Oh, and Sang Chul Park (2010Discovery of Novel Sources of Vitamin B12 in Traditional Korean Foods from Nutritional Surveys of Centenarians Curr Gerontol Geriatr Res. 2010; 2010: 374897.

11. A. O Ijabadeniyi, 2007Microbiological safety of gari, lafun and ogiri in Akure metropolis, Nigeria. African Journal of Biotechnology 626332635

12. S. E Katz, 2003Wild Fermentation: The Flavor, Nutrition, and Craft of Live-Culture Foods.White River Junction, VT.: Chelsea Green Publishing Company.

13. Moslehi-JenabianS; Pedersen,LL and Jespersen,L(2010Beneficial Effects of Probiotic and Food Borne Yeasts on Human Health.Nutrients2449473

14. R. E Ohenhen, M. J Ikenemoh, 2007shelf stability and Enzyme Activity Studies of Ogi: A corn meal fermented product. J. Am. Sci. 313842

15. N Okafor, B Lioma, and C Oylu, 1984Studies on the microbiology of cassava retting for "foo-foo"production. J. Appl. Bacteriol., 56113

16. R. I Okechukwu, N Ewelike, A. A Ukaoma, A. A Emejulu, And Azuwike C. O (2012Changes in the nutrient composition of the African oil bean meal "ugba" (Pentaclethre macrophylla Benth) subjected to solid state natural fermentation. Journal of Applied Biosciences 5135913595

17. F Oluwafemi, and M. C Ikeowa, 2005Fate of Aflatoxin B1 During Fermentation of Maize into "Ogi" Nigerian Food Journal,23www.ajol.info/journals/nifoj)

18. F Oluwafemi, and V. O Taiwo, 2003Reversal of toxigenic effects of aflatoxin B1 on cockerels by alcoholic extract of African nutmeg Monodora myristica.J. Sci. Food Agric 84333340

19. I Onyesom, P. N Okoh, and O. V Okpokunu, 2008Levels of Cyanide in Cassava Fermented with Lemon Grass (Cymbopogon citratus) and the Organoleptic Assessment of its Food Products. World Applied Sciences Journal 46860863

20. A Osho, O. O Mabekoje, and O. O Bello, 2010Comparative study on the microbial load of Gari, Elubo- isu and Iru in Nigeria. African Journal of Food Science 4646649

21. B. O Osuntogun, S. R. A Adewusi, J. O Ogundiwin, and C. C Nwasike, 1989Effect of cultivar, steeping, and malting on tannin, total polyphenol, and cyanide content of Nigerian sorghum. Cereal Chem. 668789

22. G Reid, A. W Bruce, N Fraser, C Heinemann, J Owen, and B Henning, 2001bOral probiotics can resolve urogenital infections. FEMS Immunol Med Microbiol; 304952

23. G Reid, 2002aProbiotics for urogenital healthNutr. Clin. Care; 538

24. G Reid, D Beuerman, C Heinemann, and A. W Bruce, 2001aProbiotic Lactobacillus dose required to restore and maintain a normal vaginal flora.FEMS Immunol Med Nlicrobiol; 323741

25. K. H Steinkraus, Ed. (1995Handbook of Indigenous Fermented FoodsNew York, Marcel Dekker, Inc., 776 pp.

26. UdohJ.M; Cardwell, K.F; and Ikotun, T. (2000Storage structure and aflatoxin content of maize in five agroecological zones of Nigeria,J. Stored Prod. Res 36187201

Chapter 10

SHUIDOUCHI (FERMENTED SOYBEAN) FERMENTED IN DIFFERENT VESSELS ATTENUATES HCL/ETHANOL-INDUCED GASTRIC MUCOSAL INJURY

Huayi Suo [1,2], Xia Feng [3,4], Kai Zhu [3,4], Cun Wang [3,4], Xin Zhao [3,4] and Jianquan Kan [1,2]

[1]College of Food Science, Southwest University, Chongqing 400715, China

[2]Chongqing Engineering Research Center of Regional Food, Chongqing 400715, China

[3]Department of Biological and Chemical Engineering, Chongqing University of Education, Chongqing 400067, China

[4]Chongqing Collaborative Innovation Center of Functional Food, Chongqing University of Education, Chongqing 400067, China

ABSTRACT

Shuidouchi (Natto) is a fermented soy product showing *in vivo* gastric injury preventive effects. The treatment effects of Shuidouchi fermented in different vessels on HCl/ethanol-induced gastric mucosal injury mice through their antioxidant effect was determined. Shuidouchi contained isoflavones (daidzein and genistein), and GVFS (glass vessel fermented Shuidouchi) had the highest isoflavone levels among Shuidouchi samples fermented in different vessels. After treatment with GVFS, the gastric mucosal injury was reduced as compared to the control mice. The gastric secretion volume (0.47 mL) and pH of gastric juice (3.1) of GVFS treated gastric mucosal injury mice were close to those of ranitidine-treated mice and normal mice. Shuidouchi could decrease serum motilin (MTL), gastrin (Gas) level and increase somatostatin (SS), vasoactive intestinal peptide (VIP) level, and GVFS showed the strongest effects. GVFS showed lower IL-6, IL-12, TNF-α and IFN-γ cytokine levels than other vessel fermented Shuidouchi samples, and these levels were higher than those of ranitidine-treated mice and normal mice. GVFS also had higher superoxide dismutase (SOD), nitric oxide (NO) and malonaldehyde (MDA) contents in gastric tissues than other Shuidouchi samples. Shuidouchi could raise IκB-α,

EGF, EGFR, nNOS, eNOS, Mn-SOD, Gu/Zn-SOD, CAT mRNA expressions and reduce NF-κB, COX-2, iNOS expressions as compared to the control mice. GVFS showed the best treatment effects for gastric mucosal injuries, suggesting that glass vessels could be used for Shuidouchi fermentation in functional food manufacturing.

INTRODUCTION

Soybeans fermented in water for a short time (Shuidouchi) is a traditional Chinese fermented soybean product, whose process is similar to Chungjukjang from South Korea and Natto from Japan [1]. It rich in nutrients, including proteins, vitamins and minerals, and its content of vitamin E is especially noteworthy [2]. Fermented soybeans not only have high nutritive value, but are also used as a drug in Traditional Chinese Medicine [3]. According to Traditional Chinese Medicine, fermented soybeans can clear heat and detoxify, which can treat headaches due to pathogenic wind-heat, chest distress and vomiting. Numerous soy oligosaccharides in fermented soybeans can improve the body's immunity and reduce the intestinal levels of toxic substances, which can prevent intestinal tumors [4]. Scholars in Japan and South Korea found that Chungjukjang and Natto have many health benefits, including anti-oxidation, anti-inflammatory and anti-cancer activities. As a soy product, the most active ingredient of soybeans fermented in water is soy isoflavones. Isoflavones in fermented soybean products are more active than that in raw soybeans, which have very strong anti-tumor and anti-aging effects and prevent embrittlement of blood capillaries [5].

Shuidouchi is a soy product fermented for a short time [6]. In addition to factories, it can also be homemade in many areas. Similarly, in Japan and South Korea, they often make Natto and Chungkukjang at home. Different kinds of containers are often used in fermentation of Natto and Chungkukjang both in the factories and at home. Japan has even developed an automatic Natto fermentation machine with a metal tank. South Korean scholars have studied the sensory, physical and chemical properties as well as antioxidant effects of Chungkukjang fermentation with jars and glass, they found that jars were much better for the production of Chungkukjang [7]. In China, metal, glass, plastic and ceramic containers are often used in Shuidouchi production. This research aims to study the anti-gastric mucosal injury effects of Shuidouchi fermented in different containers and explain the mechanism of the anti-oxidation effects of Shuidouchi.

The stomach is in a protected anatomical position in abdominal cavity and can move within some limits, so it is not easy to injure it with outside violence [8]. Gastric mucosal damage is very common and caused by many factors, including chemical factors such as smoking, drinking strong tea, coffee, and drugs stimulating the gastric mucosa such as aspirin and indomethacin, physical factors such as too much cold or heat, eating rough food, bacteria or their toxins [9]. The alcohol in wine can greatly stimulate the gastric mucosa, and taking in too much alcohol can lead to gastric mucosa damage and congestion in the stomach [10]. Most bean products can protect the stomach, as they take advantage of their alkaline characteristics to neutralize the amount of hydrochloric acid due to gastric damage and alleviate stomach injuries. In addition to their alkalinity, soy isoflavones in soybean products may play a key role in alleviating stomach damage [2]. Soybean foods contain many isoflavones, such as daidzein, genistein, glycitein, *etc.*, but these isolflavones cannot be immediate absorbed in the human body, and they must be hydrolysed to absorbable aglycones by β-glucosidase from the intestinal microbiota. Shuidouchi is produced by microorganisms and these microorganism could make these isoflavones change into functional compounds which could be readily absorbed by humans.

A mice model of gastric mucosal injury induced by hydrochloric acid and alcohol can determine the health effects of functional food components. Alcohol is the main factor causing gastric mucosal injury, and a certain concentration of hydrochloric acid can promote and increase gastric mucosa lesions caused by alcohol. Based on this animal model, mice are gavaged with soybean isoflavones extracted from fermented soybeans to test biochemical indicators of serum levels (MTL, Gas, SS, VIP, IL-6, IL-12, TNF-α and IFN-γ), tissue levels (SOD, NO and MDA) and mRNA expression (NF-κB, IκB-α, EGF, EGFR, nNOS, eNOS, iNOS, COX-2, Mn-SOD, Gu/Zn-SOD and CAT) in gastric tissues. The experimental results prove the gastric mucosa damage prevention effects of isoflavones in fermented soybeans and help elucidate its possible mechanism. This study also aimed to know the physicochemical properties of Shuidouchi produced by fermentation in different vessels, and the relationship between the physicochemical properties discrepancies (isoflavones content, moisture content, fermentation temperature, acidity and total bacterial count) and anti-gastric mucosa damage effects.

RESULTS AND DISCUSSION

Isoflavone Contents of Shuidouchi

In this study, the isoflavone contents of Shuidouchi was determined by a spectrophotometric method, whereby isoflavone standard solutions (daidzein and genistein) were measured, and the regression equation of contents of daidzein and genistein were made, $Y_{daidzein}$ (daidzein regression equation) = $148338 + 1.70 \times 10^8 \times (r = 0.999)$, $Y_{genistein}$ (genistein regression equation) = $-316706 + 4.20 \times 10^8 \times (r = 0.995)$.

Compared with these regression equations, the results showed that GVFS has the highest contents of daidzein and genistein (Table 1), and MVFS had more daidzein and genistein content than PVFS and CVFS ($p < 0.05$). Soybean isoflavones can prevent and cure many diseases [11]. Isoflavones are limited in Nature and soybean is the only nutritionally-meaningful food source of isoflavones [12]. Many studies had shown that soybean isoflavones have strong antioxidant effects, especially the soybean isoflavones in the human body, which have strong antioxidant and anti-inflammatory effects [13,14,15]. Recent research showed that the content of soybean isoflavones is higher in Shuidouchi, and the content was much higher in fermented soybeans because of the fermentation effect [2].

During the fermentation process, many factors could affect the contents of isoflavones in Shuidouchi, including the fermentation container type. Temperature and moisture are important factors in fermentations. By temperature control, the inside and outside of the container could remain unobstructed, which was advantageous for natural fermentation, while adequate moisture was also helpful for fermentation [5]. Diathermancy and moisture retention of glass containers were better than the other vessels, which were more advantageous to Shuidouchi fermentation [16]. They could produce more soybean isoflavones, which might inhibit gastric lesions.

Table 1: Contents of soybean isoflavones in different vessels fermented Shuidouchi.

Group	Daidzein (mg/g)	Genistein (mg/g)
CVFS	0.45 ± 0.03 [c]	0.72 ± 0.04 [c]
PVFS	0.48 ± 0.04 [c]	0.73 ± 0.03 [c]
MVFS	0.67 ± 0.02 [b]	0.96 ± 0.04 [b]
GVFS	0.84 ± 0.03 [a]	1.22 ± 0.05 [a]

[a-c] Mean values with different letters in the same column are significantly different ($p < 0.05$) according to Duncan's multiple-range test. CVFS, ceramic vessel fermented Shuidouchi; PVFS, plastic vessel fermented Shuidouchi; MVFS, metal vessel fermented Shuidouchi; GVFS, glass vessel fermented Shuidouchi.

Physicochemical Properties of Shuidouchi

After 72 h fermentation, the moisture content, temperature, acidity and total bacterial counts of Shuidouchi samples fermented in different vessels were determined. GVFS had the highest moisture content and total bacterial counts, but it had a lower temperature than CVFS, PVFS, MVFS and had lower acidity than CVFS, PVFS (Table 2).

Table 2: Physicochemical properties of Shuidouchi fermented in different vessels.

Group	Moisture Content (%)	Temperature (°C)	Acidity (%)	Total Bacterial Counts (×10⁹ CFU/g)
CVFS	53.82 ± 0.03 [d]	38.84 ± 0.08 [b]	0.92 ± 0.07 [a]	1.03 ± 0.07 [d]
PVFS	54.10 ± 0.02 [c]	39.72 ± 0.21 [a]	0.89 ± 0.06 [a]	1.24 ± 0.06 [c]
MVFS	56.39 ± 0.04 [b]	37.36 ± 0.12 [c]	0.81 ± 0.09 [ab]	1.56 ± 0.06 [b]
GVFS	57.37 ± 0.01 [a]	37.42 ± 0.13 [c]	0.72 ± 0.11 [b]	1.78 ± 0.12 [a]

[a-d] Mean values with different letters in the same column are significantly different ($p < 0.05$) according to Duncan's multiple-range test. CVFS, ceramic vessel fermented Shuidouchi; PVFS, plastic vessel fermented Shuidouchi; MVFS, metal vessel fermented Shuidouchi; GVFS, glass vessel fermented Shuidouchi.

Moisture is a essential factor for bacterial growth, as the richness of moisture favors the proliferation of bacteria, so rich moisture could help Shuidouchi to ferment [17]. Glass vessels could retain the moisture during the fermentation of Shuidouchi, which might promote this fermentation and make GVFS a high quality soybean fermented food. Shuidouchi was fermented at 36 °C in this study, and the temperatures of Shuidouchi changed in the different vessels. Under the same temperature conditions, these changes result from the heat conductivities of different vessels. The heat conductivities of glass and metal are better than those of ceramic and plastic. Thirty six (36) °C is a suitable temperature for bacterial growth [18], the fermentation temperature of glass and metal vessels fermented Shuidouchi were close to 36 °C, and glass and metal vessels could maintain a suitable temperature for Shuidouchi fermentation. The high acidity could inhibit the bacteria growth, and maintaining a convenient acidity would help bacterial growth [19]. Glass vessels can keep a convenient acidity for fermentation of Shuidouchi. From the results, glass vessels provide a better fermentation environment for the fermentation of Shuidouchi than other vessels, as glass vessels could help the Shuidouchi have more moisture content, total bacterial counts and lower temperature, acidity, and these conditions could make GVFS produce more daidzein and genistein than Shuidouchi fermented in other vessels.

Stomach Appearances of Mice

After Shuidouchi treatment, the gastric mucosal injury areas were reduced as compared to the control mice (Table 3, Figure 1), and the gastric mucosal injury area of GVFS-treated mice was significantly ($p < 0.05$) lower from other vessel fermented Shuidouchi-treated mice. The area was close to that of the ranitidine-treated mice. The inhibitory rate of GVFS-treated mice was also higher than that of other Shuidouchi-treated mice.

Table 3: Stomach appearance of HCl/ethanol induce gastric mucosal injury mice treated with Shuidouchi fermented in different vessels.

Group	Gastric Mucosal Injury	
	Gastric Mucosal Injury Area (mm²)	Inhibitory Rate (%)
Normal	0.0 ± 0.0 [f]	100 ± 0.0 [a]
Control	7.21 ± 0.62 [a]	0.0 ± 0.0 [d]
CVFS (500 mg/kg)	2.88 ± 0.36 [b]	60.1 ± 4.6 [e]
PVFS (500 mg/kg)	2.75 ± 0.31 [b]	61.9 ± 4.1 [b]
MVFS (500 mg/kg)	2.15 ± 0.18 [c]	70.2 ± 2.5 [d]
GVFS (500 mg/kg)	1.84 ± 0.21 [d]	74.5 ± 2.7 [c]
Ranitidine (50 mg/kg)	1.05 ± 0.17 [e]	85.4 ± 2.4 [b]

[a-f] Mean values with different letters in the same column are significantly different ($p < 0.05$) according to Duncan's multiple-range test. CVFS, ceramic vessel fermented Shuidouchi; PVFS, plastic vessel fermented Shuidouchi; MVFS, metal vessel fermented Shuidouchi; GVFS, glass vessel fermented Shuidouchi.

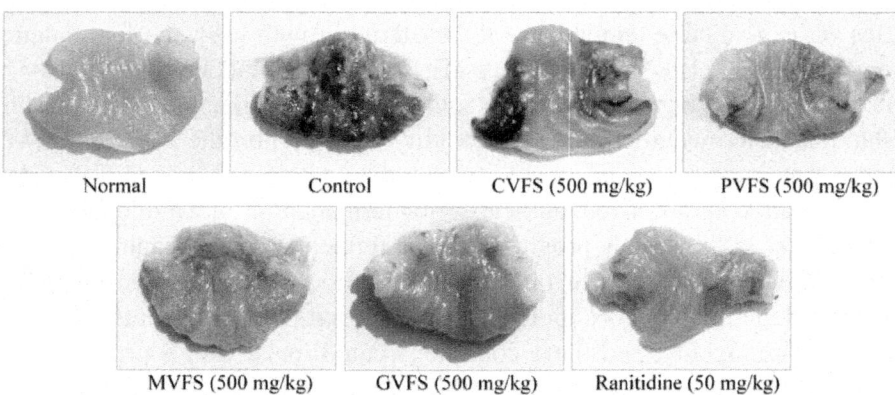

Normal   Control   CVFS (500 mg/kg)   PVFS (500 mg/kg)

MVFS (500 mg/kg)   GVFS (500 mg/kg)   Ranitidine (50 mg/kg)

Figure 1: Stomachs with HCl/ethanol-induced gastric mucosal injury in mice treated with Shuidouchi fermented in different vessels. CVFS—ceramic vessel fermented Shuidouchi; PVFS—plastic vessel fermented Shuidouchi; MVFS—metal vessel fermented Shuidouchi; GVFS—glass vessel fermented Shuidouchi.

After administering HCl/ethanol, the gastric mucosal injury area, an important index for gastric mucosal injury determination, could be increased, [20]. From this study, GVFS could decrease the HCl/ethanol induced gastric mucosal injury, and its effects seem stronger than those of Shuidouchi fermented in other vessels.

Gastric Secretion Volume and pH of Gastric Juice of Mice

GVFS-treated mice had the lower gastric secretion volume but higher than that of normal and drug (raniticine)-treated mice (Table 4). GVFS-treated mice also had higher gastric juice pH than mice treated with Shuidouchi fermented in other vessels, close to the normal and ranitidine-treated mice.

Table 4: Gastric secretion volume and pH of gastric juice of HCl/ethanol induced gastric mucosal injury mice treated with Shuidouchi fermented in different vessels.

Group	Gastric Secretion Volume (mL)	pH of the Gastric Juice
Normal	0.28 ± 0.03 [f]	3.6 ± 0.1 [a]
Control	1.12 ± 0.17 [a]	1.0 ± 0.2 [f]
CVFS (500 mg/kg)	0.71 ± 0.11 [b]	2.4 ± 0.4 [e]
PVFS (500 mg/kg)	0.68 ± 0.12 [b]	2.5 ± 0.3 [e]
MVFS (500 mg/kg)	0.57 ± 0.05 [c]	2.8 ± 0.2 [d]
GVFS (500 mg/kg)	0.47 ± 0.04 [d]	3.1 ± 0.1 [c]
Ranitidine (50 mg/kg)	0.38 ± 0.04 [e]	3.3 ± 0.2 [b]

[a-f] Mean values with different letters in the same column are significantly different ($p < 0.05$) according to Duncan's multiple-range test. CVFS, ceramic vessel fermented Shuidouchi; PVFS, plastic vessel fermented Shuidouchi; MVFS, metal vessel fermented Shuidouchi; GVFS, glass vessel fermented Shuidouchi.

After determination of gastric secretion volume and pH of gastric juice, the gastric mucosal injuries could be checked out. High gastric secretion volume and low pH of the gastric juice cause severe gastric mucosal injury [21]. Shuidouchi could reduce the gastric secretion volume and raise the pH of the gastric juice, and the GVFS had the most obvious effects.

Serum Motilin (MTL), Gastrin (Gas), Somatostatin (SS) and Vasoactive Intestinal Peptide (VIP) Levels in Mice

GVFS could significantly ($p < 0.05$) increase the serum SS, VIP levels and decrease MTL, Gas levels compared to the control mice (Table 5). The serum MTL, Gas levels of GVFS treated mice were lower than with Shuidouchi fermented in other vessels, and serum SS, VIP levels of GVFS-treated mice were higher than those of animals treated with Shuidouchi fermented in other vessels. These levels were close to those of the ranitidine-treated mice and normal mice.

MTL is the gastrointestinal hormone of excitability. After stimulation, its content increases and this case abundant secretion of hydrochloric acid, which makes the stomach acidic and worsens the degree of gastric ulceration

[22]. Stimulated by certain substances, Gas would be released into the blood and stimulate the parietal cells to secrete hydrochloric acid. The abnormal secretion of gastric acid worsens gastric mucosal injuries [23]. SS was not only a kind of neural hormone, but also a kind of neuromodulator, which could inhibit the secretion of many gastrointestinal hormones, reduce gastrointestinal peristalsis, blood flow in the viscera and portal veins, as well as release of inflammatory mediators, thus inhibiting gastric mucosal injuries [24]. VIP is a gastrointestinal inhibitory hormone, which could inhibit the secretion of stomach acid. Meanwhile, VIP could activate the D cells on the stomach wall. D cells release somatostatin, which could inhibit the secretion of gastrin by G cells on the stomach wall and play a key role of alleviating gastric mucosa injuries [25].

Table 5: Serum MTL, Gas, SS and VIP levels of HCl/ethanol induce gastric mucosal injury mice treated with Shuidouchi fermented in different vessels. Define all in table caption.

Group	MTL (μg/L)	Gas (μg/L)	SS (μg/L)	VIP (μg/L)
Normal	42.0 ± 2.3 [f]	67.6 ± 5.3 [f]	128.0 ± 5.3 [a]	99.8 ± 4.3 [a]
Control	108.3 ± 6.9 [a]	137.6 ± 7.9 [a]	74.1 ± 3.3 [f]	51.2 ± 2.9 [f]
CVFS (500 mg/kg)	73.2 ± 5.2 [b]	95.4 ± 4.8 [b]	95.4 ± 7.8 [e]	70.3 ± 3.6 [e]
PVFS (500 mg/kg)	70.6 ± 4.7 [b]	90.6 ± 6.1 [b]	97.6 ± 4.6 [e]	74.1 ± 3.3 [e]
MVFS (500 mg/kg)	59.7 ± 4.5 [c]	85.1 ± 3.6 [c]	105.3 ± 3.0 [d]	80.1 ± 2.5 [d]
GVFS (500 mg/kg)	52.6 ± 3.1 [d]	77.3 ± 3.2 [d]	114.1 ± 3.8 [c]	86.7 ± 2.8 [c]
Ranitidine (50 mg/kg)	47.6 ± 2.0 [e]	72.5 ± 2.2 [e]	120.6 ± 4.8 [b]	92.5 ± 3.1 [b]

[a-f] Mean values with different letters in the same column are significantly different ($p < 0.05$) according to Duncan's multiple-range test. CVFS, ceramic vessel fermented Shuidouchi; PVFS, plastic vessel fermented Shuidouchi; MVFS, metal vessel fermented Shuidouchi; GVFS, glass vessel fermented Shuidouchi.

Cytokine IL-6, IL-12, TNF-α and IFN-γ Levels in Mice

Cytokine IL-6, IL-12, TNF-α and IFN-γ levels in normal mice were the lowest, and these levels in control mice were the highest (Table 6). GVFS-treated mice showed a lower level than control mice and mice treated with Shuidouchi fermented in other vessels, and only higher than ranitidine-treated mice and normal mice.

Table 6: Cytokine IL-6, IL-12, TNF-α and IFN-γ levels of HCl/ethanol induce gastric mucosal injury mice treated with Shuidouchi fermented in different vessels.

Group	IL-6 (pg/mL)	IL-12 (pg/mL)	TNF-α (pg/mL)	IFN-γ (pg/mL)
Normal	41.2 ± 2.6 [g]	278.6 ± 33.5 [g]	43.5 ± 1.7 [f]	44.2 ± 1.5 [f]
Control	119.3 ± 7.3 [a]	942.1 ± 42.1 [a]	125.6 ± 8.3 [a]	99.2 ± 5.1 [a]
CVFS (500 mg/kg)	71.6 ± 4.8 [b]	651.2 ± 32.5 [b]	75.1 ± 6.2 [b]	68.7 ± 3.3 [b]
PVFS (500 mg/kg)	65.1 ± 2.5 [c]	608.6 ± 28.7 [c]	73.5 ± 6.8 [b]	65.3 ± 4.1 [b]
MVFS (500 mg/kg)	57.5 ± 3.3 [d]	506.7 ± 35.7 [d]	60.6 ± 4.9 [c]	57.1 ± 2.6 [c]
GVFS (500 mg/kg)	50.3 ± 2.3 [e]	425.6 ± 40.4 [e]	55.6 ± 3.8 [d]	51.2 ± 1.9 [d]
Ranitidine (50 mg/kg)	45.3 ± 1.9 [f]	361.5 ± 31.8 [f]	49.2 ± 2.8 [e]	47.6 ± 1.6 [e]

[a-g] Mean values with different letters in the same column are significantly different ($p <$ 0.05) according to Duncan's multiple-range test. CVFS, ceramic vessel fermented Shuidouchi; PVFS, plastic vessel fermented Shuidouchi; MVFS, metal vessel fermented Shuidouchi; GVFS, glass vessel fermented Shuidouchi.

IL-6 is a cytokine secreted by T cells, B cells and mononuclear macrophages, which becomes abnormal in many autoimmune diseases and is related to the pathological process and severity of these diseases [26]. A variety of abnormal antibodies and immune complexes in patients with gastric injuries could stimulate the monocyte-macrophages to produce and release TNF-α into blood circulation through different ways, causing increases of TNF-α levels in blood. The interaction between increased TNF-α and inflammatory cells worsens the inflammation and promotes gastric mucosa damage. IL-12 and IFN-γ are pro-inflammatory cytokines [27].

IL-12 could promote the growth and proliferation of T cells and NK cells and stimulate these cells to produce IFN-γ, which could increase the secretion of IL-12 [28]. Research had shown that IFN-γ and IL-12 could take part in gastric mucosa injuries and treatment through different interactions [29]. Under laboratory conditions, reducing the level of IL-6, IL-12, TNF-α and IFN-γ could reduce the degree of gastric injury in mice. Soy isoflavones in fermented soybeans water alleviated the effect of gastric mucosa injury by lowering the level of IL-6, IL-12, TNF-α and IFN-γ in serum [25].

Gastric Tissue SOD, NO and MDA Activities of Mice

After the gastric tissues determination, GVFS-treated mice showed higher SOD, NO contents than PVFS-, MVFS-, and CVFS-treated mice and control mice (Table 7). Control mice had the highest MDA content in gastric tissue, Shuidouchi could reduce the MDA content in gastric tissue, and GVFS decreased the MDA content the most as compared to Shuidouchi fermented in other vessels.

Table 7: Gastric tissues SOD, NO and MDA activities of HCl/ethanol induce gastric mucosal injury mice treated with Shuidouchi fermented in different vessels.

Group	SOD (kU/L)	NO (µmol/L)	MDA (µmol/L)
Normal	379.4 ± 41.2 [a]	14.5 ± 0.3 [a]	20.5 ± 1.7 [f]
Control	107.9 ± 23.5 [g]	3.1 ± 0.2 [f]	84.1 ± 4.3 [a]
CVFS (500 mg/kg)	210.6 ± 34.3 [f]	7.5 ± 0.4 [e]	48.7 ± 4.6 [b]
PVFS (500 mg/kg)	237.6 ± 33.1 [e]	7.9 ± 0.5 [e]	47.9 ± 4.4 [b]
MVFS (500 mg/kg)	271.3 ± 31.0 [d]	9.3 ± 0.4 [d]	35.1 ± 2.9 [c]
GVFS (500 mg/kg)	303.5 ± 27.6 [c]	11.2 ± 0.5 [c]	30.1 ± 2.5 [d]
Ranitidine (50 mg/kg)	341.2 ± 25.6 [b]	12.7 ± 0.3 [b]	26.4 ± 2.3 [e]

[a-g] Mean values with different letters in the same column are significantly different ($p < 0.05$) according to Duncan's multiple-range test. CVFS, ceramic vessel fermented Shuidouchi; PVFS, plastic vessel fermented Shuidouchi; MVFS, metal vessel fermented Shuidouchi; GVFS, glass vessel fermented Shuidouchi.

NO in the body is generated by NOS catalysis, which has cytotoxic effects and is involved in mediating immune reactions [30]. In recent years, studies have shown that many pathological stomach disease processes are associated with abnormal changes of NO levels [30,31]. As the only synthetase of NO, the distribution and function of NOS in the stomach is a hotspot of current research. NO could protect the gastric mucosa, which is one of its main functions in the stomach. It was widely accepted that NO mediates gastric mucosa to produce prostaglandins and increase the blood flow of gastric mucosa to protect it [32]. It was generally agreed that the effect of NO on gastric motility is an inhibitory process, and NO has a diastolic function and could delay gastric emptying ability. It has been found that a large number of NOS exist in gastric smooth cells and between the gastric muscle nerve plexus, which shows that NO plays an important role in regulating gastric motility [33]. When gastritis happens, the content of NO in stomach tissues would drop significantly, while this study also showed the same result [34].

Oxygen free radicals are a kind of oxygenic gene with high chemical activity and produced by oxygen metabolism. As inflammatory mediators, free radicals were closely related with gastritis and also a very important initiation factor and independent pathogenic factor of gastric mucosal injury [35]. Under normal circumstances, SOD is a protection factor of gastric mucosal cells, which could remove oxygen free radicals and resist lipid peroxidation of the gastric mucosa epithelium in order to keep oxygen free radicals at a low level and avoid gastric mucosal epithelial cell damage. However, under pathological conditions, the body produces a large number of oxygen free radicals by enzyme and/or non-enzyme systems [36]. These free radicals could attack polyunsaturated fatty acids in the phospholipids of biological membranes and can cause lipid peroxidation, which might lead to biological membrane damage, protein denaturation, DNA damage, gene aberration and cell necrosis [37]. MDA is the final product of lipid peroxides, so the level of MDA can reflect the level of oxygen free radicals. Under normal circumstances, the free radical removing system in the human body could effectively decompose the free radicals to avoid too much MDA and cause no harm to the body, but if the stomach is injured, there are too many free radicals, or the defense system for removing free radicals fails, this could cause free radical damage to the body and too much MDA [38].

mRNA Expression of NF-κB, IκB-α, EGF and EGFR in Gastric Tissues of Mice

By the RT-PCR assay, the results showed that Shuidouchi could decrease the NF-κB mRNA expressions while increase the IκB-α and EGF. EGFR expressions of gastric tissues as compared to the control mice (Figure 2). In the gastric tissues of GVFS-treated mice, the NF-κB, IκB-α, EGF and EGFR were at 0.57, 3.95, 9.46 and 2.37 fold the levels of control mice, close to those of drug (ranitidine)-treated mice and normal mice.

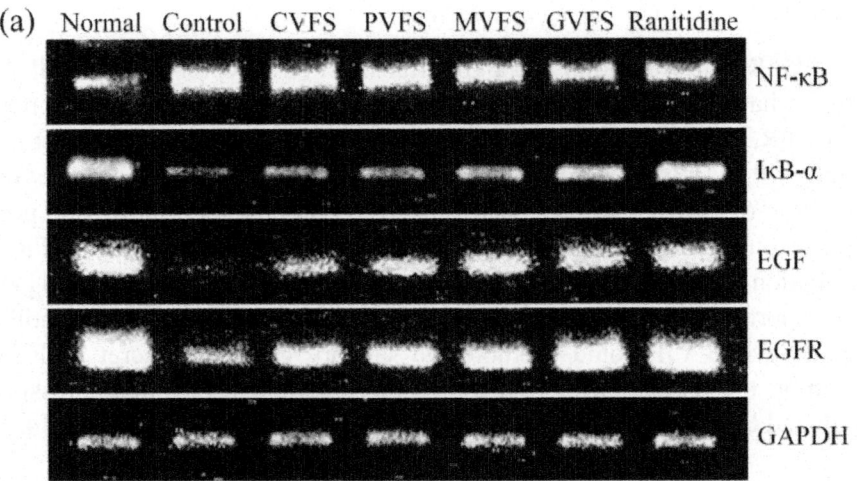

(a)   Normal   Control   CVFS   PVFS   MVFS   GVFS   Ranitidine

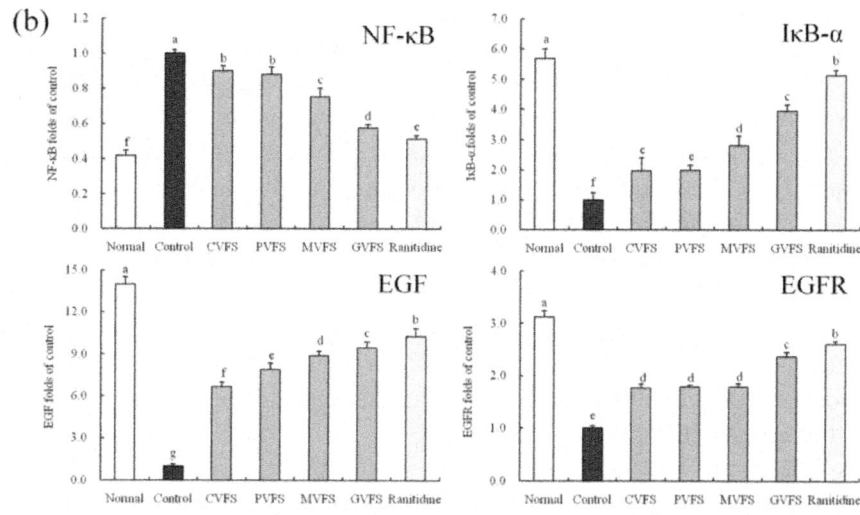

Figure 2: Effect of Shuidouchi fermented in different vessels on the NF-κB, IκB-α, EGF and EGFR mRNA expression in HCl/ethanol induced gastric mucosal injury mice. **(a)** bands of mRNA gene expression; **(b)** quantitative and ratio analysis. Fold-ratio: gene expression/GAPDH × control numerical value (control fold ratio: 1). [a–g] Mean values with different letters over the bars are significantly different ($p < 0.05$) according to Duncan's multiple range test. CVFS, 500 mg/kg ceramic vessel fermented Shuidouchi; PVFS, 500 mg/kg plastic vessel fermented Shuidouchi; MVFS, 500 mg/kg metal vessel fermented Shuidouchi; GVFS, 500 mg/kg glass vessel fermented Shuidouchi; Ranitidine, 50 mg/kg ranitidine.

EGFR is proliferation and signal transduction receptor of EGF cells [39]. Studies have proven that many solid tumors show high or abnormal expression of EGFR, while appropriate EGFR and EGF is helpful to alleviate stomach injury [40,41]. Besides, EGF could inhibit gastric acid secretion, increase the secretion of gastric juice, increase mucous membrane blood flow, and protect the mucous membrane. All these functions are mainly mediated by EGFR. The regulation mechanism of EGF synthesis and release is very complicated [42]. It was reported that androgen can increase EGF synthesis in mice submandibular glands, while VIP could promote the salivary glands to secret EGF. As a receptor, the regulation of EGFR was more complex, but the expression of EGF and EGFR are positively correlated when the stomach is injured [43].

mRNA Expression of nNOS, eNOS, iNOS and COX-2 in Gastric Tissues of Mice

GVFS had higher nNOS (2.36-fold the control group), eNOS (2.05-fold the control group) mRNA expressions and lower iNOS (0.24-fold the control group), COX-2 (0.16-fold the control group) expressions than Shuidouchi fermented in other vessels in gastric tissues of mice (Figure 3). The nNOS, eNOS expressions of GVFS-treated mice were also significantly ($p < 0.05$) higher than control mice, while the iNOS, COX-2 expressions were lower than in control mice.

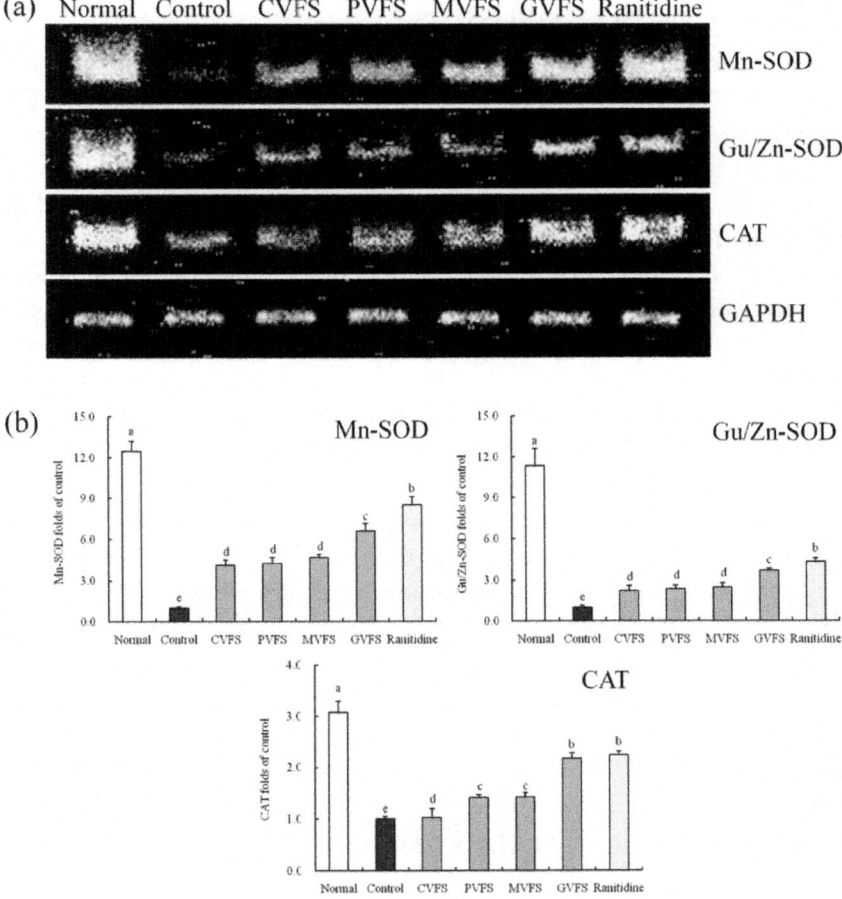

Figure 3: Effect of Shuidouchi fermented in different vessels on the nNOS, eNOS, iNOS and COX-2 mRNA expression in HCl/ethanol induced gastric mucosal injury mice. (a) bands of mRNA gene expression; (b) quantitative and ratio analysis. Fold-ratio: gene expression/GAPDH × control numerical value (control fold ratio: 1). [a-e] Mean values with different letters over the bars are significantly different ($p < 0.05$) according to Duncan's multiple range test. CVFS, 500 mg/kg ceramic vessel fermented Shuidouchi; PVFS, 500 mg/kg plastic vessel fermented Shuidouchi; MVFS, 500 mg/kg metal vessel fermented Shuidouchi; GVFS, 500 mg/kg glass vessel fermented Shuidouchi; Ranitidine, 50 mg/kg ranitidine.

Normal endothelial cells have anticoagulant, anti-inflammatory, and angiogenesis inhibiting functions and also promote vasodilatation by generating NO, prostacyclin and other vasodilator substances [44]. Endothelial dysfunction could cause stomach inflammation, and lower bioavailability of NO is an important factor of endothelial dysfunction. NO is synthesized by NOS catalysis. NOS has three different subtypes, including NOS1 (nNOS), NOS2 (iNOS) and NOS3 (eNOS).

Under normal physiological conditions, NO in vascular endothelial cells mainly came from eNOS to regulate normal physiological functions [45], but under some pathological conditions, eNOS shows dysfunction, generating O_2^- instead of NO, which decreases the bioavailability of NO and increases the oxidative stress, causing or aggravating endothelial dysfunction [46]. In a rest state, iNOS doesn't express, but under pathological conditions, a large amount of iNOS and NO are produced. NO plays a dual role in inflammation. After activation for 4 to 6 h, iNOS could produce a lot of NO. In this study, the content of NO fell in the early stages of stomach injury and might increase continuously. In the early stages of inflammation, iNOS increases sharply and turns into NO after 4 to 6 h [47]. Excessive NO could intensify any gastric mucosa damage and activate COX-2 by combining with COX-2 at its active site, resulting in aggravated inflammation [48]. eNOS and nNOS are both Ca^{2+}-dependent NOS, whose expression always shows a positive correlation in stomach tissues. When the gastric mucosa is damaged, the expression of NOS decreases, so does the expression of nNOS [49]. iNOS and COX-2 which are also important expressors of inflammation. Inflammation caused by stomach mucosa damage increases the expression of iNOS and COX-2, which is observed in both human clinical experiments and animal experiments. Stomach tissues of control group mice with gastric damage also showed the same expression [50]. Soy isoflavones from soybeans fermented in glass vessels significantly alleviate the phenomenon, and with the increase of its concentration, the alleviating effect was enhanced.

mRNA Expression of Mn-SOD, Gu/Zn-SOD and CAT in Gastric Tissues of Mice

After inducing the gastric mucosal injury (control group), the mRNA expressions of Mn-SOD, Gu/Zn-SOD and CAT were significantly ($p < 0.05$) reduced as compared to normal mice (Figure 4). Shuidouchi could retard these reductions, and GVFS had the best effects. The Mn-SOD (6.60-fold of control group), Gu/Zn-SOD (3.68-fold of control group) and CAT (2.17-fold of control group) expressions of gastric tissues in GVFS-treated mice were the highest among the Shuidouchi-treated mice, and these expressions were close to those of ranitidine-treated mice and normal mice.

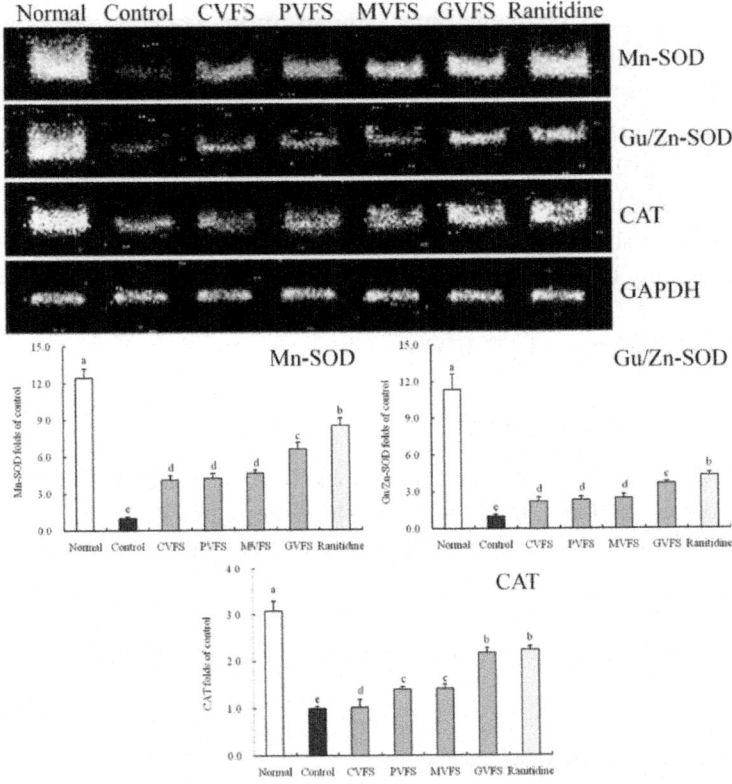

Figure 4: Effect of Shuidouchi fermented in different vessels on the Mn-SOD, Gu/Zn-SOD and CAT mRNA expression in HCl/ethanol induced gastric mucosal injury mice. (a) bands of mRNA gene expression; (b) quantitative and ratio analysis. Fold-ratio: gene expression/GAPDH × control numerical value (control fold ratio: 1). [a–e] Mean values with different letters over the bars are significantly different ($p < 0.05$) according to Duncan's multiple range test. CVFS, 500 mg/kg ceramic vessel fermented Shuidouchi; VFS, 500 mg/kg plastic vessel fermented Shuidouchi; MVFS, 500 mg/kg metal vessel fermented Shuidouchi; GVFS, 500 mg/kg glass vessel fermented Shuidouchi; Ranitidine, 50 mg/kg ranitidine. Before sending to publish, try your best to move the caption in one page.

SOD has three kinds of isomers in animals, including Cu/Zn-SOD, Mn-SOD and EC-SOD (SOD3), while Cu/Zn-SOD and Mn-SOD are the two main types of SODs, which are antioxidant enzymes in the body [51]. Extreme decreases of Cu/Zn-SOD and Mn-SOD in the body imply the production of large numbers of free radicals, which could lead to inflammation and put the body in a pathological state. Maintaining Cu/Zn-SOD and Mn-SOD in the body at normal levels it is an important way of controlling stomach injuries [52]. As one of the key enzymes in the biological defense system, CAT antioxidant enzymes could remove oxygen free radicals, promote decomposition of H_2O_2 to molecular oxygen and water and also remove hydrogen peroxide in body in order to avoid the damage of H_2O_2 to cells and reduce tissue injuries.

Ethanol gastric mucosal injury is closely related to a decrease of gastric mucosal blood flow and lipid peroxidation induced by oxygen free radicals [53]. By strengthening the removal of free radicals, keeping the activity of active enzymes such as Cu/Zn-SOD, Mn-SOD and CAT in body is an important way of avoiding ethanol gastric mucosal injury [54].

EXPERIMENTAL SECTION

Shuidouchi Fermentation

Five kilograms of dry soybeans were soaked in 12.5 L distilled water for 12 h, and cooked under 120 °C for 1 h. then the water was removed and the cooked soybeans were equally divide into five parts, which were placed in ceramic, plastic, metal and glass vessels covered by gauze in a constant temperature incubator at 45 °C so natural fermentation could occur for 72 h. After fermentation, soybeans were cooled, dried and crushed to extract the soybean isoflavones.

Isoflavone Determination

Two kilograms of freeze-dried fermented soybean powder were Soxhlet extracted with 32 L 70% aqueous ethanol solution for 6 h, and then the distilled ethanol was added with 5 mol/L HCl and N-503. The mixture was kept in a water bath at 30 °C for hydrolysis and extraction of soybean isoflavones for 2 h. Stratification in the separating funnel also takes up the N-503 layer. NaOH (5 mol/L) solution was added for back extraction, then the water layer was added with hydrochloric acid after stratification to precipitate soy isoflavones. After centrifugation, the sediment was washed to a neutral state and freeze-dried to extract soybean isoflavones. Taking daidzein and genistein as standards, the absorbance of different concentrations of daidzein and genistein were determined at 260 nm, and a standard curve of soybean isoflavone concentration was plotted for the content of isoflavones in soybeans [4].

Mice Experiment

Mice for this experiment were divided into seven groups, including normal group, control group, ceramic vessel fermented Shuidouchi (CVFS) group, plastic vessel fermented Shuidouchi (PVFS) group, metal vessel fermented Shuidouchi (MVFS) group, glass vessel fermented Shuidouchi (GVFS) group and ranitidine group, having 10 mice in each group. During the first 14 d, mice in the normal group and the control group were gavaged with 0.2 mL distilled water once a day, while mice in the other groups were gavaged 0.2 mL of Shuidouchi extract with the concentration of 500 mg/kg once. Mice in the drug treatment group were gavaged 0.2 mL ranitidine with a concentration of 50 mg/kg. From the 14th day, all mice were cut off food, but allowed to drink water freely. In addition to mice in the normal group, all mice in other group were gavaged a stomach injury inducer (0.1 mL HCl/ethanol/10 g body weight, 60% in 150 mM HCl) after 24 h and then killed after 1 h [8]. Heart blood was taken for centrifugal separation (4000 r/min, 10 min), where the upper serum was kept and the stomach was anatomized for further use. The experiments were performed following a protocol approved by the Animal Ethics Committee of Chongqing Medical University (Chongqing, China).

Mice Gastric Mucosal Injury Evaluation

The gastric secretion volume of mice were determined with a 10 mL measuring cylinder, and the pH of gastric juice of mice were determined using a SevenEasy pH meter (Mettler Toledo, Schwerzenbach, Switzerland). The isolated stomachs were inflated by injecting 1% formalin solution (10 mL) for 10 min to fix the tissues, and opened along the greater curvature. The area (mm^2) of hemorrhagic lesions that had developed in the stomach was measured under a Leica MZ7.5 dissecting microscope (Leica, Bensheim, Germany) with a square grid.

Mice Serum Levels Measurement

Serum MTL, Gas, SS and VIP levels were determined with radioimmunoassay kits (Beijing Puer Weiye Biotechnology Co., Ltd., Beijing, China) according to the manufacturer's protocols.

Mice Cytokine IL-6, IL-12, TNF-α and IFN-γ Levels Measurement

Serum IL-6, IL-12, TNF-α and IFN-γ levels were measured with a commercial ELISA kit (ELISA MAX, Biolegend, San Diego, CA, USA) according to the manufacturer's protocol.

Gastric Tissues SOD, NO and MDA Activities Measurement

Gastric tissues SOD, NO and MDA activities were determined with appropriate assay kits (Nanjing Jiancheng Bioengineering Institute, Nanjing, Jiangsu, China) according to the manufacturer's protocols.

mRNA Expression Determination (RT-PCR Assay)

Stomach tissues in mice were shattered by an ultrasonic pulverizer and RNA was extracted using RNAzol reagent. RNA extract of stomach tissues was diluted to 1 μg/μL. oligodT18 (1 μL), RNase, dNTP with MLV enzymes and 10 μL 5 × buffer were added into 2 μL RNA extraction of stomach tissues to synthesize cDNA under the conditions of 37 °C for 120 min, 99 °C for 4 min, 4 °C for 3 min. Then the expressions were amplified by the reverse transcription-polymerase chain reaction method, while house-keeping gene GAPDH was taken as reference. Agarose gel (1%) with ethidium bromide was used for electrophoresis to check the PCR amplification products [8].

Statistical Analysis

Experimental data were presented as mean ± standard deviation (SD). Differences between the mean values for individual groups were assessed by one-way analysis of variance (ANOVA) with Duncan's multiple range test. $p < 0.05$ was considered to indicate a statistically significant difference. SAS version 9.2 (SAS Institute Inc., Cary, NC, USA) was used to conduct the statistical analyses.

CONCLUSIONS

Through molecular biology methods, this research built stomach injury animal models to study the effect on inhibition of stomach injuries when Shuidouchi fermented in different vessels was administered to mice. By analyzing the content of soybean isoflavones, the results showed that the glass vessel was more advantageous for fermentation, producing more soybean isoflavones. These isoflavones had functional effects, which could cause molecular changes in mice bodies, as inflammation and oxidation factors were changed by Shuidouchi. Shuidouchi could inhibit the inflammation factors and increase the antioxidation factors. By further analyzing animal serum and tissues, using glass vessel to ferment Shuidouchi could decrease the MTL, Gas serum levels and increase the SS, VIP serum levels compared with Shuidouchi fermented in other vessels and no Shuidouchi treatment for control mice. The Shuidouchi fermented in glass vessel could also better lower cytokine levels (IL-6, IL-12, TNF-α and IFN-γ) in stomach injury of mice, increase the content of SOD, NO and reduce the content of MDA in mice gastric tissues. By further analyzing mRNA in related genes in stomach tissues with RT-PCR experimental technology at the molecular level, it had been found out that Shuidouchi could improve the strength of expression of IκB-α, EGF, EGFR, nNOS, eNOS, Mn-SOD, Gu/Zn-SOD, CAT in stomach-injured mice tissues and reduce the expression strength of NF-κB, COX-2, iNOS. The effect of Shuidouchi fermentation in glass vessels was more intense, which was significantly different from that of Shuidouchi produced in other kinds of vessels.

ACKNOWLEDGMENTS

This work was partly supported by National Natural Science Foundation of China (Grant No. 31201411), Ability Improvement Project of Chongqing Engineering Research Center of Regional Food (cstc2014pt-gc8001) and Scientific and Technological Research Program of Chongqing Municipal Education Commission (Grant No. KJ1401415), China.

AUTHOR CONTRIBUTIONS

H. Suo and X. Zhao designed the study, analyzed the data and wrote the manuscript. H. Suo, X. Feng and K. Zhu did the animal experiments. C. Wang did the statistical data analysis. J. Kan edited and revised the manuscript.

CONFLICTS OF INTEREST

The authors declare no conflict of interest.

REFERENCES

1.   Zhao, X.; Song, J.L.; Wang, Q.; Qian, Y.; Li, G.J.; Pang, L. Comparisons of Shuidouchi, Natto and Chungkukjang in their physicochemical properties, antimutagenic and anticancer effects. *Food Sci. Biotechnol.* **2013**, *22*, 1077–1084.

2.   Zhao, X.; Qian, Y.; Li, G.J. Fermentation period influences on gastric injury preventive effects of Shuidouchi. *J. Chin. Cereals Oils Assoc.* **2014**, *29*, 14–17.

3.   Zhao, X.; Qian, Y. Fermentation period influences on physicochemical properties and *in vitro*MCF-7 human breast adenocarcinoma cells anticancer effects of Shuidouchi. *J. Chin. Cereals Oils Assoc.* **2015**, *30*, 27–32.

4.   Zhao, X.; Wang, Q. Study on comparisons of different ripening fermentation periods fermented Shuidouchi in their physicochemical

properties. *Sci. Technol. Food Ind.* **2014**, *35*, 346–349.

5. Zhao, X.; Bak, S.S.; Rhee, S.H.; Park, L.Y. Fermentation period influences on antimutagenic and *in vitro* anticancer effects of Shuidouchi. *J. Cancer Prev.* **2008**, *13*, 40–46.

6. Zhao, X.; Li, G.J. Influences of soaked water volume on physicochemical properties of Shuidouchi. *Sci. Technol. Food Ind.* **2013**, *34*, 217–220.

7. Ryu, K.J.; Zhao, X.; Bak, S.S.; Kim, B.K.; Jeon, J.T.; Park, K.Y. *In vitro* anticancer effect of Chungkukjangs from folk villages of Sunchang region in HT-29 human colon cancer cells. *J. Cancer Prev.* **2008**, *13*, 62–67.

8. Zhao, X.; Wang, Q.; Qian, Y. *Ilex kudingcha* CJ Tseng (Kudingcha) prevents HCl/ethanol-induced gastric injury in Sprague-Dawley rats. *Mol. Med. Rep.* **2013**, *7*, 1613–1616.

9. Zhao, X.; Song, J.L.; Jung, O.S.; Lim, Y.I.; Park, K.Y. Chemical properties and *in vivo* gastric protective effects of bamboo salt. *Food Sci. Biotechnol.* **2014**, *23*, 895–902.

10. Loguercio, C.; Tuccillo, C.; Federico, A.; Fogliano, V.; Del Vecchio Blanco, C.; Romano, M. Alcoholic beverages and gastric epithelial cell viability: Effect on oxidative stress-induced damage. *J. Physiol. Pharmacol.* **2009**, *60*, 87–92.

11. Messina, M. Soy foods, isoflavones, and the health of postmenopausal women. *Am. J. Clin. Nutr.* **2014**, *100*, 423S–430S.

12. Takagi, A.; Kano, M.; Kaga, C. Possibility of breast cancer prevention: use of soy isoflavones and fermented soy beverage produced using probiotics. *Int. J. Mol. Sci.* **2015**, *16*, 10907–10920.

13. Zhou, C.; Lin, H.; Ge, X.; Niu, J.; Wang, J.; Wang, Y.; Chen, L.; Huang, Z.; Yu, W.; Tan, X. The effects of dietary soybean isoflavones on growth, innate immune responses, hepatic antioxidant abilities and disease resistance of juvenile golden pompano *Trachinotus ovatus.Fish Shellfish Immunol.* **2015**, *43*, 158–166.

14. Ferguson, J.F.; Ryan, M.F.; Gibney, E.R.; Brennan, L.; Roche, H.M.; Reilly, M.P. Dietary isoflavone intake is associated with evoked responses to inflammatory cardiometabolic stimuli and improved glucose homeostasis in healthy volunteers. *Nutr. Metab. Cardiovasc. Dis.* **2014**, *24*, 996–1003.

15. Malardé, L.; Groussard, C.; Lefeuvre-Orfila, L.; Vincent, S.; Efstathiou, T.; Gratas-Delamarche, A. Fermented soy permeate reduces cytokine level and oxidative stress in streptozotocin-induced diabetic rats. *J. Med.*

Food **2015**, *18*, 67–75.

16. Yoo, S.M.; Kim, J.S.; Shin, D.H. Quality changes of traditional *Doenjang* fermented in different vessels. *J. Korean Soc. Agric. Chem. Biotechnol.* **2001**, *44*, 230–234.

17. Yin, S.; Wang, X.J.; Ai, J.W.; Liu, J.H.; Tian, D.; Wang, J.H.; Yang, Z.B. Research on post-fermentation and flavoring technology of Guizhou Natto. *China Condiment* **2015**, *40*, 54–56.

18. Huang, Z.W.; Shangguan, X.C.; Cheng, J.H.; Wu, C.S.; Wang, F. Studies on the mixed fermentation technique of Natto. *J. Chin. Inst. Food Sci. Technol.* **2005**, *5*, 70–73.

19. Tienungoon, S.; Ratkowsky, D.A.; McMeekin, T.A.; Ross, T. Growth limits of Listeria monocytogenes as a function of temperature, pH, NaCl, and lactic acid. *Appl. Environ. Microbiol.* **2000**, *66*, 4979–4987.

20. Song, J.L.; Zhou, Y.L.; Feng, X.; Zhao, X. White tea (*Camellia sinenesis* (L.)) ethanol extracts attenuate reserpine-induced gastric ulcers in mice. *Food Sci. Biotechnol.* **2015**, *24*, 1159–1165.

21. Yi, R.K.; Wang, R.; Sun, P.; Zhao, X. Antioxidant-mediated preventative effect of Dragon-pearl tea crude polyphenol extract on reserpine-induced gastric ulcers. *Exp. Ther. Med.* **2015**,*10*, 338–344.

22. Depoortere, I.; Thijs, T.; Janssen, S.; de Smet, B.; Tack, J. Colitis affects the smooth muscle and neural response to motilin in the rabbit antrum. *Br. J. Pharmacol.* **2010**, *159*, 384–393.

23. Wakim, S.A.; Ahmed, M.A.; Ali, R.H. Gastric acid secretion in experimental acute uremia.*Can. J. Physiol. Pharmacol.* **2013**, *91*, 693–699.

24. Peng, L.; Liu, M.; Chang, X.; Yang, Z.; Yi, S.; Yan, J.; Peng, Y. Role of the nucleus tractus solitarii in the protection of pre-moxibustion on gastric mucosal lesions. *Neural Regen. Res.***2014**, *9*, 198–204.

25. Chen, S.; Zhu, K.; Wang, R.; Zhao, X. Preventive effect of polysaccharides from the large yellow croaker swim bladder on HCl/ethanol induced gastric injury in mice. *Exp. Ther. Med.***2014**, *8*, 316–322.

26. Liang, J.; Li, Y.; Liu, X.; Xu, X.; Zhao, Y. Relationship between cytokine levels and clinical classification of gastric cancer. *Asian Pac. J. Cancer Prev.* **2011**, *12*, 1803–1806.

27. Wang, F.Y.; Liu, J.M.; Luo, H.H.; Liu, A.H.; Jiang, Y. Potential protective effects of Clostridium butyricum on experimental gastric ulcers in mice. *World J. Gastroenterol.* **2015**,*21*, 8340–8351.

28. Sun, L.; Wu, Q.; Han, B.; Li, G.; Sun, Z.; Zhang, J.; An, L. Mechanisms

of immune injury and heterogeneity of bone marrow hematopoietic cells island in patients with auto-immuno-related hematocytopenia. *J. Immunoassay Immunochem.* **2014**, *35*, 378–387.

29. Lindgren, Å.; Yun, C.H.; Sjöling, Å.; Berggren, C.; Sun, J.B.; Jonsson, E.; Holmgren, J.; Svennerholm, A.M.; Lundin, S.B. Impaired IFN-γ production after stimulation with bacterial components by natural killer cells from gastric cancer patients. *Exp. Cell Res.* **2011**, *317*, 849–858.

30. Silva, R.O.; Lucetti, L.T.; Wong, D.V.; Aragão, K.S.; Junior, E.M.; Soares, P.M.; Barbosa, A.L.; Ribeiro, R.A.; Souza, M.H.; Medeiros, J.V. Alendronate induces gastric damage by reducing nitric oxide synthase expression and NO/cGMP/K(ATP) signaling pathway. *Nitric Oxide* **2014**, *40*, 22–30.

31. Liu, Y.; Gou, L.; Fu, X.; Li, S.; Lan, N.; Yin, X. Protective effect of rutin against acute gastric mucosal lesions induced by ischemia-reperfusion. *Pharm. Biol.* **2013**, *51*, 914–919.

32. Yandrapu, H.; Sarosiek, J. Protective factors of the gastric and duodenal mucosa: an overview. *Curr. Gastroenterol. Rep.* **2015**, *17*, 24.

33. Rocha, B.S.; Gago, B.; Barbosa, R.M.; Cavaleiro, C.; Laranjinha, J. Ethyl nitrite is produced in the human stomach from dietary nitrate and ethanol, releasing nitric oxide at physiological pH: Potential impact on gastric motility. *Free Radic. Biol. Med.* **2015**, *82*, 160–166.

34. Rouhollahi, E.; Moghadamtousi, S.Z.; Hamdi, O.A.; Fadaeinasab, M.; Hajrezaie, M.; Awang, K.; Looi, C.Y.; Abdulla, M.A.; Mohamed, Z. Evaluation of acute toxicity and gastroprotective activity of curcuma purpurascens BI. rhizome against ethanol-induced gastric mucosal injury in rats. *BMC Complement. Altern. Med.* **2014**, *14*, 378.

35. Kwiecien, S.; Jasnos, K.; Magierowski, M.; Sliwowski, Z.; Pajdo, R.; Brzozowski, B.; Mach, T.; Wojcik, D.; Brzozowski, T. Lipid peroxidation, reactive oxygen species and antioxidative factors in the pathogenesis of gastric mucosal lesions and mechanism of protection against oxidative stress-induced gastric injury. *J. Physiol. Pharmacol.* **2014**, *65*, 613–622.

36. Martínez Aranzales, J.R.; Cândido de Andrade, B.S.; Silveira Alves, G.E. Orally administered phenylbutazone causes oxidative stress in the equine gastric mucosa. *J. Vet. Pharmacol. Ther.* **2015**, *38*, 257–264.

37. Strickertsson, J.A.; Desler, C.; Martin-Bertelsen, T.; Machado, A.M.; Wadstrøm, T.; Winther, O.; Rasmussen, L.J.; Friis-Hansen, L. Enterococcus faecalis infection causes inflammation, intracellular oxphos-independent ROS production, and DNA damage in human gastric cancer cells. *PLoS ONE* **2013**, *8*, e63147.

38. Zhao, Z.; Gong, S.; Wang, S.; Ma, C. Effect and mechanism of evodiamine against ethanol-induced gastric ulcer in mice by suppressing Rho/NF-κB pathway. *Int. Immunopharmacol.***2015**, *28*, 588–595.

39. Moschini, I.; Dell'Anna, C.; Losardo, P.L.; Bordi, P.; D'Abbiero, N.; Tiseo, M. Radiotherapy of non-small-cell lung cancer in the era of EGFR gene mutations and EGF receptor tyrosine kinase inhibitors. *Future Oncol.* **2015**, *11*, 2329–2342.

40. Khan, A.J.; Fligiel, S.E.; Liu, L.; Jaszewski, R.; Chandok, A.; Majumdar, A.P. Induction of EGFR tyrosine kinase in the gastric mucosa of diabetic rats. *Proc. Soc. Exp. Biol. Med.* **1999**,*221*, 105–110.

41. Liu, M.; Chang, X.R.; Yan, J.; Yi, S.X.; Lin, Y.P.; Yue, Z.H.; Peng, Y.; Zhang, H. Effects of moxibustion on gastric mucosal EGF and TGF-alpha contents and epidermal growth factor receptor expression in rats with gastric mucosal lesion. *Zhen Ci Yan Jiu* **2011**, *6*, 403–408.

42. Xu, J.J.; Huang, P.; Wu, Q.H.; Cao, H.Y.; Wen, S.; Liu, J. Study on efficacy and mechanism of weiyangning pills against experimental gastric ulcer. *Zhongguo Zhong Yao Za Zhi* **2013**, *38*, 736–739.

43. De Heuvel, E.; Wallace, L.; Sharkey, K.A.; Sigalet, D.L. Glucagon-like peptide 2 induces vasoactive intestinal polypeptide expression in enteric neurons via phophatidylinositol 3-kinase-γ signaling. *Am. J. Physiol. Endocrinol. Metab.* **2012**, *303*, E994–E1005.

44. Zhao, J.; Suyama, A.; Tanaka, M.; Matsui, T. Ferulic acid enhances the vasorelaxant effect of epigallocatechin gallate in tumor necrosis factor-alpha-induced inflammatory rat aorta. *J. Nutr. Biochem.* **2014**, *25*, 807–814.

45. Bogdan, C. Nitric oxide synthase in innate and adaptive immunity: An update. *Trend. Immunol.* **2015**, *36*, 161–178.

46. Banerjee, M.; Vats, P. Reactive metabolites and antioxidant gene polymorphisms in Type 2 diabetes mellitus. *Redox. Biol.* **2013**, *2C*, 170–177.

47. Rashed, E.; Lizano, P.; Dai, H.; Thomas, A.; Suzuki, C.K.; Depre, C.; Qiu, H. Heat shock protein 22 (Hsp22) regulates oxidative phosphorylation upon its mitochondrial translocation with the inducible nitric oxide synthase in mammalian heart. *PLoS ONE* **2015**, *10*, e0119537.

48. Lee, Y.Y.; Yang, Y.P.; Huang, P.I.; Li, W.C.; Huang, M.C.; Kao, C.L.; Chen, Y.J.; Chen, M.T. Exercise suppresses COX-2 pro-inflammatory pathway in vestibular migraine. *Brain Res. Bull.* **2015**, *116*, 98–105.

49. Darra, E.; Rungatscher, A.; Carcereri de Prati, A.; Podesser, B.K.;

Faggian, G.; Scarabelli, T.; Mazzucco, A.; Hallström, S.; Suzuki, H. Dual modulation of nitric oxide production in the heart during ischaemia/ reperfusion injury and inflammation. *Thromb. Haemost.* **2010**, *104*, 200–206.

50. Zheng, C.; Lei, C.; Chen, Z.; Zheng, S.; Yang, H.; Qiu, Y.; Lei, B. Topical administration of diminazene aceturate decreases inflammation in endotoxin-induced uveitis. *Mol. Vis.* **2015**, *21*, 403–411.

51. Gottfredsen, R.H.; Larsen, U.G.; Enghild, J.J.; Petersen, S.V. Hydrogen peroxide induce modifications of human extracellular superoxide dismutase that results in enzyme inhibition. *Redox. Biol.* **2013**, *1*, 24–31.

52. Othman, A.I.; El-Missiry, M.A.; Amer, M.A. The protective action of melatonin on indomethacin-induced gastric and testicular oxidative stress in rats. *Redox. Rep.* **2001**, *6*, 173–177.

53. Ligumsky, M.; Sestieri, M.; Okon, E.; Ginsburg, I. Antioxidants inhibit ethanol-induced gastric injury in the rat. Role of manganese, glycine, and carotene. *Scand. J. Gastroenterol.* **1995**, *30*, 854–860.

54. Yoshikawa, T.; Minamiyama, Y.; Ichikawa, H.; Takahashi, S.; Naito, Y.; Kondo, M. Role of lipid peroxidation and antioxidants in gastric mucosal injury induced by the hypoxanthine-xanthine oxidase system in rats. *Free Radic. Biol. Med.* **1997**, *23*, 243–250.

Chapter 11

A GLUTAMIC ACID-PRODUCING LACTIC ACID BACTERIA ISOLATED FROM MALAYSIAN FERMENTED FOODS

Mohsen Zareian, Afshin Ebrahimpour, Fatimah Abu Bakar, Abdul Karim Sabo Mohamed, Bita Forghani, Mohd Safuan B. Ab-Kadir and Nazamid Saari

Faculty of Food Science and Technology, Universiti Putra Malaysia, Serdang 43400, Selangor, Malaysia

ABSTRACT

L-glutamaic acid is the principal excitatory neurotransmitter in the brain and an important intermediate in metabolism. In the present study, lactic acid bacteria (218) were isolated from six different fermented foods as potent sources of glutamic acid producers. The presumptive bacteria were tested for their ability to synthesize glutamic acid. Out of the 35 strains showing this capability, strain MNZ was determined as the highest glutamic-acid producer. Identification tests including 16S rRNA gene sequencing and sugar assimilation ability identified the strain MNZ as *Lactobacillus plantarum*. The characteristics of this microorganism related to its glutamic acid-producing ability, growth rate, glucose consumption and pH profile were studied. Results revealed that glutamic acid was formed inside the cell and excreted into the extracellular medium. Glutamic acid production was found to be growth-associated and glucose significantly enhanced glutamic acid production (1.032 mmol/L) compared to other carbon sources. A concentration of 0.7% ammonium nitrate as a nitrogen source effectively enhanced glutamic acid production. To the best of our knowledge this is the first report of glutamic acid production by lactic acid bacteria. The results of this study can be further applied for developing functional foods enriched in glutamic acid and subsequently γ-amino butyric acid (GABA) as a bioactive compound.

INTRODUCTION

Glutamic acid is a multifunctional amino acid involved in taste perception, excitatory neurotransmission and intermediary metabolism [1]. It plays an important role in gastric phase digestion with multiplicity effects in the gastrointestinal tract when consumed with nutrients by enhancing gastric exocrine secretion [2]. Glutamic acid is a specific precursor for other amino acids*i.e.*, arginine and proline as well as for bioactive molecules such as γ-amino butyric acid (GABA) and glutathione. GABA possesses several well-known physiological functions (*i.e.*, anti-hypertension [3] and anti-diabetic [4]) and glutathione plays a key role in the protection of the mucosa from peroxide damage and from dietary toxins [5]. Furthermore, a number of studies have shown the possible usefulness of glutamic acid in enhancing nourishment in the elderly and in patients with poor nutrition [6,7]. At the present time, glutamic acid is largely produced through microbial fermentation because the chemical method produces a racemic mixture of glutamic acid (D- and L-glutamic acid) [8]. Numerous studies have reported glutamic acid excretion by various micro-organisms; however, most of them were not food-grade micro-organisms. Lactic acid bacteria (LAB) are well known to produce a variety of primary metabolites. The existence of the *gdh*gene in LAB which is responsible for the production of glutamic acid has been proven [9]. In addition, LAB are essential in the processing of food materials and they have been applied extensively in the food industry [10]. LAB can enhance the shelf-life and safety of foods, improve food texture, and contribute to the nutritional value of food products and pleasant sensory profile of the end-use products [11]. Employing LAB, on the other hand, with the potential to produce glutamic acid can facilitate production of functional foods rich in bioactive molecules such as GABA. The key advantage of the glutamic acid production by LAB is that the amino acid produced in this way is biologically active (L-glutamic acid) and the production process is considered to be safe and eco-friendly. This can be achieved through selection of appropriate LAB strains from indigenous micro-organisms which are well-adapted to a particular product, more competitive and with elevated metabolic capacities. Thus, research of this technological potential is of highest interest to the industry.

As a result, selection of the most efficient glutamic acid-producing LAB strains may contribute to new fermented products with improved general standards with respect to the naturally-biosynthesized glutamic acid. This could lead to the development of fermented foods rich in glutamic acid and consequently GABA. Therefore, it is hypothesized that screening various LAB capable of producing glutamic acid may explore new avenues for mass production of functional foods rich in GABA.

RESULTS AND DISCUSSION

In this study, a number of indigenous fermented foods available in Malaysia were used for the isolation of lactic acid bacteria. These isolated LAB strains were individually examined based on their potential to produce glutamic acid. Table 1 summarizes the characteristics of such LAB exhibiting the capacity to produce glutamic acid. It shows that 14 strains of 35 that produced glutamic acid were isolated from fermented soybean. Higher concentrations of glutamic acid produced by LAB strains isolated from fermented soybean showed that these strains were more efficient in biosynthesizing glutamic acid as compared to LAB strains isolated from other food samples. Among fermented food-derived LAB strains, only one strain showed a superior glutamic acid-producing potential with a contribution of 489.46 μmol/L glutamic acid. This strain assigned as MNZ, was subjected to further phenotypic and genotypic identification tests.

Table 1: Characteristics of the LAB isolates which biosynthesized glutamic acid in MRS broth.

Type of Sample	Total LAB isolates	Total LAB with glutamic acid production	Glutamic acid production range (μmol/L)
Fermented soybean	53	14	20–489
Fermented durian flesh	42	5	3.2–20
Fermented tapioca	21	3	34–59
Fermented glutinous rice	26	3	18–65
Fermented shrimp sauce	27	4	2–11
Fermented fish sauce	49	6	22–106

Glutamic acid production was previously reported by Zalán *et al.* (2010) [12] and Tarek and Mostafa (2009) [13] for some of the LAB species such as *Lactobacillus paracasei* and *Lactobacillus* spp.; however, in this study, *Lactobacillus plantarum* was reported as a glutamic acid producer. The level of glutamic acid produced by LAB in other studies was reported to be 68.7 mg/L [13] and <25 mmol/L [12]. Gram-positive micro-organisms other than LAB were also shown to produce glutamic acid. For example, *Brevibacterium* spp. were found to produce this amino acid between 10 to 46 mmol/L [14].

Phenotypic and Genotypic Identification

In order to identify the LAB strain MNZ, the API carbohydrate fermentation kit was utilized. The results showed that this strain was able to assimilate the following carbon sources: Ribose, galactose, D-glucose, fructose, mannose, manitol, sorbitol, α-methyl-D-mannoside, *N*-acetyl glucosamine, arbutin, cellobiose, maltose, lactose, melibiose, sucrose, gentiobiose, turanose, salicin and aesculin; while this ability was negative for the rest of carbon sources. The pattern of assimilation ability of this isolate for carbohydrates was in agreement with the metabolic characteristics of *Lactobacillus plantarum* described by Bergey's Manual of Systemic Bacteriology [15]. Taxonomic identification of the strain MNZ at species level was performed by 16S rRNA gene sequencing. The sequence was deposited on NCBI under accession number HM175883 (*Lactobacillus plantarum* MZ-02).

Effect of Various Carbon Sources on Glutamic Acid Production

The effect of different carbon sources on the production of glutamic acid by the LAB strain MNZ was investigated in this study. It was found that the following carbohydrates were fermented by*Lactobacillus plantarum* MNZ: Ribose, sorbitol, manitol, fructose, glucose, sucrose and lactose. Three concentrations (6, 12 and 24% *w/v*) were tested for each carbon source and the production of glutamic acid was determined in MRS medium supplemented with each carbon source individually at 30 °C for 120 h. Based on a preliminary test (data not shown), the range of 6–24% (*w/v*) was found to be the most suitable for assessment of various carbon sources. The results presented in Figure 1 indicate that glucose was the best source of carbohydrate for the production of glutamic acid among the tested sources where 12% was found to be the optimum point for glutamic acid production. Since 12% and 24% glucose concentrations did not show any significant differences ($p < 0.05$) regarding the production of glutamic acid by the LAB strain MNZ, the lowest amount was chosen for further studies

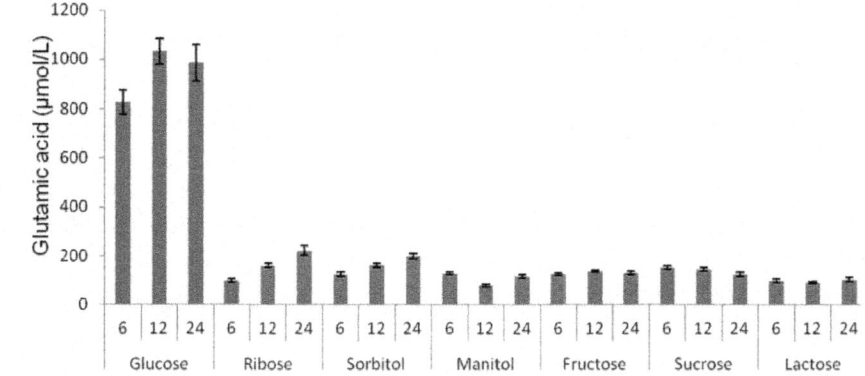

Figure 1: Effect of different carbon sources with different concentrations on the production of glutamic acid by the LAB strain MNZ cultured in MRS medium at 30 °C for 120 h. Values are mean ± standard deviation of two independent experiments.

In the present study, a number of various carbon sources was tested to select the most suitable one for the purpose of glutamic acid production by the LAB strain MNZ. Media containing various carbon sources have already been used for the production of amino acids by the bacteria [16]. In case of the microbial glutamic acid production, a number of different carbon sources such as glucose, sucrose, fructose, ribose, *etc.* have been tested and similar results were reported compared to the results obtained in this study. For example, Roy and Chatterjee [17] reported that fructose and glucose in order of effectiveness were the best carbon sources for the glutamic acid production in a synthetic media. Other carbon sources including sucrose and manitol showed far less effect on glutamic acid production. The best glucose concentration (8%) reported by Roy and Chatterjee [17] was lower than the one found in this study (12%).

This study demonstrates the effectiveness of glucose as an appropriate carbon source in glutamic acid production compared to other sugars, especially those that are not hexoses, such as ribose, lactose and sucrose. The higher growth rate of lactic acid bacteria on glucose as a hexose sugar compared to non-hexoses [18] could be the reason for such an elevated glutamic acid production from glucose. On the other hand, glutamic acid was found to be produced in bacteria from glucose via Krebs cycle intermediates [19]. The carbon-energy source of glucose can be converted into pyruvic acid by the glycolysis pathway, a preface to the TCA cycle and the electron-transport chain [20]. Therefore, glucose was found to be the most appropriate sugar for glutamic acid production by the LAB strain MNZ in this study.

The result of the present study concerning the effects of different carbon sources on glutamic acid biosynthesis revealed that optimum levels of glucose can effectively promote glutamic acid production possibly through the Krebs cycle. Such findings can be further used for appropriate selection of other substrates containing glucose at higher levels for enhanced production of glutamic acid.

Study of Glucose Consumption

In order to investigate the glucose consumption pattern of the LAB strain, the strain MNZ was cultivated in MRS broth at 30 °C for 144 h. The broth was supplemented with 12% glucose in order to examine the effects of glucose concentration on the production of glutamic acid. As shown inFigure 2, there was a sharp decrease in glucose content between 12 and 72 h of fermentation. This might be due to the microbial growth with the highest $\log_{10}$ CFU/mL of 8.3 after 72 h of fermentation. In other investigations using other species of glutamic acid-producing bacteria, the glucose consumption pattern during glutamic acid production was found to be similar to the present study [21]. A continuing decrease in the sugar concentration which is consumed by the bacterium corresponds with the growth of the culture [22].

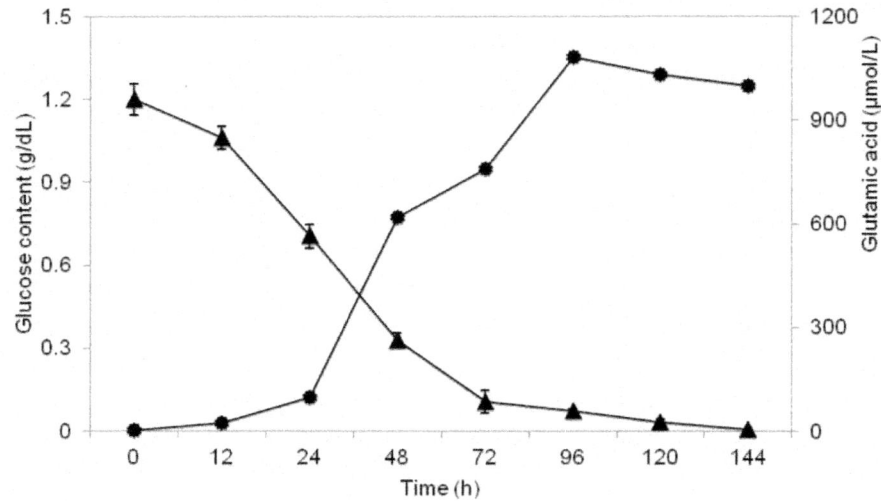

Figure 2: Consumption of glucose (g/dL) by the LAB strain MNZ cultured in MRS medium supplemented with 12% glucose at 30 °C. Symbols used were: (▲), Glucose consumption profile of the bacterium; (●), glutamic acid production.

Effect of Different Concentrations of Ammonium Nitrate on Glutamic Acid Production

Ammonium nitrate as a nitrogen source was chosen in this study and the effect of its different concentrations on the glutamic acid production was tested. The results presented in Figure 3 show that 0.7% (*w/v*) was the best concentration of ammonium nitrate for *Lactobacillus plantarum* to produce glutamic acid. The mechanism by which ammonia enhances glutamic acid production is associated with nitrogen as an essential component for amino acid production. Nitrogen plays an important role in fermentative cultivation of glutamic acid-producing bacteria. Therefore, nitrogen is taken up by cells, and thereafter assimilated to accomplish their metabolism [23].

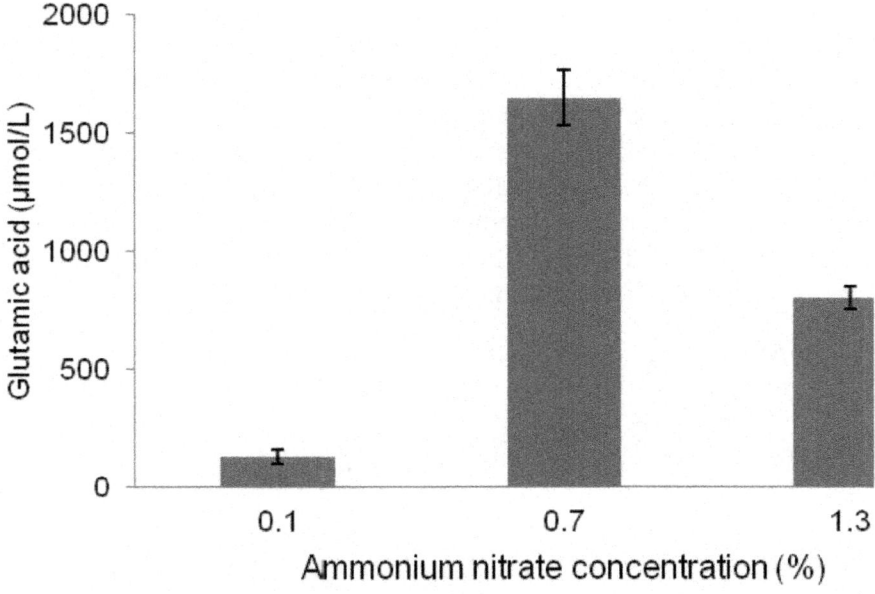

Figure 3: Effect of different concentrations of ammonium nitrate on glutamic acid production by the LAB strain MNZ cultured at 30 °C in MRS.

It is well-known that the uptake of nitrogen sources into the cells is mediated either by passive diffusion (ammonium) or active transport [24]. Generally the bacteria follow two different pathways for ammonium assimilation to form glutamic acid. When the ammonia concentration is low and the diffusion into cells becomes restricted, a unique ammonium transporter (AmtB) encoded by the *amt* gene is activated to cope with the nitrogen starvation, and ammonium is assimilated to glutamine by glutamine synthetase [25]. In contrast, in the presence of high concentrations of ammonium, the diffusion of uncharged ammonia (NH_3) occurs through the cytoplasmic membrane. This would promote the growth of the bacterial cells, and thereby ammonium is assimilated by glutamate dehydrogenase to form glutamic acid. Similarly, Tesch *et al.* [26] showed that most of the ammonium (72%) is assimilated by the glutamate dehydrogenase to form glutamic acid which has been proven to demonstrate high activity in *Lactobacillus plantarum*[27].

pH Association with Glutamic Acid Production

Effect of various initial pH on glutamic acid production by the LAB strain MNZ was investigated in this study. Results show that an initial pH value of 4.5 was the best compared to other pH values (Figure 4). pH plays an important role in biological processes and the pH of the medium is important for the glutamic acid production [28]. *Lactobacillus plantarum* prefers a moderately acidic pH 6.5 for optimal growth [29]. However, it was noted that the maximum glutamic acid production in this study occurred at a lower pH (4.5). The pH of the culture medium can influence the growth rate of *Lactobacilli* [29]. An initial pH value of lower than 6.5 decreases the growth rate of*Lactobacillus plantarum* in the medium [29]. According to Krämer [30], glutamic acid secretion occurs by an overflow metabolism whenever growth is limited. This could cause a redirection of 2-oxoglutarate efflux towards glutamic acid production which leads to an increase of the glutamic acid excretion rate [31].

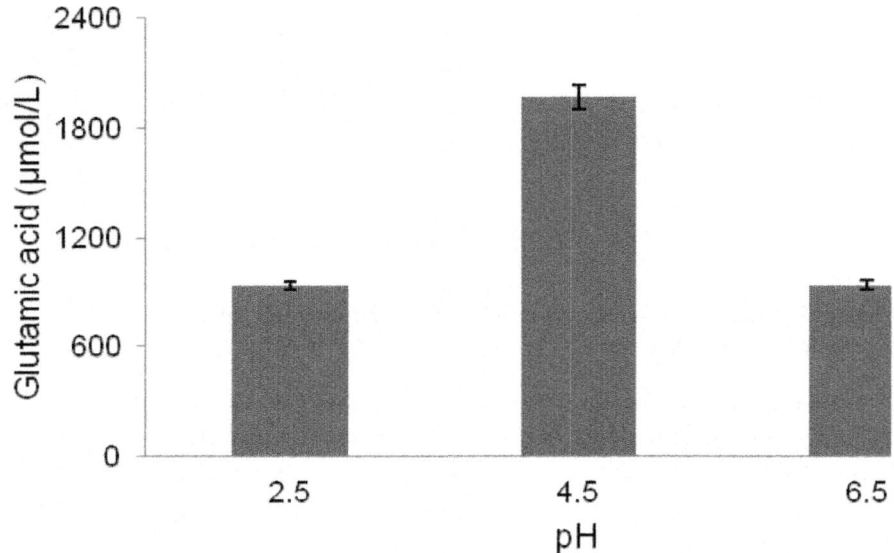

Figure 4: Effect of initial pH on glutamic acid production by the LAB strain MNZ cultured at 30 °C in MRS medium.

It was also proved that *Lactobacillus plantarum* produces ammonia in an acidic environment, which contributes to pH homeostasis and thereby survival of the micro-organism through neutralizing the pH [32]. As a result, the ammonia produced this way can be utilized in glutamic acid production. Thus, the ability of *Lactobacillus plantarum* to decrease the pH not only is considered as a food safety factor, but also improves glutamic acid production in this bacterium through redirecting metabolic afflux of 2-oxoglutarate towards glutamic acid production and producing ammonia which results in an enhanced glutamic acid production.

In order to understand the pH association with glutamic acid production, the pH profile of the fermentation medium was examined and the results presented in Figure 5. It was revealed that between 6 to 72 h of fermentation, the pH immediately decreased from 6.2 to 4.2 followed by a slow increase throughout the remaining fermentation period (Figure 5). Such a reduction in pH value during fermentation might reflect an acid-producing ability of the bacterium, an important feature for the production of quality-fermented foods. The production of organic acids such as lactic and acetic acids is the result of bacterial metabolic activities. The decline in pH is an important characteristic required by the starter strains for acidifying their environment rapidly. The acid production and the accompanying pH decrease are well-known to extend the lag phase of sensitive organisms including food-borne pathogens [33]. Acid production during fermentation, resulting in acidification to pH levels lower than 4.2, constitutes a major food safety factor [34].

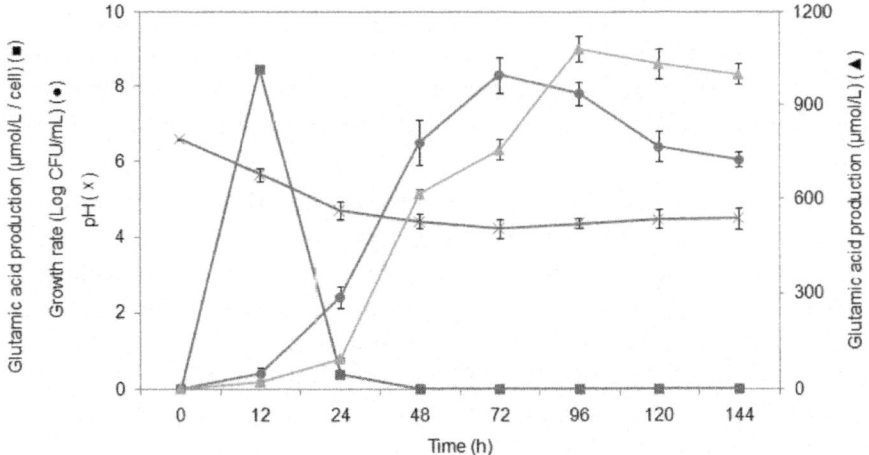

Figure 5: pH, growth rate, glutamic acid production/cell and glutamic acid production profile of *Lactobacillus plantarum*MNZ. Symbols used were: (×), pH profile; (●), growth rate (log CFU/mL); (■), glutamic acid production (μmol/L/cell); (▲), glutamic acid produced (μmol/L).

Lactobacillus plantarum as a facultative hetero-fermentative lactic acid bacterium ferments pentoses through the phosphoketolase pathway and hexoses via glycolysis[18]. In addition, this lactic acid bacterium is capable of mixed acid fermentation which ferments hexoses under specific conditions to various inorganic acids such as acetatic, lactatic and formic acids [35]. The glucose content of the MRS medium can go through the above-mentioned pathways resulting in the production of inorganic acids by the LAB strain MNZ which results in a decrease in pH.

Study of Growth Profile and Association with Glutamic Acid Production

Growth characteristics for LAB strain MNZ were monitored and are presented in Figure 5. The purpose of such a growth characteristic test was to obtain a better understanding about the mechanism of the glutamic acid production in LAB strain MNZ and to find out at which phase of the microbial growth glutamic acid can be produced. Figure 5 shows that *Lactobacillus* cell growth increased exponentially between 18 and 72 h of fermentation in the MRS broth. This stage is recognized as the log phase or exponential phase. It is obvious that the exponential growth phase cannot go on indefinitely. This is due to the fact that the medium is soon depleted of nutrients and enriched with other metabolites. After 72 h of fermentation, a decline in bacterium cell growth was noted and a stationary phase reached. During the stationary phase, the microbial growth rate decreased as a result of nutrient depletion and accumulation of products that may suppress glutamic acid production. This phase is reached as the bacteria begin to exhaust the resources that are available to them.

On the other hand, it was noted that the ratio of glutamic acid production/cell during the first 12 h of fermentation increased to 8.4 μmol/L/cell (Figure 5). Such an elevation of the ratio of glutamic acid production/cell may be associated with the induction of the *gdh* gene. After 12 h of fermentation, the ratio of glutamic acid production/cell drastically decreased to 0.38, 1.9 × 10^{-4}, 3.8 × 10^{-5}, 1.7 × 10^{-5}, 4.1 × 10^{-4} and 8.9 × 10^{-4} μmol/L/cell for 24, 48, 72, 96, 120 and 144 h of fermentation time, respectively. Total glutamic acid production increased between 24 and 96 h although the ratio of glutamic acid production/cell decreased. The increase in glutamic acid production especially after 48 h was due to increasing numbers of bacterial cells in the log phase, which caused an elevation in the glutamic acid yield. The metabolic activity of the bacteria that produced acids resulted in a fall of pH especially for the first 24 h of fermentation. An acidic condition, therefore, could be responsible for triggering the *gdh* gene resulting in elevated glutamic acid production/cell ratio. In this study, a decrease in pH of the fermentation medium was found to be associated with cell growth owing to the glutamic acid production, which was elevated during the log phase of the bacterial growth profile. A similar result was also found in a study by Nakamura *et al.* [21] using *Corynebacterium glutamicum* for glutamic acid production. However, other studies showed that despite the fact that bacterial cells grew well in the synthetic medium, glutamic acid production was depressed during the log phase and only started to increase during the stationary phase [36]. In contrast to our results, it was suggested that glutamic acid production was not associated with growth. Bacterial strain differences and various media containing other factors might be responsible for the differences in the obtained results.

Study of Glutamic Acid-Production Profile

The glutamic acid-producing ability of the isolate MNZ was assessed by conducting a time course analysis of intra-cellular and extra-cellular glutamic acid contents in this strain in a culture medium (MRS broth supplemented with 12% glucose). The results are shown in Figure 6. The highest intra-cellular glutamic acid content (502 μmol/L) was achieved after 48 h of cultivation. While the highest extra-cellular glutamic acid biosynthesis, reaching 933 μmol/L, was attained after 96 h of cultivation. Total glutamic acid production (intra- and extra-cellular glutamic acid) was found to be at a maximum (1082 μmol/L) at 96 h of cultivation.

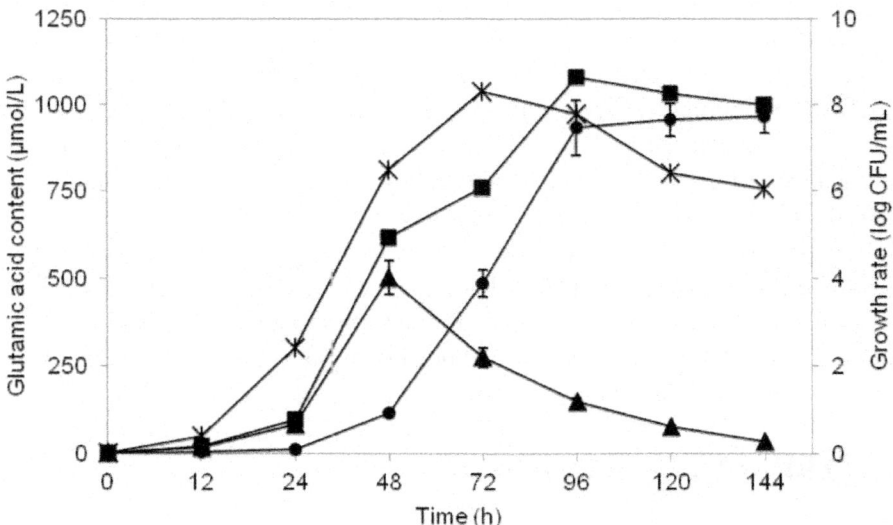

Figure 6: Changes in extra-cellular and intra-cellular glutamic acid content produced by *Lactobacillus plantarum* MNZ cultured in MRS medium supplemented with 12% glucose at 30 °C. Symbols used were: (▲) Intra-cellular glutamic acid; (●) extra-cellular glutamic acid; (■) total glutamic acid; (×) growth profile. Values are mean ± standard deviation of three independent experiments.

Production of glutamic acid in bacteria is mostly dependent on the cytoplasmic glutamate dehydrogenase (GDH) which catalyzes this amino acid formation from α-ketoglutarate [37]. This enzyme has been demonstrated to occur in high frequency and activity in *Lactobacillus plantarum* with the GDH-encoding gene (*gdh*) implying such an elevated activity [27]. On the other hand, the intra-cellular concentration of glutamic acid in this study regularly decreased after 48 h to such an extent—owing to its secretion into the extra-cellular medium—that the feedback inhibition which regulates the internal glutamic acid pool is abolished [38]. Recently, a glutamic acid exporter in a bacterium was found responsible for excreting glutamic acid to the extra-cellular medium [21].

The mutual effect of a feedback inhibition system in the bacterium along with the glutamic acid permease caused the decline of the intra-cellular glutamic acid in *Lactobacillus plantarum* after 48 h of fermentation process. This research finding along with the reported localization of glutamic acid dehydrogenase in cytoplasm suggests that glutamic acid was synthesized in the cytoplasm of*Lactobacillus plantarum* and then secreted into the culture medium.

To gain a better insight of the mechanism by which glutamic acid is secreted into the extra-cellular medium, three models have been described over the past several years as the crucial steps for glutamic acid efflux. These included functional inversion of uptake systems, diffusion and the existence of particular excretion systems [30]. Recently it was suggested that the mechanism of glutamic acid production in bacteria is not mostly related to the cell membrane structure, but that the production of this amino acid is caused by an alteration in metabolic efflux which leads to glutamic acid biosynthesis [31]. Furthermore, two genes in bacteria were shown to be involved in the glutamic acid efflux properties [39] although the functions of these genes still remain unclear [40]. Consequently, appropriate investigations are required to discover the mechanisms involved in glutamic acid production in Lacobacilli.

EXPERIMENTAL SECTION

Samples

Six locally available fermented foods including *tempoyak* (fermented durian flesh), *tempeh*(fermented soybean), *tapai ubi* (fermented tapioca), *tapai pulut* (fermented glutinous rice), *budu* (fish sauce) and *cincalok* (fermented shrimp sauce) were purchased from wet markets in Perlis, Kelantan and Selangor states, peninsular Malaysia as LAB-strain local sources.

Chemicals and Media

HPLC-grade solvents were purchased from Sigma (Sigma-Aldrich, St. Louis, MO, USA). HPLC-grade water was prepared using a Sartorius apparatus

(Arium 611 UV, Sartorius Stedim Biotech, Göttingen, Germany). MRS medium was purchased from Difco (Detroit, MI, USA).

Isolation of Lactic Acid Bacteria

Isolation of lactic acid bacteria (LAB) from fermented food products was performed according to the method described by Adnan and Tan [41] with minor modifications. Each sample (10 g) was separately blended with 90 mL of 0.85% NaCl solution for 2 min (Lab Blender Seward, Stomacher 400). This blended food (10 mL) was mixed with MRS broth (90 mL) in a 250 mL Erlenmeyer flask. The broths containing the food samples were enriched with glucose (2% w/v). The flasks were incubated at 30 °C, 100 rpm, for 7 days. Aliquots of the culture from each of the flasks were serially diluted from 10_1 to 10_{12} times and 0.1 mL of each dilution was spread evenly on MRS agar plates. Colonies of LAB were counted on MRS agar plates after anaerobic incubation for 72 h at 30 °C in GasPaks jars (GasPaks System, BBL) and colonies were reported as $\log_{10}$ CFU/mL. Colonies with distinct morphological differences such as color, size and shape were randomly picked from countable MRS agar plates and subcultured on fresh MRS agar plates. Pure colonies were maintained in 20% v/v glycerol in MRS broth for storage at −80 °C. Each of the isolates was first tested for catalase reaction based on bubble formation after applying 3% hydrogen peroxide solution on the cells. The isolates with catalase-negative results were Gram-stained, and those with Gram-positive activity were considered as lactic acid bacteria for further experiments and analyses.

Quantification of Glutamic Acid-Producing LAB

LAB isolates obtained from various fermented foods were assessed for their potential to produce glutamic acid in a basal medium. The LAB isolates were grown in MRS broth in 15 mL test tubes with yellow caps at 30 °C for 7 days under anaerobic condition (GasPaks System, BBL). After anaerobic incubation, each test tube was quantified for extra-cellular glutamic acid and intra-cellular glutamic acid (inside the bacterial cells).

Extraction of Glutamic Acid in MRS Broth

The contents of extra-cellular glutamic acid accumulated in the culture medium were extracted according to the method of Yang [42] with minor modifications. First, the culture broth was separated from cells by centrifugation (8000× g for 15 min at 4 °C) and the supernatant was diluted 50-fold with 7% (v/v) of glacial acetic acid. The diluted sample was centrifuged at 8000× g for 15 min at 4 °C, and the supernatant was filtered using Nylon 0.22-μm-pore-size membrane and then collected for further analysis.

Intra-cellular glutamic acid was determined following the method described by Komatsuzaki [43] with minor modifications. First, the cells cultured in MRS broth at 30 °C for 7 days, were separated from the culture broth by centrifugation (8000× g for 15 min at 4 °C); the cells were washed with 0.9% NaCl three times, and re-suspended in 20 mL of phosphate buffer saline (PBS, pH 7.0) which consisted of 8.0 g NaCl, 0.2 g KCl, 0.91 g Na$_2$HPO$_4$, 0.12 g KH$_2$PO$_4$. The cells were suspended in 1.0 mL of 75% (v/v) ethanol, the homogenate was centrifuged at 8000× g for 15 min at 4 °C, and the supernatant was filtered using a Nylon 0.22 μm/L pore-size filter. A 100 mL proportion of the filtered supernatant was collected for quantitative analysis of glutamic acid concentration.

Quantitative Analysis of Glutamic Acid

Glutamic acid extracted from MRS broth was subjected to derivatization according to the method described by Rossetti and Lombard [44]. Quantitative measurement of the glutamic acid was performed by running an HPLC analysis of the glutamic acid according to the method described by Yang [42]. The derivatized samples were dissolved in 200 mL of initial mobile phase, consisting of a mixture of 60% solution A (aqueous solution of 10.254 g sodium acetate, 0.5 mL tri ethylamine and 0.7 mL acetic acid in 1000 mL, final pH 5.8), 12% solution B (acetonitrile) and 28% solution C (deionized water). Gradient HPLC separation was performed on a Shimadzu (Kyoto, Japan) LC 20AT apparatus, consisting of a pump system, a CTO-10ASVP model oven with a 20 μL injection loop injector, and a model SPD-M20A PDA (Photo Diode Array) Detector, in conjunction with a DELL Optiplex integrator. A Prevail C18 column (250 mm × 4.6 mm I.D., particle size 5 μm/L, Alltech, IL, USA) was used during the analysis. The mobile phase for the gradient elution was pumped at 0.6 mL/min flow rate and 27 °C temperature, and glutamic acid detection was performed at 254 nm.

Identification of LAB Isolates

Phenotypic Identification

Microscopic and conventional biochemical and physiological techniques were used to initially characterize all the LAB isolates. The isolates were individually propagated in MRS broth. Each strain was primarily subjected to catalase and Gram-reaction tests. Based upon the phenotypic characteristics, only the catalase-negative and Gram-positive isolates were selected. The selected LAB isolates were also tested for colony formation, cell morphologies, and cell grouping using a light microscope (NIKON YS 100). Further identification of the superior glutamic-acid-producing isolate was carried out by employing API 50 CHL fermentation strips (BioMérieux, Marcy l'Etoile, France) at 30 °C according to the manufacturer's instructions.

Genotypic Identification

The LAB strain with superior glutamic acid production was identified by 16S rRNA gene sequencing analysis. Genomic DNA from the LAB selected was extracted according to the method described by Sambrook [45]. The 16S rRNA gene amplification was performed using a genomic DNA sample as template and following universal primer pairs; 27F (5' AGAGTTTGATCCTGGCTCAG 3') and 1492R (5' GTTTACCTTGTTACGACTT 3'). The following thermal cycle was used: 95 °C for 3 min; 40 cycles of: 95 °C for 30 s (denaturation), 55 °C for 55 s (annealing), and 72 °C for 60 s (extension); and one cycle final primer extension 72 °C for 10 min. The purified PCR products using QIAquick Gel Extraction Kit (QIAGEN, Germany) were sequenced by First BASE Laboratories Sdn. Bhd. (Shah Alam, Selangor, Malaysia). The 16S rRNA sequences were aligned and compared with other 16S rRNA genes in the GenBank by using the NCBI Basic Local Alignment Search Tools, nucleotide (BLASTn) program [46]. The 16S rRNA gene sequence described in this study was deposited into the Genbank Data Library and assigned the accession number HM175883.

CONCLUSION

A new glutamic acid-producing strain of Lactobacilli isolated from a traditional fermented food locally available in Malaysia was identified as *Lactobacillus plantarum* MNZ according to the phenotypic and genotypic tests. Evaluation of the intra- and extra-cellular glutamic acid content confirmed that this amino acid was produced inside the cell and excreted into the extra-cellular medium. Among various carbon sources tested, the finding showed that glucose not only supported *Lactobacillus plantarum* growth, but also significantly enhanced glutamic acid production by this LAB strain. Studying the influence of other factors showed that a 7% (w/v) concentration of ammonium nitrate and a pH 4.5 improved glutamic acid production. The physiological characteristics of *Lactobacillus plantarum* including pH profile as investigated in this study support its potential use as starter culture in fermented foods while improving glutamic acid production as well. It is essential to select LAB strains with a suitable acid production profile to develop fermented foods with optimum levels of glutamic acid as a precursor of GABA. This new glutamic acid-producing LAB strain could be utilized for mass production of GABA-rich products, thus accelerating the development of functional fermented foods to benefit the consumers.

ACKNOWLEDGEMENT

This research was fully supported by the Fundandamental Grant (05-10-07-380 FR) from the Ministry of Higher Education, Malaysia, which was awarded to Nazamid Saari.

REFERENCES

1.   Kondoh, T.; Mallick, H.N.; Torii, K. Activation of the gut-brain axis by dietary glutamate and physiologic significance in energy homeostasis. *Am. J. Clin. Nutr* **2009**, *90*, 832S–837S.

2.   Zolotarev, V.; Khropycheva, R.; Uneyama, H.; Torii, K. Effect of free dietary glutamate on gastric secretion in dogs. *Ann. N. Y. Acad. Sci* **2009**, *1170*, 87–90.

3.   Inoue, K.; Shirai, T.; Ochiai, H.; Kasao, M.; Hayakawa, K.; Kimura, M.; Sansawa, H. Blood pressure lowering effect of a novel fermented milk containing g amino butyric acid (GABA) in mild hypertensives. *Eur. J. Clin. Nutr* **2003**, *27*, 490–495.

4.   Hagiwara, H.; Seki, T.; Ariga, T. The effect of pre-germinated brown rice intake on blood glucose and PAI-1 levels in streptozotocin-induced diabetic rats. *Biosci. Biotechnol. Biochem* **2004**, *68*, 444–447.

5.   Beyreuther, K.; Biesalski, H.; Fernstrom, J.; Grimm, P.; Hammes, W.; Heinemann, U.; Kempski, O.; Stehle, P.; Steinhart, H.; Walker, R. Consensus meeting: Monosodium glutamate—An update. *Eur. J. Clin. Nutr* **2006**, *61*, 304–313.

6.   Tomoe, M.; Inoue, Y.; Sanbe, A.; Toyama, K.; Yamamoto, S.; Komatsu, T. Clinical trial of glutamate for the improvement of nutrition and health in the elderly. *Ann. N. Y. Acad. Sci* **2009**, *1170*, 82–86.

7.   Yamamoto, S.; Tomoe, M.; Toyama, K.; Kawai, M.; Uneyama, H. Can dietary supplementation of monosodium glutamate improve the health of the elderly? *Am. J. Clin. Nutr* **2009**, *90*, 844S–849S.

8.   Sano, C. History of glutamate production. *Am. J. Clin. Nutr* **2009**, *90*, 728S–732S.

9. Tanous, C.; Chambellon, E.; Sepulchre, A.M.; Yvon, M. The gene encoding the glutamate dehydrogenase in Lactococcus lactis is part of a remnant Tn3 transposon carried by a large plasmid. *J. Bacteriol* **2005**, *187*, 5019–5022.

10. Leroy, F.; de Vuyst, L. Lactic acid bacteria as functional starter cultures for the food fermentation industry. *Trends Food Sci. Technol* **2004**, *15*, 67–78.

11. Lücke, F.K. Utilization of microbes to process and preserve meat. *Meat Sci* **2000**, *56*, 105–115.

12. Zalán, Z.; Hudáček, J.; Štětina, J.; Chumchalová, J.; Halász, A. Production of organic acids by Lactobacillus strains in three different media. *Eur. Food Res. Technol* **2010**, *230*, 395–404.

13. Tarek, M.; Mostafa, H.E. Screening of potential infants' lactobacilli isolates for amino acids production. *Afr. J. Microbiol. Res* **2010**, *4*, 226–232.

14. Nampoothiri, K.M.; Pandey, A. Effect of different carbon sources on growth and glutamic acid fermentation by *Brevibacterium* sp. *J. Basic Microbiol* **1995**, *35*, 249–254.

15. Sneath, P.H.A.; Mair, N.S.; Sharpe, M.E.; Holt, J.G. *Bergey's Manual of Systematic Bacteriology*; Williams and Wilkins: Baltimore, MD, USA, 1986.

16. Kiefer, P.; Heinzle, E.; Wittmann, C. Influence of glucose, fructose and sucrose as carbon sources on kinetics and stoichiometry of lysine production by Corynebacterium glutamicum. *J. Ind. Microbiol. Biotechnol* **2002**, *28*, 338–343.

17. Roy, D.K.; Chatterjee, S.P. Production of glutamic acid by Arthrobacter globiformis: Influence of cultural conditions. *Folia Microbiol* **1989**, *34*, 11–24.

18. Zaunmüller, T.; Eichert, M.; Richter, H.; Unden, G. Variations in the energy metabolism of biotechnologically relevant heterofermentative lactic acid bacteria during growth on sugars and organic acids. *Appl. Microbiol. Biotechnol* **2006**, *72*, 421–429.

19. Yoneda, Y.; Roberts, E.; Dietz, G.W., Jr. A new synaptosomal biosynthetic pathway of glutamate and GABA from ornithine and its negative feedback inhibition by GABA. *J. Neurochem* **1982**, *38*, 1686–1694.

20. Williams, A.G.; Withers, S.E.; Brechany, E.Y.; Banks, J.M. Glutamic acid dehydrogenase activity in lactobacilli and the use of glutamic acid

dehydrogenase-producing adjunct*Lactobacillus* spp. cultures in the manufacture of cheddar cheese. *J. Appl. Microbiol* **2006**, 1062–1075.

21. Nakamura, J.; Hirano, S.; Ito, H.; Wachi, M. Mutations of the Corynebacterium glutamicum NCgl1221 gene, encoding a mechanosensitive channel homolog, induce l-glutamic acid production. *Appl. Environ. Microbiol* **2007**, *73*, 4491–4498.

22. Nampoothiri, K.M.; Pandey, A. Urease activity in a glutamate producing *Brevibacterium* sp.*Process Biochem* **1996**, *31*, 471–475.

23. Burkovski, A. Ammonium assimilation and nitrogen control in *Corynebacterium glutamicum*and its relatives: An example for new regulatory mechanisms in actinomycetes. *FEMS Microbiol. Rev* **2003**, *27*, 617–628.

24. Meier-Wagner, J.; Nolden, L.; Jakoby, M.; Siewe, R.; Krämer, R.; Burkovski, A. Multiplicity of ammonium uptake systems in *Corynebacterium glutamicum*: Role of Amt and AmtB.*Microbiology* **2001**, *147*, 135–143.

25. Jakoby, M.; Nolden, L.; Meier-Wagner, J.; Krämer, R.; Burkovski, A. AmtR, a global repressor in the nitrogen regulation system of *Corynebacterium glutamicum. Mol. Microbiol* **2000**, *37*, 964–977.

26. Tesch, M.; de Graaf, A.; Sahm, H. *In vivo* fluxes in the ammonium-assimilatory pathways in*Corynebacterium glutamicum* studied by15N nuclear magnetic resonance. *Appl. Environ. Microbiol* **1999**, *65*, 1099–1109.

27. De Angelis, M.; Calasso, M.; di Cagno, R.; Siragusa, S.; Minervini, F.; Gobbetti, M. NADP-glutamate dehydrogenase activity in non-starter lactic acid bacteria: Effects of temperature, pH and NaCl on enzyme activity and expression. *J. Appl. Microbiol* **2010**, *109*, 1763–1774.

28. Eggeling, L.; Bott, M. *Handbook of Corynebacterium Glutamicum*; CRC Press: Boca Raton, FL USA, 2005.

29. Zacharof, M.; Lovitt, R. Development of an optimised growth strategy for intensive propagation, lactic acid and bacteriocin production of selected strains of Lactobacilli genus.*Int. J. Chem. Eng. Appl* **2010**, *1*, 55–63.

30. Krämer, R. Secretion of amino acids by bacteria: Physiology and mechanism. *FEMS Microbiol. Rev* **1994**, *13*, 75–93.

31. Asakura, Y.; Kimura, E.; Usuda, Y.; Kawahara, Y.; Matsui, K.; Osumi, T.; Nakamatsu, T. Altered metabolic flux due to deletion of odhA causes l-glutamate overproduction in Corynebacterium glutamicum. *Appl. Environ. Microbiol* **2007**, *73*, 1308–1319.

32. Jaichumjai, P.; Valyasevi, R.; Assavanig, A.; Kurdi, P. Isolation and characterization of acid-sensitive *Lactobacillus plantarum* with application as starter culture for Nham production.*Food Microbiol* **2010**, *27*, 741–748.

33. Smulders, F.; Barendsen, P.; van Logtestijn, J.; Mossel, D.; van der Marel, G. Review: Lactic acid: Considerations in favour of its acceptance as a meat decontamininant. *Int. J. Food Sci. Technol* **1986**, *21*, 419–436.

34. Holzapfel, W. Use of starter cultures in fermentation on a household scale. *Food Control* **1997**,*8*, 241–258.

35. Boekhorst, J.; Siezen, R.J.; Zwahlen, M.C.; Vilanova, D.; Pridmore, R.D.; Mercenier, A.; Kleerebezem, M.; de Vos, W.M.; Brüssow, H.; Desiere, F. The complete genomes of*Lactobacillus plantarum* and *Lactobacillus johnsonii* reveal extensive differences in chromosome organization and gene content. *Microbiology* **2004**, *150*, 3601–3611.

36. Cocaign-Bousquet, M.; Guyonvarch, A.; Lindley, N.D. Growth rate-dependent modulation of carbon flux through central metabolism and the kinetic consequences for glucose-limited chemostat cultures of *Corynebacterium glutamicum*. *Appl. Environ. Microbiol* **1996**, *62*, 429–436.

37. Börmann, E.; Eikmanns, B.; Sahm, H. Molecular analysis of the *Corynebacterium glutamicum*gdh gene encoding glutamate dehydrogenase. *Mol. Microbiol* **1992**, *6*, 317–326.

38. Nampoothiri, K.; Hoischen, C.; Bathe, B.; Mockel, B.; Pfefferle, W.; Krumbach, K.; Sahm, H.; Eggeling, L. Expression of genes of lipid synthesis and altered lipid composition modulates l-glutamate efflux of *Corynebacterium glutamicum*. *Appl. Microbiol. Biotechnol* **2002**, *58*, 89–96.

39. Kimura, E. Triggering mechanism of-glutamate overproduction by DtsR1 in coryneform bacteria. *J. Biosci. Bioeng* **2002**, *94*, 545–551.

40. Eggeling, L.; Krumbach, K.; Sahm, H. l-glutamate efflux with *Corynebacterium glutamicum*: Why is penicillin treatment or Tween addition doing the same? *J. Mol. Microbiol. Biotechnol***2001**, *3*, 67–68.

41. Mohd Adnan, A.F.; Tan, I.K.P. Isolation of lactic acid bacteria from Malaysian foods and assessment of the isolates for industrial potential. *Bioresour. Technol* **2007**, *98*, 1380–1385.

42. Yang, S.Y.; Lu, F.X.; Lu, Z.X.; Bie, X.M.; Jiao, Y.; Sun, L.J.; Yu, B. Production of γ-amino butyric acid by *Streptococcus salivarius* subsp. *thermophilus* Y2 under submerged fermentation. *Amino Acids* **2008**, *34*, 473–478.

43. Komatsuzaki, N.; Shima, J.; Kawamoto, S.; Momose, H.; Kimura, K. Production of γ amino butyric acid (GABA) by *Lactobacillus paracasei* isolated from traditional fermented foods. *Food Microbiol* **2005**, *22*, 497–504.

44. Rossetti, V.; Lombard, A. Determination of glutamic acid decarboxylase by high-performance liquid chromatography. *J. Chromatogr. B* **1996**, *681*, 63–67.

45. Sambrook, J.; Fritsch, E.F.; Maniatis, T. *Molecular Cloning: A Laboratory Manual*, 5th ed; Cold Spring Harbor Laboratory Press: New York, NY, USA, 1989.

46. *BLAST Home Page*, Available online: http://www.ncbi.nlm.nih.gov/BLAST/ accessed on 3 May 2012.

CITATION

CHAPTER 1

Stephanie N. Chilton, Jeremy P. Burton and Gregor Reid, Inclusion of Fermented Foods in Food Guides around the World, doi:10.3390/nu7010390

CHAPTER 2

Charina Gracia B. Banaay, Marilen P. Balolong and Francisco B. Elegado (2013). Lactic Acid Bacteria in Philippine Traditional Fermented Foods, Lactic Acid Bacteria - R & D for Food, Health and Livestock Purposes, Dr. J. Marcelino Kongo (Ed.), ISBN: 978-953-51-0955-6, InTech, DOI: 10.5772/50582.

CHAPTER 3

M. Puttananjaiah, M. Dhale and V. Govindaswamy, "Non-Toxic Effect of Monascus purpureus Extract on Lactic Acid Bacteria Suggested Their Application in Fermented Foods," Food and Nutrition Sciences, Vol. 2 No. 8, 2011, pp. 837-843. doi: 10.4236/fns.2011.28115.

CHAPTER 4

O. Oyedeji, S. Ogunbanwo and A. Onilude, "Predominant Lactic Acid Bacteria Involved in the Traditional Fermentation of Fufu and Ogi, Two Nigerian Fermented Food Products," Food and Nutrition Sciences, Vol. 4 No. 11A, 2013, pp. 40-46. doi: 10.4236/fns.2013.411A006.

CHAPTER 5

Chiara Devirgiliis, Paola Zinno, Mariarita Stirpe, Simona Barile, and Giuditta Perozzi, "Functional Screening of Antibiotic Resistance Genes from a Representative Metagenomic Library of Food Fermenting Microbiota," BioMed Research International, vol. 2014, Article ID 290967, 9 pages, 2014. doi:10.1155/2014/290967

CHAPTER 6

Grazina Juodeikiene, Elena Bartkiene, Pranas Viskelis, Dalia Urbonaviciene, Dalia Eidukonyte and Ceslovas Bobinas (2012). Fermentation Processes Using Lactic Acid Bacteria Producing Bacteriocins for Preservation and Improving Functional Properties of Food Products, Advances in Applied Biotechnology, Prof. Marian Petre (Ed.), ISBN: 978-953-307-820-5, InTech, DOI: 10.5772/30692.

CHAPTER 7

Lavinia Claudia Buruleanu, Magda Gabriela Bratu, Iuliana Manea, Daniela Avram and Carmen Leane Nicolescu (2013). Fermentation of Vegetable Juices by Lactobacillus Acidophilus LA-5, Lactic Acid Bacteria - R & D for Food, Health and Livestock Purposes, Dr. J. Marcelino Kongo (Ed.), ISBN: 978-953-51-0955-6, InTech, DOI: 10.5772/51309.

CHAPTER 8

Diana I. Serrazanetti, Davide Gottardi, Chiara Montanari and Andrea Gianotti (2013). Dynamic Stresses of Lactic Acid Bacteria Associated to Fermentation Processes, Lactic Acid Bacteria - R & D for Food, Health and Livestock Purposes, Dr. J. Marcelino Kongo (Ed.), ISBN: 978-953-51-0955-6, InTech, DOI: 10.5772/51049.

CHAPTER 9

Egwim Evans, Amanabo Musa, Yahaya Abubakar and Bello Mainuna (2013). Nigerian Indigenous Fermented Foods: Processes and Prospects, Mycotoxin and Food Safety in Developing Countries, Dr. Hussaini Makun (Ed.), ISBN: 978-953-51-1096-5, InTech, DOI: 10.5772/52877.

CHAPTER 10

Huayi Suo, Xia Feng, Kai Zhu, Cun Wang, Xin Zhao and Jianquan Kan, Shuidouchi (Fermented Soybean) Fermented in Different Vessels Attenuates HCl/Ethanol-Induced Gastric Mucosal Injury, doi:10.3390/molecules201119654

CHAPTER 11

Mohsen Zareian, Afshin Ebrahimpour, Fatimah Abu Bakar, Abdul Karim Sabo Mohamed, Bita Forghani, Mohd Safuan B. Ab-Kadir and Nazamid Saari, A Glutamic Acid-Producing Lactic Acid Bacteria Isolated from Malaysian Fermented Foods, doi:10.3390/ijms13055482

INDEX